Intelligent Algorithms in Ambient and Biomedical Computing

T0181582

Philips Research

VOLUME 7

Editor-in-Chief
Dr. Frank Toolenaar
Philips Research Laboratories, Eindhoven, The Netherlands

SCOPE TO THE *'PHILIPS RESEARCH BOOK SERIES'*

As one of the largest private sector research establishments in the world, Philips Research is shaping the future with technology inventions that meet peoples' needs and desires in the digital age. While the ultimate user benefits of these inventions end up on the high-street shelves, the often pioneering scientific and technological basis usually remains less visible.

This 'Philips Research Book Series' has been set up as a way for Philips researchers to contribute to the scientific community by publishing their comprehensive results and theories in book form.

Dr. Rick Harwig

Intelligent Algorithms in Ambient and Biomedical Computing

Edited by

Wim Verhaegh
Philips Research Laboratories, Eindhoven,
The Netherlands

Emile Aarts
Philips Research Laboratories, Eindhoven,
The Netherlands

and

Jan Korst
Philips Research Laboratories, Eindhoven,
The Netherlands

 Springer

A C.I.P. Catalogue record for this book is available from the Library of Congress.

ISBN-10 94-007-8728-6
ISBN-13 978-94-007-8728-5
ISBN-10 1-4020-4995-1 (eBook)
ISBN-13 978-1-4020-4995-8 (eBook)

Published by Springer,
P.O. Box 17, 3300 AA Dordrecht, The Netherlands.

www.springer.com

Printed on acid-free paper

Contents

Part III Technology

Contributing Authors

Emile Aarts
Philips Research Laboratories Eindhoven
High Tech Campus 36, 5656 AE Eindhoven, The Netherlands
emile.aarts@philips.com

Lalitha Agnihotri
Philips Research Laboratories USA
345 Scarborough Rd., Briarcliff Manor, NY 10510, USA
lalitha.agnihotri@philips.com

Zharko Aleksovski
Philips Research Laboratories Eindhoven
High Tech Campus 31, 5656 AE Eindhoven, The Netherlands
zharko@cs.vu.nl

Mauro Barbieri
Philips Research Laboratories Eindhoven
High Tech Campus 34, 5656 AE Eindhoven, The Netherlands
mauro.barbieri@philips.com

Richard Y. Chen
Philips Research Laboratories USA
345 Scarborough Rd., Briarcliff Manor, NY 10510, USA
richard.chen@philips.com

Chun-Ting Chou
University of Michigan
1301 Beal Avenue, Ann Arbor, MI 48109, USA
choujt@umich.edu

Christopher D. Clack
University College London
Gower Street, London WC1E 6BT, United Kingdom
clack@cs.ucl.ac.uk

Nevenka Dimitrova
Philips Research Laboratories USA
345 Scarborough Rd., Briarcliff Manor, NY 10510, USA
nevenka.dimitrova@philips.com

Aukje E.M. van Duijnhoven
Numerando Groep
Amersfoortseweg 10e, 3705 GJ Zeist, The Netherlands
aukjevanduijnhoven@hotmail.com

Gijs Geleijnse
Philips Research Laboratories Eindhoven
High Tech Campus 34, 5656 AE Eindhoven, The Netherlands
gijs.geleijnse@philips.com

Frank van Harmelen
Vrije Universiteit Amsterdam
De Boelelaan 1081a, 1081 HV Amsterdam, The Netherlands
frank.van.harmelen@cs.vu.nl

Herman J. ter Horst
Philips Research Laboratories Eindhoven
High Tech Campus 31, 5656 AE Eindhoven, The Netherlands
herman.ter.horst@philips.com

Nick de Jong
Tiobe Software B.V.
De Zaale 11, 5612 AJ Eindhoven, The Netherlands
nick.de.jong@tiobe.com

Warner ten Kate
Philips Research Laboratories Eindhoven
High Tech Campus 31, 5656 AE Eindhoven, The Netherlands
warner.ten.kate@philips.com

Martin L. Kersten
Centrum voor Wiskunde en Informatica
Kruislaan 413, 1098 SJ Amsterdam, The Netherlands
martin.kersten@cwi.nl

Jan Korst
Philips Research Laboratories Eindhoven
High Tech Campus 34, 5656 AE Eindhoven, The Netherlands
jan.korst@philips.com

Ben J.A. Kröse
University of Amsterdam
Kruislaan 403, 1098 SJ Amsterdam, The Netherlands
krose@science.uva.nl

Akash Kumar
Philips Research Laboratories Eindhoven
High Tech Campus 45, 5656 AE Eindhoven, The Netherlands
akakumar@natlab.research.philips.com

Martin McKinney
Philips Research Laboratories Eindhoven
High Tech Campus 36, 5656 AE Eindhoven, The Netherlands
martin.mckinney@philips.com

Steffen Pauws
Philips Research Laboratories Eindhoven
High Tech Campus 34, 5656 AE Eindhoven, The Netherlands
steffen.pauws@philips.com

Sai Shankar N.
Qualcomm Standards Engineering Dept.
5775 Morehouse Drive, San Diego, CA 92121, USA
nsai@qualcomm.com

Sergei Sawitzki
Philips Research Laboratories Eindhoven
High Tech Campus 31, 5656 AE Eindhoven, The Netherlands
sergei.sawitzki@philips.com

Ruediger Schmitt
Philips Research Laboratories USA
345 Scarborough Rd., Briarcliff Manor, NY 10510, USA
ruediger.schmitt@philips.com

Berry Schoenmakers
Technische Universiteit Eindhoven
Den Dolech 2, 5612 AZ Eindhoven, The Netherlands
berry@win.tue.nl

Kang G. Shin
University of Michigan
1301 Beal Avenue, Ann Arbor, MI 48109, USA
kgshin@umich.edu

Arno P.J.M. Siebes
Universiteit Utrecht
Padualaan 14, 3584 CH Utrecht, The Netherlands
arno.siebes@cs.uu.nl

Janto Skowronek
Philips Research Laboratories Eindhoven
High Tech Campus 36, 5656 AE Eindhoven, The Netherlands
janto.skowronek@philips.com

Pim Tuyls
Philips Research Laboratories Eindhoven
High Tech Campus 34, 5656 AE Eindhoven, The Netherlands
pim.tuyls@philips.com

Wim F.J. Verhaegh
Philips Research Laboratories Eindhoven
High Tech Campus 34, 5656 AE Eindhoven, The Netherlands
wim.verhaegh@philips.com

Michael Verschoor
Technische Universiteit Eindhoven
Den Dolech 2, 5612 AZ Eindhoven, The Netherlands
m.p.f.verschoor@tm.tue.nl

Nikos Vlassis
University of Amsterdam
Kruislaan 403, 1098 SJ Amsterdam, The Netherlands
vlassis@science.uva.nl

Wojciech Zajdel
University of Amsterdam
Kruislaan 403, 1098 SJ Amsterdam, The Netherlands
wzajdel@science.uva.nl

Kees van Zon
Philips Research Laboratories USA
345 Scarborough Rd., Briarcliff Manor, NY 10510, USA
kees.van.zon@philips.com

Preface

The rapid growth in electronic systems in the past decade has boosted research in the area of computational intelligence. As it has become increasingly easy to generate, collect, transport, process, and store huge amounts of data, the role of intelligent algorithms has become prominent in order to visualize, manipulate, retrieve, and interpret the data. For instance, intelligent search techniques have been developed to search for relevant items in huge collections of web pages, and data mining and interpretation techniques play a very important role in making sense out of huge amounts of biomolecular measurements. As a result, the added value of many modern systems is no longer determined by hardware only, but increasingly by the intelligent software that supports and facilitates the user in realizing his or her objectives.

Over the past years, considerable progress has been made in the area of computational intelligence, which can be positioned at the intersection of computer science, discrete mathematics, and cognitive science. This has led to a growing community of practitioners within Philips Research that develop, analyze, and apply intelligent algorithms. The Symposium on Intelligent Algorithms (SOIA) intends to provide this community of practitioners with a platform to exchange information. The first edition of SOIA, held in 2002, addressed the topic of intelligent algorithms in ambient intelligence. To share the output of the symposium with a larger audience, a selection of papers was edited and published by Kluwer in the Philips Research Book Series under the title "Algorithms in Ambient Intelligence." For the second edition, held in 2004, the scope of the symposium was broadened so as to comply with the three main topics of the Philips company strategy, i.e., Healthcare, Lifestyle and Technology. Again a selection of papers was edited, resulting in the present book. It consists of 17 chapters, divided over three parts corresponding to the strategic topics mentioned above. The main topic in Healthcare is the understanding of biological processes, for Lifestyle the main topic is content retrieval and manipulation, and finally for Technology most contributions relate to media processing. Below we present more detailed information about the individual chapters.

Part I consists of four chapters. In Chapter 1, Chris Clack discusses the topic of modeling biological systems, thus allowing to perform in-silico experiments by means of computer simulation, to formulate hypotheses. In Chapter 2, Nevenka Dimitrova gives an overview of the reverse approach, where one does not use computers to simulate biological processes, but where one uses biology to perform computations, in DNA computing and synthetic biology. In Chapter 3, Martin Kersten and Arno Siebes discuss data management inspired by biology, resulting in an organic database system. In Chapter 4, Kees van Zon discusses how to achieve machine consciousness, and how it can be applied.

Par II consists of eight chapters, addressing problems from the area of content management and retrieval. In Chapter 5, Wim Verhaegh discusses the problem of making a schedule of preferred TV programs, while at the same time selecting TV programs for recording, under the assumption of a limited number of tuners. In Chapter 6, Mauro Barbieri, Nevenka Dimitrova, and Lalitha Agnihotri present a technique to automatically summarize video into a condensed preview, allowing one to quickly browse and access large amounts of stored programs. Chapters 7–9 concerns audio applications. First, Janto Skowronek and Martin McKinney discuss in Chapter 7 the topic of automatic classification of audio and music, for which they developed the automatic extraction of the higher-level feature of percussiveness. In Chapter 8, Steffen Pauws presents a technique to automatically extract the key from a piece of music, providing an emotional connotation to it, and making it possible to build well-sounding music mixes. In Chapter 9, Zharko Aleksovski, Warner ten Kate, and Frank van Harmelen address the problem of combining multiple databases of music data in a semantic way, by approximating matches of music classes. Next, Jan Korst, Gijs Geleijnse, Nick de Jong, and Michael Verschoor discuss in Chapter 10 the possibilities to fill a knowledge database, using an ontology to collect and structure data from web pages. In the last chapter of part II, which Wim Verhaegh, Aukje van Duijnhoven, Pim Tuyls, and Jan Korst resolve the privacy issue of population-based recommenders by encrypting the users' profiles and performing the required algorithms on encrypted data.

Part III consists of six chapters, focusing on the technology underlying intelligent algorithms and intelligent systems. The first two chapters discuss theoretical aspects of intelligent algorithms. In Chapter 12, Peter Grünwald gives an overview on the minimum description length principle to resolve the problem of model selection, based on the fundamental idea to see learning as a form of data compression. In Chapter 13, Herman ter Horst discusses the computational complexity of reasoning with semantic web ontologies, such as RDF Schema and OWL. Next, Wojciech Zajdel, Ben Kröse, and Nikos Vlassis present in Chapter 14 an introduction to dynamic Bayesian networks, and show their application in robot localization and multiple-person tracking. In

Chapter 15, Berry Schoenmaker and Pim Tuyls discuss efficient protocols for securely matching two user profiles, without leaking information on the details of the profiles. Finally, Chapters 16 and 17 address resource issues in intelligent systems. In Chapter 16, Sai Shankar N., Richard Chen, Ruediger Schmitt, Chun-Ting Chou, and Kang Shin revisit fairness in multi-rate wireless networks, and present a solution to fairly schedule airtime. Finally, in Chapter 17, Akash Kumar and Sergei Sawitzki discuss the design alternatives of Reed Solomon decoders, and address the problem of making optimal design decisions to obtain a high-throughput, low-power solution.

We are convinced that the chapters presented in this book comprise an interesting collection of examples of the use of intelligent algorithms in different settings, and that the book reconfirms that the area of computational intelligence is a truly challenging field of research.

WIM F.J. VERHAEGH, EMILE AARTS, AND JAN KORST
Philips Research Laboratories Eindhoven

Acknowledgments

We would like to thank the following people who helped us to review the contributed chapters: Dee Denteneer, Nevenka Dimitrova, Hans van Gageldonk, Srinivas Gutta, Herman ter Horst, Jan Nesvadba, Dave Schaffer, and Peter van der Stok.

Part I

HEALTHCARE

Chapter 1

BIOSCIENCE COMPUTING AND THE ROLE OF COMPUTATIONAL SIMULATION IN BIOLOGY

Christopher D. Clack

Abstract   Bioscience computing exploits the synergy of challenges facing both computer science and biology, drawing inspiration from biology to solve computer science challenges and simultaneously using new bio-inspired adaptive software to model and simulate biological systems. This chapter first provides an introduction to bioscience computing — discussing the role of computational simulation in terms of hypothesis formulation and prototyping for biologists and medics, and explaining how bioscience computing is both timely and well-suited to systems biology. A concrete example of computational simulation is then provided — the artificial cytoskeleton, which utilises swarm agents and a cellular automaton to model cell morphogenesis. Morphological adaptation for tasks such as chemotaxis and phagocytosis are presented, and the role of the artificial cytoskeleton and its swarm-based techniques in both computer science and biology is explained.

Keywords   Bioscience computing, systems biology, computational simulation, morphogenesis, adaptive systems, agent based modelling, swarm agents.

1.1   Introduction to bioscience computing

Bioscience computing exploits the synergy of challenges facing both computer science and biology, drawing inspiration from biology to solve problems in computer science and simultaneously using new bio-inspired adaptive software to model and simulate self-organising, adaptive, biological systems.

There has recently been a substantial increase in inter-disciplinary research interactions between computer science and the life sciences. From the biologist's perspective, the post-genomic era is characterised by huge amounts of data but little understanding of how genes map to physiological functions, and there is an urgent need for the application of intelligent computing techniques to gain increased understanding. From the computer scientist's perspective, the new biological data and expanding understanding of biological processes

3

provide both an excellent driver for new methods in bioinformatics and an increasing source of ideas for new computational techniques in areas such as intelligent systems and artificial life.

The purpose of the first part of this chapter is to provide an introduction to the biological context and to explain the role of bioscience computing within that context.

1.1.1   A change of focus in biology and medicine

The traditional reductionist view of biology is rooted in analysis and biophysics; it is based on a hierarchical perspective where the functioning of the physiome[1] is the deterministic product of a 'one-way upward causation from genes to cells, organs, system and whole organisms' [Noble, 2002], and has been remarkably successful with fundamental achievements such as discovering the structure of DNA and mapping the genome for not one but several organisms. The traditional role of computer science in biology (e.g. of bioinformatics) has been to support this endeavour by providing data-handling, data visualisation, numerical simulation and data-mining services.

However, in the post-genomic era the super-abundance of data and relative paucity of understanding, coupled with a clearer perspective of the complexity of living organisms, are causing biologists to question whether the traditional view is sufficient as a basis for a full understanding of nature. The traditional view is giving way to a new biology, often referred to as systems biology.

The rise of systems biology has caused a much closer relationship to develop between biologists and computer scientists. In systems biology, the computer science techniques are no longer merely a data service to the biologists, but are intimately involved in the formulation of biological hypotheses as biologists embrace the process-oriented world of the computer scientist. systems biology considers an organism as a self-organising, adaptive, complex, dynamic system providing an information framework with global constraints and multiple feedback and regulation paths between high and low levels (e.g. controlling gene expression); the sub-modules are too inextricably connected, there are too many interactions between levels, for a one-way hierarchy to be possible [Noble, 2002]. Biologists now experiment not just in-vivo and in-vitro, but increasingly *in-silico*. These in-silico experiments are the basis for what we term bioscience computing.

1.1.2   Modelling and simulation

The primary aims of modelling and simulation in biology are to improve understanding of a process or hypothesis, to highlight gaps in knowledge, and

---

[1]A glossary of biological terms is provided in Table 1.1 at the end of this chapter.

to make clear, testable predictions [Kirkwood et al., 2003]. Note, however, that an in-silico experiment itself can never truly be used to *test* a biological hypothesis — rather, computational simulation in biology should be viewed as a process of *prototyping to assist hypothesis formulation.*

Wet-lab experimental techniques tend to focus analytic attention on single mechanisms. By contrast, computational simulation can contribute to the activity of synthesis, of integrating many separate elements that form a network of activity. The resultant interaction and synergy can provide a qualitatively much improved experimental framework. These in-silico results may then guide the choice of (more expensive) subsequent wet-lab experiments.

Techniques. There is a wide spectrum of techniques available to support modelling and simulation, ranging from high-level phenomenological approaches which generally represent qualitative features of a system, to low-level mechanistic simulations which typically represent quantitative aspects (though abstraction and quantification need not be mutually exclusive concepts [Ideker & Lauffenburger, 2003]). Examples of available techniques include statistical data-mining, clustering and classification (e.g. support vector machines), Bayesian networks, Markov chains, fractal theory, Boolean logic, and fuzzy logic. At the mechanistic extreme there are cellular automata and agent-based simulations. Differential equations are widely used and capable of capturing detail at varying levels of abstraction. See Figure 1.1.

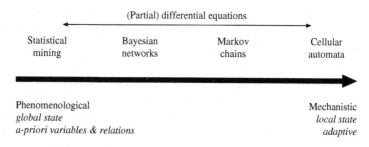

Figure 1.1. Comparative spectrum of available techniques.

Phenomenological models tend to focus on the *global* state of a system. Often they describe an a-priori given set of relations between an a-priori given set of variables [Giavitto et al., 2002]; the two sets cannot evolve jointly with the running system, and very few of these models successfully capture a rich enough semantics to be able to predict complex behaviour [Anderson & Chaplain, 1998]. By contrast, mechanistic models provide local interaction modelling, where cells react (often adaptively) to a *local* environment, not to the state of the system as a whole (thereby supporting heterogeneity). This leads to a rich model of spatiotemporal dynamics, and offers insights into the parameters and mechanisms responsible for system dynamics [Gatenby & Maini, 2003] and for collective organisational behaviour at the microscopic level [Patel, 2004].

Differential equations and partial differential equations provide an excellent mechanism for detailed expression of behaviours of many kinds, but are unsatisfactory for some highly detailed spatiotemporal behaviours [Araujo & McElwain, 2004]. For example, where precise local effects due to intermolecular interactions and random molecular movement are required, a great number of equations must be generated and solved [Succi et al., 2002]. In practice, the computational limits on solving a large number of related partial differential equations leads to the technique normally being applied only to *abstractions* of internal mechanisms and processes.

An interesting mechanistic approach is the use of cellular automata — e.g. Scalerandi's 2D model of cardiac growth dynamics [Scalerandi et al., 2002]. When coupled with agent-based modelling, using a 'swarm' of thousands of tiny agents (a mechanism itself inspired by nature) each representing a separate macromolecule, this method has the advantages of both mathematical simplicity and that the spatiotemporal fates of individual components (cell, proteins etc.) can be tracked in minute detail. The resulting system is very good at representing spatiotemporal dynamics and organisational behaviour, particularly for the simulation of adaptive behaviour.

Objects and processes. The specific attraction of computational simulation is that the computational approach corresponds more naturally to the way that biologists think about their subject. Biologists (in particular molecular biologists) naturally focus on *objects*, *interactions* and *processes*.

Computational simulation permits biologists to express biological systems in terms of computational objects, interactions and processes that relate directly to their biological counterparts and are therefore far easier to understand and easier to manipulate than differential equations. Computational simulations can be expressed in terms of information networks and can use interaction-centric models (e.g. local-neighbourhood operations within a cellular automaton grid), all of which naturally map onto (for example) cell structure and the interaction of macromolecules.

The experience of systems biology has been that biologists have increasingly adopted the computational systems concepts of computer scientists. This should not come as a surprise, since computer scientists have extensive experience of building, modelling, and simulating complex systems that require analysis and synthesis at many different levels of abstraction.

1.1.3 A computational approach to biological complexity

The computational approach to biology enables simulations as dynamic emergent hierarchies of biological complexity, with interactions and feedback between the levels, for example as illustrated in Figure 1.2. At the lowest level, system components are lightweight agents governed by local-neighbourhood rules. The rules provide the system of dynamic interaction between agents, and from this comes the self-organising properties of the simulated organism (threshold parameters may need to be derived via automatic search methods). The emergent behaviour of the system is dependent on a combination of the competitive and co-operative interactions of the underlying local-neighbourhood rules, the regulatory effects that arise from the self-organising properties of those rules, and sets of global constraints (which may be derived from experimental observation). The result is a complex, dynamic system, which can itself be considered as an agent in a larger network of agents of similar complexity, each undergoing interactions according to local-neighbourhood rules at a higher level, and from which yet more complex behaviour emerges.

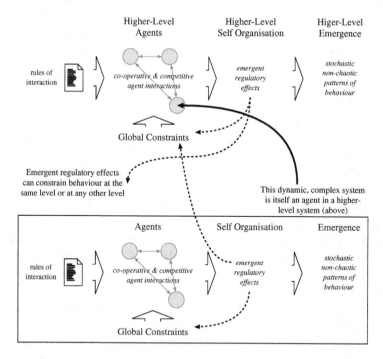

Figure 1.2. The dynamic emergence of hierarchies of biological complexity.

While emergent behaviour has the potential for chaotic results, in a hierarchy of levels each can constrain the realisable solutions of the other levels — thus, an understanding of dynamic emergence in complex hierarchies is a fundamental step in understanding the underlying mechanisms of biology.

1.1.4   Summary: The role of bioscience computing

The first part of this chapter has explored the role of bioscience computing in biology and argued that it is both timely and well-suited to the emergence of systems biology: it provides *in-silico* experiments; focuses on interactions and integration of concurrent mechanisms; is intimately involved in the formulation of biological hypotheses; manipulates objects, processes and interactions; is mathematically straightforward, with a low barrier to uptake; and captures rich spatiotemporal detail at low computational cost.

1.2   Simulating adaptive behaviour

This second part provides a concrete example of the bioscience computing techniques discussed in the first part of this chapter, and presents the *artificial cytoskeleton*, a computational simulation of the development and adaptation of the shape and form of an organism: *morphogenesis*. The work is more fully described by Bentley & Clack [2004; 2005].

Organisms in nature exhibit complex adaptive behaviour that far surpasses the ability of current state-of-the-art autonomous software and robotics. Our research focuses on morphological adaptation, the continuous lifetime reconfiguration of phenotypic form (shape) exhibited by natural systems in order to continue to survive in a changing environment. Many unicell organisms exhibit complex adaptations of their shape in rapid response to environmental changes — e.g. fibroblast cells change shape to assist movement during wound healing, and immune system cells change shape to eat invading bacteria — even though they have no centralized control system. We aim to understand the underlying mechanisms and principles that govern this adaptive behaviour, to explore the concept of morphological adaptation as a mapping from environment to phenotype rather than merely from genotype to phenotype, and to draw inspiration from those mechanisms to improve the adaptive behaviour of artificial systems.

The detailed spatiotemporal aspects of morphogenesis are difficult to compute using partial differential equations and so we turned to a bioscience computing technique; a cellular automaton and agent-based computing using a very large number of simple agents ('swarm' agents).

1.2.1   The artificial cytoskeleton

Our mechanism, the 'artificial cytoskeleton', is closely modelled on the eukaryotic cytoskeleton, a complex, dynamic network of protein filaments which extends throughout the cytoplasm and which gives the cell dynamic structure and function. In particular *actin* cytoskeleton microfilaments are involved in rapid changes to membrane shape in response to environmental signals [Alberts et al., 1994]. We use agent-based swarm techniques combined with 3D cellular automaton (CA) rules to allow proteins to exist and interact with their 26 nearest neighbours in a 3D voxellated environment. The agent-based swarm technique permits the modelling and tracking of individual components and their interactions. The CA simplifies visualization, supports 3D spatial placement and movement, and reduces system complexity. The combination of the two techniques (agent-based swarm and CA) provides opportunities for optimizing computational overhead (e.g. it is not always necessary to compute interactions for all cells in the CA — only those that contain or abut an agent). The CA rules for chemical diffusion and agent interactions can be checked against current understanding of the biology.

The artificial cytoskeleton resides within a membrane-bound 'cell' and receptors (sensors) in the membrane relay external signals to the artificial cytoskeleton via a pathway of protein reactions: the transduction pathway (TP). See Figure 1.3. For efficiency, the artificial cytoskeleton and transduction pathway comprise only a small selection of proteins — just those necessary for a particular experiment. The artificial cytoskeleton's non-rigid form permits it to disassemble rapidly and re-form in a more advantageous distribution; it constantly responds to environmental cues by reorganizing, i.e. altering the cell's internal topography and the membrane morphology.

The underlying mechanism. The artificial cytoskeleton consists of *structural* proteins (actin and a nucleator), which make up the filaments, and several *accessory* proteins, which regulate a filament's behaviour (e.g. inhibiting, activating, severing, bundling). Environmental signals filter into the cell via the transduction pathway, affecting concentrations of accessory proteins and structural protein behaviour. The cooperative and competitive interactions of these structural and accessory proteins can dramatically alter the cytoskeleton's filamentous structure, affecting the shape and structure of the cell as a whole, and resulting in rich diversity in cell shape [Alberts et al., 1994].

The protein interactions are defined by a set of functions; these functions encapsulate the complete mapping from environmental cues to cell morphology (which in turn may affect the environment). We call this function set the 'environment-phenotype map' (or 'E-P map'). Different cell behaviours may require different E-P maps. The following explanation of the underlying mechanism will focus on the E-P map for chemotaxis; see Figures 1.3 and 1.4.

Each voxel in the cellular automaton contains one of the following units:

1. environment which may contain concentrations of a chemoattractant '*C*'.

2. cytoplasm which may contain concentrations of the protein profilin;

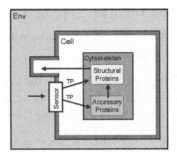

Figure 1.3. A generalized environment-phenotyope map. The cytoskeleton is affected by input from the environment (Env) via the transduction pathway (TP) and can affect the shape of the cell, and thereby also the environment.

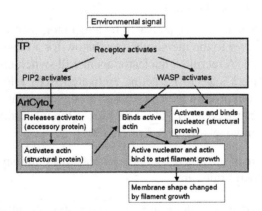

Figure 1.4. The environment-phenotype map as by Bentley & Clack [2004] abstracted from the biological pathway for fibroblast chemotaxis. The simplified transduction pathway (TP) contains a receptor and two macromolecules PIP2 and WASP, which convey information to the artificial cytoskeleton (ArtCyto).

3. an agent which may be either:

> **actin:** which may be in the states S-actin (inactive), P1, P2, or F-actin (in a filament) and which has 2 opposing binding sites ('+','−'); or
>
> **a nucleator:** the protein complex 'Arp 2/3', which may be switched on or off and has one binding site.

The interactions of these two agents drive the creation, growth and disassociation of actin filaments. The growth of actin filaments forces local membrane shape changes, therefore altering the cell's overall shape.

4. cell membrane which may contain a receptor and/or the two transduction pathway proteins WASP and PIP_2.

> The membrane separates the cell from the environment. Initially, no membrane units contain WASP or PIP_2 but each has a probability of containing a receptor.
>
> Cell surface receptors are embedded in the membrane and mediate signals from the external environment to the cytoskeleton. Membrane units containing receptors sum the concentration of C in their adjacent environment voxels. If the sum exceeds a threshold, a cascade reaction inside the cell is triggered; WASP and PIP2 are activated for the receptor and for its adjacent membrane voxels. If the receptor deactivates, WASP and PIP_2 deactivate. See Figure 1.5.
>
> The **WASP** proteins, when activated by a receptor, recruit agents *nucleator* and *P1 actin* to the membrane (see below for a further explanation of recruitment). A recruited nucleator agent will switch on and recruited P1 actin changes state to P2 actin. Activated **PIP_2** releases a one-off plume of protein profilin which diffuses through cytoplasm units. Deactivated PIP_2 causes *removal* of all profilin in the membrane unit's adjacent cytoplasm voxels [Holt & Koffer, 2001].

Protein behaviour is governed by both general rules and specific rules of interaction. The general rules are:

1. *Diffusion:* accessory proteins are represented as concentration gradients which diffuse through cytoplasm voxels. Diffusion is calculated as by Glazier & Graner [1993]; each cytoplasm voxel has a protein threshold, the excess being evenly distributed to its cytoplasm neighbours.

2. *Random movement:* when not bound or stuck, an agent moves randomly. When it moves to a new position, the protein concentration currently in that position is diffused away and the voxel acquires the agent's identifier; the agent's previous voxel becomes cytoplasm.

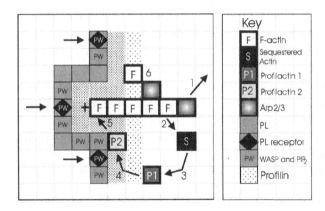

Figure 1.5. Artificial cytoskeleton interactions. Receptors detect chemoattractant, WASP and PIP$_2$ activate and cause the cytoskeletal behaviours shown in stages 1–6, see text for details.

3. *Recruitment:* the biological concept of recruitment of proteins, to a specific protein S, is modelled as follows: an agent follows random movement until it encounters an S in its nearest neighbours. It then can only move such that an S is still in its nearest neighbours. Recruitment stops if there is no S nearest neighbour.

The specific rules of interaction for the chemotaxis environment-phenotype map consist of rules governing actin filament formation (and destruction) and rules governing modifications to the shape of the cell membrane. These are illustrated in Figure 1.5 and the stages are described in detail below:

An actin filament (AF) is created when a nucleator agent combines with an actin agent. Figure 1.5 illustrates a chain of F-actin agents 'F' and a nucleator ('Arp2/3'). Each F-actin agent has two binding sites ('+'/'−'): filament growth occurs at the end with the exposed '+' binding site. Subsequently other actin agents may join the filament by attaching to an actin agent already in the filament. Over time the nucleator disassociates (and un-sticks) from its AF and deactivates (stage 1). Similarly actin in a filament (F-actin) loses affinity for the filament allowing cofilin (a severing protein) to disassociate it; it then gets sequestered and changes to the inactive S-actin state (stage 2). Disassociation always occurs at the filament's '−' end. The actin or nucleator agent disassociates with a probability that increases with time spent in a filament. As the '+' end of the filament grows, the '−' end shrinks and the filament, as a higher level entity, moves towards the membrane.

Actin agents are initiated in the inactive state S-actin; S-actin units sum the concentration of profilin in their nearest neighbours — if it exceeds the threshold then the actin binds to profilin and changes to state P1, removing an amount of profilin from the surrounding cytoplasm (stage 3). P1 actin is recruited to active WASP to form P2 (stage 4). After recruited movement, if P2 actin has an actin filament '+' site in its nearest neighbours, it binds to it, changes state to F-actin, releases profilin to the surrounding cytoplasm, and moves to the nearest neighbour cytoplasm voxel that permits its '−' site to directly abut the actin filament '+' site (stage 5).

A nucleator agent activates when recruited by WASP and then can nucleate (start) actin filaments and set their orientation by binding to a P2 actin agent in its nearest neighbours (also see push-out rule below). If there is a fully bound F-actin nearest neighbour, then a nucleator can also 'stick' to it and nucleate a *branch* actin filament (stage 6 [Alberts et al., 1994]).

There are three interactions affecting the cell membrane:

1. A gap must exist or be created between the AF's '+' end and the membrane to allow P2 actin to bind either to F-actin or a nucleator. Adjacent membrane is 'pushed-out' — the membrane voxels become cytoplasm and the adjacent environment voxels become membrane (C is diffused away first).[2] The precise biology for this process is unclear [Condeelis, 2001].

2. To keep cytoplasm volume constant following 'push-out', the cytoplasm or agent (but not F-actin) voxel within the cell that is furthest from the newly created cytoplasm is replaced with a membrane voxel (any affected profilin or agent is displaced).

3. If a membrane unit has no contact with inner cellular units, it is removed (becomes an environment unit); this ensures there are no doubled-up layers of membrane.

The combination of the above three interactions contracts the cell at the opposite side to a leading edge and allows the cell's centre of mass to move.

Experiments. *Chemotaxis experiment.* The artificial cytoskeleton was tested in a simple experiment based on animal cell chemotaxis, requiring the cell to undergo transformations in form in response to an external chemical stimulus. The specific E-P map for our chemotaxis experiment [Bentley & Clack, 2004] is given in Figure 1.4. The artificial cytoskeleton's response to the stimulus mimicked that of a real fibroblast cell (Figure 1.6), forming a leading edge with protrusions. It moved towards the chemical source purely by lifetime adapation of shape: see Figure 1.7.

Phagocytosis experiment. In nature, a single adaptive mechanism is able to provide different morphologies in response to different environmental stimuli.

---

[2]After implementing this rule, a nucleator would switch off as it would no longer have WASP nearest neighbours, so we permit a nucleator to remain switched on if any of its 26 nearest neighbours or any of their surrounding 98 voxels contain WASP.

For example, compare chemotaxis (movement morphology) with phagocytosis (ingestion morphology): these two examples are distinct both topologically and functionally, yet are known to be controlled by the same underlying biological mechanism. In chemotaxis, a cell detects a chemical gradient and transforms its morphology in order to follow it to the source. By contrast, phagocytosis is the process of engulfment of a foreign particle for degradation or ingestion [Castellano et al., 2001]; a fairly universal cell function relying on profound rearrangements of the cell membrane.

In phagocytosis, cell surface receptors trigger and bind to the particle, tethering it; this causes reactions involving the same proteins downstream as in chemotaxis, but leading to a different morphology — in this case, internal structure change near the edge touching the particle causes an enclosing concave morphology called a 'phagocytic cup'; see Figure 1.8. The simulated morphology is shown in Figure 1.9 — by comparing the medial axis functions for chemotaxis and phagocytosis shapes, we were able to demonstrate a bifurcation of morphology based on a single E-P map; this is a clear demonstration of the potential of E-P maps, and of our ability to reproduce the multifunctionality of nature in our artificial simulations.

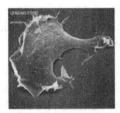

Figure 1.6. Leading edge morphology (top left) during chemotaxis movement.

Figure 1.7. Simulation of chemotaxis leading edge morphology.

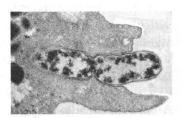

Figure 1.8. Phagocytic cup morphology the cup forms around the particle.

Figure 1.9. Simulation of phagocytic cup morphology (particle not shown).

1.3 Impact and future directions for bioscience computing

In the computer sciences. An improved understanding of the internal mechanisms and organisational principles of adaptive behaviour and lifetime plasticity, especially adaptation of morphology, will provide foundational results that are applicable to many forms of adaptive response. The improved understanding of plasticity will provide the basis for a new breed of software systems that are adaptable to continuously changing, dynamic and unpredictable environments, are robust in the face of unexpected change, and are efficient. This includes improvements to synthetic systems such as autonomous software agents (e.g. as used for trading and fund optimisation in the financial markets, which need to adapt to a constantly changing environment — note that 'shape' may be remapped into an analogous concept such as asset allocation), automatic design systems for physical artefacts, some automatic systems in clinical medicine, some control systems in the automotive and aeronautical industries, and embodied robots (which currently either have a fixed shape or comprise several modular units that may be dynamically reconfigured).

As in nature, when a software agent alters its 'shape', it alters its exposure to and interaction with its environment. Further benefit will be in the area of mechanisms for the distributed control of adaptive response. Overall, the impact of our work will be better designed, more robust, more reliable, more adaptable, more efficient and more effective systems.

In the life sciences. There is much current interest in our computational simulation techniques from our collaborators at the Natural History Museum (NHM), at the UCL Department of Oncology, and elsewhere. For example, with the UCL Department of Oncology we are currently using agent-based swarm techniques to simulate the transport of antibody-based drugs through the extra-cellular matrix during cancer therapy, aiming to improve understanding and increase the efficacy of therapy. A particular attraction is the use of computational simulation as a biologist's 'hypothesis prototyping tool', and the fact that the simulation permits the fate of individal molecules to be tracked.

Collaboration with the Natural History Museum (NHM). Our initial work with the NHM was a study of the morphogenesis of diatoms (single-celled algae), whose patterned cell walls are thought to be an adaptive response to their environment. Diatoms are one of the most important groups of primary producers on the planet, which have thousands of forms and behaviours, each adapted to a different environment. If their adaptive response to environmental pressure were better understood, they would be a good bio-indicator of changes occurring in the natural environment [Davey & Crawford, 1986].

The observable diatom cell wall morphologies are not explicable by the physics of diffusion alone; electron microscopy studies reveal that the cytoskeleton is intimately involved in the patterning of the cell wall and may also incorporate the use of cytoplasmic organelles as moulds for different cell wall components. Our simulation of cell wall morphogenesis used the artificial cytoskeleton to represent the physical position of cell wall and cytoskeletal components, and a genetic algorithm to evolve the cytoskeletal control mechanism. Our model generated representations of diatom cell walls that were, at each stage of development, consistent with empirical observations and exhibited some of the functions of diatom cell walls. More importantly, understanding of the mechanism of the cytoskeleton during morphogenesis was improved; e.g. the need to consider not only genetics but also environmental effects and developmental genetic encoding. This was a significant advance in understanding for the NHM and a fruitful collaboration for both parties, for example leading to three research publications: [Bentley et al., 2005], [Bentley & Clack, 2004] and [Bentley & Clack, 2005].

We are currently seeking funding to conduct a further experiment to aid the NHM in the understanding of diatom colony behaviour. Certain species of diatom have developed a complex set of interactions during morphogenesis, which allows them to form and disband colonies, triggered by environmental cues and giving them a greater chance of survival (e.g. by altering sinking rates to optimize nutrient and light exposure). In general, a colony-forming diatom will, upon cell division, grow two new cells such that their abutting cell walls interlock and hold the cells together; this continues until the filament reaches a certain average length, then (statistically) the most central dividing cell will divide into two new cells that do not interlock, thus dividing the filament in two [Davey & Crawford, 1986]. Diatom colony formation is an explicit and interesting example of morphological adaptation to environmental changes; it is a type of cyclomorphosis (where adaptation cycles through two or more forms). There has been a large amount of speculation as to how and why certain species of diatom form colonies; it contributes to current understanding within diatom research, and also provides a good model to improve understanding of the hierarchical adaptive systems that underlie morphological plasticity.

1.4 Summary and conclusions

Bioscience computing draws inspiration from biology to solve computer science challenges and simultaneously uses new bio-inspired adaptive software to simulate biological systems. It is a rapidly emerging field for interdisciplinary research, with synergistic benefits for both computer science and the life sciences.

Biologists have a computational, process-oriented understanding of their subject — they think in terms of objects, interactions and processes: processes are more important than the end result; dynamic behaviour is more important than equilibrium; and behaviour and interactions of *individual* objects are important. Computational simulation provides in-silico experiments as a prototyping technique for hypothesis formulation that more directly maps to the biologists' understanding of their subject, and can more directly assist their thought processes. The process-oriented approach to simulating biological complexity leads to an increased understanding of dynamic emergence and regulatory interaction and control: this is a fundamental step towards a future theory of biology.

The artificial cytoskeleton has been presented as an example of the computational simulation techniques of bioscience computing, illustrating real benefits accruing to both computer science and the life sciences.

Acknowledgements

The research contributions of the following colleagues are gratefully acknowledged: Katie Bentley (artifical cytoskeleton, PhD student), Dr. Eileen Cox (diatom morphology, NHM), Dr. Sylvia Nagl (oncology and mathematical biology) and Manish Patel (model integration, PhD student). All are members of the UCL BioScience Computing Interest Group (www. cs.ucl.ac.uk/research/bioscience).

References

Alberts, B., D. Bray, J. Lewis, M. Raff, K. Roberts, and J.D. Watson [1994]. *Molecular Biology of The Cell*. Garland Publishing, 3rd edition.

Anderson, A.R.A., and M.A.J. Chaplain [1998]. Continuous and discrete mathematical models of tumour-induced angiogenesis. *Bulletin of Mathematical Biology*, 60:857–900.

Araujo, R., and D. McElwain [2004]. A history of the study of solid tumour growth: The contribution of mathematical modelling. *Bulletin of Mathematical Biology*, 66:1039–1091.

Bentley, K., and C.D. Clack [2004]. The artificial cytoskeleton for lifetime adaptation of morphology. *SODANS Workshop proceedings of the 9th Intl. Conf. on the Simulation and Synthesis of Living Systems (ALIFE IX)*, pages 13–16.

Bentley, K., and C.D. Clack [2005]. Morphological plasticity: Environmentally driven morphogenesis. *Proceedings of the 8th European Conference on Artificial Life (ECAL 2005), Lecture Notes in Artificial Intelligence*, 3630:118–127.

Bentley, K., E. Cox, and P. Bentley [2005]. Nature's batik: A computer evolution model of diatom valve morphogenesis. *Nanoscience and Nanotechnology Journal*, 5(1):25–34.

Castellano, F., P. Chavrier, and E. Caron [2001]. Actin dynamics during phagocytosis. *Seminars in Immunology*, 13:347–355.

Condeelis, J. [2001]. How is actin polymerization nucleated *in vivo*? *TRENDS in Cell Biology*, 11(7):288–293.

Davey, M.C., and R.M. Crawford [1986]. Filament formation in the diatom melosira granulata. *Journal of Phycology*, 22:144–150.

Gatenby, R.A., and P.K. Maini [2003]. Mathematical oncology: Cancer summed up. *Nature*, 421:321–324.

Giavitto, J-L., C. Godin, O. Michel, and P. Prusunkiewicz [2002]. *Modelling and Simulation of Biological Processes in the Context of Genomics, chapter Computational Models for Integrative and Developmental Biology*. Hermes.

Glazier, J.A., and F. Graner [1993]. Simulation of the differential adhesion driven rearrangement of biological cells. *Physical Review E*, 47(3):2128–2154.

Holt, M.R., and A. Koffer [2001]. Cell motility: Proline-rich proteins promote protrusions. *TRENDS in Cell Biology*, 11(1):38–46.

Ideker, T., and D. Lauffenburger [2003]. Building with a scaffold: emerging stategies for high- to low-level cellular modelling. *Trends in Biotechnology*, 21(6):255–262.

Kirkwood, T.B.L., R.J. Boys, C.S. Gillespie, C.J. Proctor, D.P. Shanley, and D.J. Wilkinson [2003]. Towards an e-biology of ageing: Integrating theory and data. *Nature Reviews, Molecular Cell Biology*, 4:243–249.

Noble, D. [2002]. The rise of computational biology. *Nature Reviews, Molecular Cell Biology*, 3:460–463.

Patel, M. [2004]. *Internal report*. Department of Clinical Oncology, UCL.

Priami C. (ed.) [2003]. *Proc. 1st Workshop Computational Methods in Systems Biology*. Springer.

Scalerandi, M., B. Capogrosso Sansone, C. Benati, and C.A. Condat [2002]. Competition effects in the dynamics of tumor cords. *Physical Review E*, 65(5 Pt 1):051918.

Succi, S., I.V. Karlin, and H. Chen [2002]. Colloquium: Role of the h theorem in lattice boltz-mann hydrodynamics simulations. *Reviews of Modern Physics*, 74:1203–1220.

Table 1.1. Glossary of biological terms.

term	description
actin	A protein that links into chains (polymers), forming microfilaments in muscle and other contractile elements in cells.
allele	The precise sequence of nucleotides for a specified gene.
cellular secretion	The escape of substances from a cell to its environment.
chemotaxis	Migration of cells along a concentration gradient of an attractant.
chromosome	The self-replicating genetic structure of cells containing the cellular DNA that contains the linear array of genes.
cytoplasm	The contents of a cell (but not including the nucleus).
cytoskeleton	A system of molecules within eukaryotic cells providing shape, internal spatial organization, and motility, and may assist in communication with other cells.
diatoms	Microscopic algae with cell walls made of silicon and of two separating halves.
eukaryote	A cell or organism with a membrane-bound nucleus (and other subcellular compartments). Includes all organisms except viruses, bacteria, and bluegreen algae.
extracellular matrix (ECM)	Any material produced by cells and secreted into the surrounding medium. The properties of the ECM determine the properties of the tissue (e.g. bone versus tendon) and can also affect the behaviour of cells.
fibroblast	A cell found in most tissues of the body, involved in wound repair and closure; they migrate towards the wound site via chemotaxis.

Table 1.1 (contd). Glossary of biological terms.

term	description
gene	The fundamental unit of heredity; a sequence of nucleotides in a particular position on a chromosome that encodes a specific functional product (e.g. a protein).
gene expression	The process by which a gene's coded information is translated into the structures present and operating in the cell (either proteins or RNA).
genome	The set of different types of gene for a specified organism, distinguished by allele type and position on the chromosome.
genotype	The set of different gene alleles existing in an organism.
leukocyte	A white blood cell, an important component of the body's immune system.
macromolecule	A molecule composed of a very large number of atoms. Includes proteins, starches and nucleic acids (e.g. DNA).
metabolic pathway	A series of chemical reactions in a cell resulting in either a metabolic product to be used/stored by the cell or the initiation of another metabolic pathway.
morphogenesis	The development and adaptation of the shape and form of an organism.
nucleotide	A subunit of DNA or RNA.
organelle	A membrane-bound structure in a eukaryotic cell that partitions the cell into regions which carry out different cellular functions.
phagocytic cup	An inward folding of the cell membrane creating an interior pocket, formed by an actin dependent process during phagocytosis.
phagocytosis	The engulfment of a particle or a microorganism by leukocytes.
phenotype	The physical characteristics of an organism.
physiome	The functional behavior of the physiological state of an individual or species, describing the physiological dynamics of the normal intact organism.
PIP2	Phosphatidylinositol [4, 5]-biphosphate; formed and broken down in the cell membrane; mediates cell motility in fibroblast chemotaxis.
protist	An organism with eukaryotic cells that is neither plant nor animal nor fungi.
protrusions	Thin fingerlike extensions from the surface of a cell.
WASP	Wiskott-Aldrich syndrome protein. Regulates the formation of actin chains.

Chapter 2

THE MANY STRANDS OF DNA COMPUTING

Nevenka Dimitrova

Abstract   Reaching the theoretical limit of Moore's law has inspired new computing paradigms. DNA computing uses properties of biomolecules and techniques from
molecular biology to perform computations, instead of using the traditional
silicon-based computer technologies. To date experiments have been performed
both in-vitro and in-vivo. In this chapter, we will give an overview and examples of the different implementations of DNA computing: molecular computing,
aqueous computing, DNA Turing machines, and the nascent field of synthetic
biology.

Keywords   DNA computing, aqueous computing, molecular computing.

2.1   Introduction

The advances in biology since the discovery of the structure of the double
helix in 1953 can be only described as big strides. New areas of biology have
been born giving rise to new approaches in widely varied fields such as agri-
culture, medicine, and forensics. Most prominently, genomics and proteomics
have greatly improved our knowledge of the components of biological systems
at the molecular level. Scientists have elucidated the complete gene sequences
of several model organisms and provided general understanding of the molecular machinery involved in gene expression. Next is the combination of dis-
parate types of data that interpret changes in genes, proteins, and metabolites
on a cellular level, to result in a set of parameters that can provide a definitive
means of diagnosis and evaluation of therapeutic intervention to alter disease
outcome.

Now, all these advances have also facilitated a change in attitude. We understand enough biology now to tinker with molecules in a predictable manner
and 'compute' the outcome. So the topic is to use biology with an engineering approach: to compute with molecules or to synthesize new reactions and
organisms with the available biological knowledge. In this chapter we will

Wim F.J. Verhaegh et al. (Eds.), Intelligent Algorithms in Ambient and Biomedical Computing, 21-35.

provide a brief overview of the advances in DNA computing and synthetic biology.

We give an overview of DNA computing in Section 2.2 and the more nascent field of synthetic biology in Section 2.3. In Section 2.4, we conclude and suggest open new directions for future research.

2.2 DNA computing

DNA computing is a form of computing that uses DNA and molecular biology, instead of the traditional silicon-based computer technologies. This field was started by Leonard Adleman of the University of Southern California [Adleman, 1994]. In 1994, Adleman demonstrated a proof-of-concept use of DNA as form of computation that was used to solve the Hamiltonian path problem. Since the initial Adleman experiments, DNA computing has made advances and has shown to have potential as a means to solve several large-scale combinatorial search problems. There has been research over one-dimensional lengths, two-dimensional tiles, and even three-dimensional DNA graphs processing, self-assembling DNA graphs [Sa-Ardyen et al., 2003]. A new term, natural computing, has been introduced to describe computing going on in nature and computing inspired by nature [Brauer et al., 2002]. The advancements in the field include computing with membranes – P systems [Paŭn, 2000]. A P system is a computing model which abstracts from the way the alive cells process chemical compounds in their compartmental structure. Benenson et al. [2003] constructed a DNA computer, coupled with an input and output module, capable of diagnosing cancerous activity within a cell, and then releasing an anti-cancer drug upon diagnosis [Benenson et al., 2004].

The field of DNA computing has been expanding greatly as evidenced by the variety of topics covered at the 11th International Meeting on DNA Computing, held in 2005 in London Ontario (see http://www.csd.uwo.ca/dna11/). Here we decided to present only a cross section of approaches to DNA computing: molecular computing, aqueous computing, and Turing machines.

2.2.1 Molecular computing

In 1994 Leonard Adleman published his paper: Molecular Computation of Solutions to Combinatorial Problems [Adleman, 1994] in which he described the experimental use of DNA as a computational system. He showed how to solve a seven-node instance of the Hamiltonian Path problem, an NP-Complete problem similar to the traveling salesman problem. While the seven-node instance is considered a toy problem, this paper is the first known example of the successful use of DNA to compute an algorithm.

In Adleman's version of the Hamiltonian Path Problem (HPP), a hypothetical salesman tries to find an optimum route through a set of cities so that he visits each city exactly once. As the number of cities increases, the solution run time grows exponentially relative to the number of cities at which point the problem requires brute force search methods. HPPs with a large number of cities quickly become computationally expensive, making them less than feasible to solve on even the latest super- or grid- computer. Adlemans demonstration only involves seven cities, making it in some sense a trivial problem. If the problem involves a large number of cities the molecular computing approach would also be very difficult because of the required mass of molecules. Nevertheless, his work is significant for a number of reasons:

- It is the first time to combine computer science, chemistry, and biology. It illustrated the possibilities of using DNA to solve a class of problems that is difficult or impossible to solve using traditional computing methods.

- It is an example of computation at a molecular level, potentially a size limit that may never be reached by the semiconductor industry.

- In an innovative way the DNA is used as a data structure to encode symbols.

- It showed the potential use of DNA as memory: We should note here that DNA at 0.34 nm spacing between the bases, produces 18 Mbits per inch (linear); or 1 million Gbits = 1 petabits/sq. inch. This is important because current hard disk drives have a capacity of 400 GB. In research, Seagate has reached densities of 50 terabits (Tb) per square. In 50 terabits we can store over 3.5 million high-resolution photos, 2800 audio CDs, 1600 hours of television, or the entire printed collection of the US Library of Congress.

- The computing machinery works at molecular levels with the use only of DNA strands and enzymes.

- It demonstrated the possibility for massively parallel computation, as many enzymes can work on many DNA molecules simultaneously.

Consider the example of Figure 2.1, where we have to find a path from Boston to Phoenix that visits each city exactly once. For this example, the molecular solution is as follows.

Step 1. Represent the cities in the graph (i.e. encode) with single stranded DNA sequences, as shown in Figure 2.2,

Step 2. Generate all possible connections (represented by edges in the graph) using DNA hybridization. As shown in Figure 2.2, a connection between two cities is encoded by taking the complement of three letters from the starting city 'tac' which is 'atg' and the complement of the first three letters of

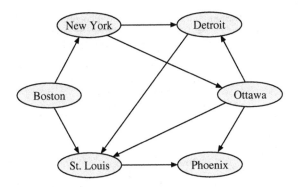

Figure 2.1.   An example HP graph.

the ending city 'atc', which is 'tag'. The connection from Boston to Phoenix is encoded as 'atgtag'. In this manner, all the different connections can be encoded.

Step 3. Select itineraries that start and end with the correct city. In this step, the goal is to copy and amplify paths that start with Boston and end with Phoenix. To achieve this, polymerase chain reaction (PCR) is used, with primers that are complimentary to Boston and Phoenix.

Step 4. Select itineraries with the correct number of cities. In order to achieve this, first the DNA is sorted by length, and then only chains with exactly six cities are selected. In this process, so-called gel electrophoresis is used to select the chains with six cities.

Step 5. Select itineraries that contain each city only once. To acheive this goal, affinity purification is used to fish out the correct ending city with a magnetic bead attached to the complement of the desired city, i.e. Phoenix, as shown in Figure 2.3. We should note here that if there is more than one solution then all the solutions will be attracted.

Step 6. The final answer is obtained by sequencing, i.e. reading, the output or by using so-called graduated PCR.

2.2.2   Aqueous computing

Aqueous computing refers to a method of recording information on DNA molecules while they are dissolved in water [Head et al., 2002b].

The resulting solution of information containing molecules is considered to constitute a 'fluid memory'. He also introduced schemes for reading information from these molecules. A simple instance of the Satisfiability Problem (SAT) of a set of boolean clauses was proposed. Given a set of boolean clauses in a number of variables, the problem is to find truth values for the variables for which the clauses are satisfied (true). A procedure for solving SAT is described

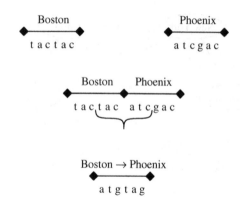

Figure 2.2. City and connection representation with DNA sequences.

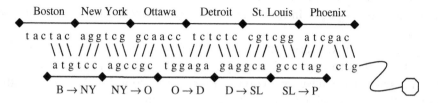

Figure 2.3. Affinity purification example.

by Head et al. [2002b] that illustrates the DNA computing method called the aqueous algorithm. The wet-lab procedure for the aqueous computing has been carried out in the laboratory of Susannah Gal. A projected transparency illustrated the reading procedure and confirmed that the computation was correctly carried out. This work is part of a world-wide search for information storage techniques and computational procedures that take advantage of the vast parallelism of biomolecular operations. All known sequential solutions of the SAT problem, which is NP-complete, require a number of steps that grows exponentially in the size of the problem instance. The number of steps required by the aqueous algorithm grows only linearly. Another application of aqueous computing is the cardinality of a maximal independent subset of a graph has been computed and reported by Head et al. [2000]. The cardinality of a minimal dominating subset of a graph has also been computed by the same group. The satisfiability of a set of four disjunctive clauses in three boolean variables has been determined and reported by Head et al. [2002b], with preliminary reports by Head et al. [1999]. The aqueous approach suggests a convenient way to carry out computations in the style introduced by Lipton [1995]. Finding the patterns in which non-attacking knights can be placed on a 3×3 chessboard has been reported by Heat et al. [2002a].

In aqueous computing there are three basic operations: partition, unite, and write.

- Partition is the action of fast replication of the entire memory, i.e. dividing the memory into n segments

- Unite is the fast action to merge separate units of fluid memory, yielding a single uniform unit containing all the information that was present in the previously separate units.

- A write into memory is the bit alteration at the same station on every word of memory in a given body of fluid. Although the words in a fluid memory cannot be addressed individually, processes are available that allow individual bit locations on each molecule to be addressed, with the provision that the same bit location on every molecule in the memory is acted on in the same way. We will call these crucial locations on the molecule the stations of the molecule.

We should note here that this kind of abstract computation can be implemented in different ways. One possibility is to use a protein or a polypeptide as the molecule to write on, if writing can be done by attachment of anti-bodies to specific sites on the peptide.

Next, we discuss how the operations Partition, Unite, and Write can be orchestrated to provide a useful computation. We will do this for the maximum independent set problem.

For example, let us consider a procedure for determining the cardinality of a maximum independent set in the graph $G = (V, E)$, with vertex set $V = \{a, b, c, d, e, f\}$ and four undirected edges $\{a, b\}, \{b, c\}, \{c, d\}, \{d, e\}$. Let m be a molecule that possesses six stations, each admitting a write operation. We designate these six stations with the vertices a, b, c, d, e, f, respectively. Also, assume that the initial condition of each of the stations is a representation of the bit '1'. A station at which a write has been performed will be regarded as a representation of the bit '0'. This entails that when a write is performed at a station, the bit '0' is written at the station. If a '0' has previously been written at a station then a second write performed at that station is '0' again, i.e. it entails no change.

With each memory molecule, m, we associate a non-negative integer $P(m)$. At each phase of the computation, the value of $P(m)$ will be the number of stations that remain in their original condition, i.e. representing a '1'. Thus, initially $P(m) = 6$ for each m. We also suppose that the molecules can be sorted by this parameter and that the value of the parameter can be determined for each of the classes resulting from the sort. Next, we introduce the procedure.

Initialize test tube T_0 to contain, in solution, a large number of the memory molecules. T_0 now contains only molecules that encode 111111 representing the set $\{a,b,c,d,e,f\}$.

1. Partition T_0 into tubes denoted as T_1 and T_2. In T_1 write at station a; in T_2 write at station b.

2. Unite T_1 & T_2 into T_0. After this step, T_0 contains molecules representing 011111 & 101111, which encode the two subsets $\{b,c,d,e,f\}$ & $\{a,c,d,e,f\}$ of the vertex set V.

3. Partition T_0 into tubes T_1 and T_2. In T_1 write at station b; in T_2 write at station c.

4. Unite the contents from T_1 & T_2 into T_0. After this step, T_0 contains molecules representing 001111, 010111, 101111 & 100111.

5. Partition the contents of T_0 into tubes T_1 and T_2. In T_1 write at station c; in T_2 write at station d.

6. Unite T_1 & T_2 into T_0. After this step, T_0 contains molecules representing 000111, 001011, 010111, 010011, 100111, 101011, 100111 & 100011.

7. Partition T_0 into tubes T_1 and T_2. In T_1 write at station d; in T_2 write at station e.

8. Unite T_1 & T_2 into T_0. After this step, T_0 contains molecules representing 000011, 000101, 001011, 001001, 010011, 010101, 010011, 010001, 100011, 100101, 101011, 101001, 100011 & 100001, each representing an independent set.

9. Sort the memory molecules m remaining in T_0 according to their values $P(m)$.

The cardinality of a maximum independent set of vertices in the graph G is $\max\{P(m)|m \in T_0\}$.

In the present case, the largest parameter value is expected to be attained by the molecules representing 101011. For each such molecule m, $P(m) = 4$ and m represents the set $\{a,c,e,f\}$, i.e., the expected maximal independent set.

One can think of the above program as follows: a FOR loop that is traversed once for each edge of the graph G; followed by a Sort treated as a single step. The number of steps required to find the cardinality of a maximal independent subset of a graph, following this abstract program, grows linearly in the number of edges of the graph. Finding such a cardinal number for a graph is, of course, one of the classical complete algorithmic problems [Garey & Johnson, 1979]. Of course one limiting factor for graphs with large number of edges would be the number of molecules that are required to provide the solution. In addition, if a solution contains a huge number of different molecules the question is whether these molecules can find and bind with each other therefore posing a question whether the write actions can be performed in constant time. Consequently, the finding of a high-speed dependable technology for implementing this abstract aqueous program would be a substantial contribution to computing.

2.2.3   DNA approach to Turing machines

One of the most exciting moments in DNA computing was the physical realization of an abstract Turing machine using DNA molecules [Benenson et al., 2001]. The paradigm is to use DNA as software, and enzymes as hardware. The way in which these molecules undergo chemical reactions with each other allows simple operations to be performed as a byproduct of the reactions. The devices can be controlled by the composition of the DNA software molecules. Of course this is a completely different approach as compared to pushing electrons around a dry circuit in a conventional computer.

Turing machines.   In the 1930's several mathematicians began to think about what it means to be able to compute a function. Alan Turing and Alonzo Church independently arrived at equivalent conclusions. As we might phrase their common definition now, a function is computable if it can be computed by a Turing machine.

In fact, a Turing machine (TM) is a very simple machine. Yet, a TM has the power of any digital computing machinery. A Turing machine processes an infinite tape. This tape is divided into squares (cells), any square of which may contain a symbol from a finite alphabet, with the restriction that there can be only finitely many non-blank squares on the tape. The TM has a read/write head positioned at some square on the tape. Furthermore, at any time, the Turing machine is in any one of a finite number of internal states. The Turing machine is further specified by a set of instructions of the following form:

$$(currentState,\ currentSymbol,\ nextState,\ newSymbol,\ left/right).$$

This instruction means: if the TM is now in *currentState*, and the symbol under the readwrite head is *currentSymbol*, change its internal state to *nextState*, replace the symbol on the tape at its current position by *newSymbol*, and move the readwrite head one square in the given direction (left or right). If a Turing machine is in a condition for which it has no instruction, it halts.

DNA implementation of Turing machines.   In 2001 Shapiro and colleagues introduced the first version of a Turing machine (biomolecular computer) created in a test tube capable of performing simple mathematical calculations [Benenson et al., 2001; Regev et al., 2002]. In this implementation they used ATP molecules for energy. An improved design that uses its input DNA molecule as its sole source of energy, was introduced in 2003 [Benenson et al., 2003].

In this approach, symbols are encoded with 5 base pairs: the symbol a is encoded by 'tggct', the symbol b is encoded by 'gcagg' and the terminal symbol t is encoded by 'gtcgg'. The sticky ends for state-symbol pairs are encoded as follows: $\langle S_1, a \rangle$ is encoded by 'tggc', $\langle S_1, b \rangle$ is encoded by 'gcag', $\langle S_1, t \rangle$ is encoded by 'gtcg', $\langle S_0, a \rangle$ is encoded by 'ggct', $\langle S_0, b \rangle$ is encoded by 'cagg', $\langle S_0, t \rangle$ is encoded by 'tcgg'. The 'hardware' is implemented using the property of the FokI enzyme and recognition site: FokI always recognizes the sequence 'ggatg' and then 'cuts' 9 and 13 nucleotides on the $5' \rightarrow 3'$ and $3' \rightarrow 5'$ strands, respectively, leaving 'sticky' (not even) ends (see Figure 2.4). DNA stores energy, available upon hybridization of complementary strands or hydrolysis of its phosphodiester backbone.

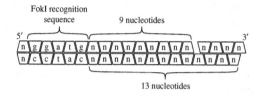

Figure 2.4. FokI enzyme recognition sequence and cutting action.

Figure 2.5 shows an example of the DNA molecules that embody the software (i.e. transition rules): each molecule realizes a different transition rule by detecting a current state and symbol and determining a next state. Each transition consists of a FokI recognition site, $\langle state, symbol \rangle$ detector molecule (four nucleotides, single stranded). A so called *spacer* – double stranded DNA of variable length that determines the FokI cutting site inside the next symbol is introduced. The spacers define the next state. There are empty spacers that realize S_1 to S_0 transition, 1-base-pair long spacers that maintain the current state, and 2-base-pair spacers that transfer S_0 to S_1. For transition T_1, the molecule contains the FokI recognition site, then one base pair spacer, followed by the ccga which is the complement of $\langle S_0, a \rangle$ (represented by 'ggct', as introduced above). The software molecules (shown in Figure 2.5) effectively operate as a family of cofactors of variable specificity to FokI enzyme, each determining a specific FokI cleavage site on the input molecule. A fixed amount of software and hardware molecules can process any input molecule of any length.

An example of an input molecule is shown in Figure 2.6 where the exposed uneven (or 'sticky') end at the $5'$ terminus of the DNA molecule encodes the

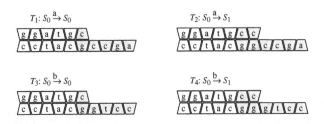

Figure 2.5. DNA sequences representing transitions starting with state S_0.

initial state and first symbol. Each symbol is encoded with 5 base pairs separated by 3-base-pair spacers.

Figure 2.6 illustrates one computational cycle of the automaton. The computation proceeds via a series of transition cycles. During each transition, the hardware-software complex cuts and scatters one input symbol. This is exemplified with the input molecule that starts with ab... in the initial state S_0 and the transition $S_0 \xrightarrow{a} S_1$. At the top of the figure there is a software molecule and an input molecule. In the middle of the figure a ligated software-input molecule is shown with the recognition site for the FokI enzyme. FokI cuts the input molecule inside the next symbol as shown in the third row of the figure. We should note here that the hardware-software molecules are recycled. Each subsequent computational step of the automaton consists of a reversible hybridization between a software molecule – and an input molecule, followed by an irreversible software-directed cleavage (cutting) of the input molecule, which drives the computation forward by increasing entropy and releasing heat. The cleavage uses the capability of the restriction enzyme FokI, which serves as the hardware, to operate on a noncovalent software-input hybrid. In Shapiro's initial implementation, the software-input ligation step consumed one software molecule and two ATP molecules per step. In this implementation there is no need for ligation, which means that a fixed amount of software and hardware molecules can, in principle, process any input molecule of any length without external energy.

Medical computer using DNA Turing machines. A novel application of the DNA turing machine is to assess concentrations of specific RNA molecules [Benenson et al., 2004] (in vitro), which may be overproduced or underproduced, in specific types of cancer. Using pre-programmed medical knowledge, the computer then makes its diagnosis based on the detected RNA levels. In response to a cancer diagnosis, the output unit of the computer can initiate the controlled release of a single-stranded DNA molecule that is known to interfere with the cancer cell's activities, causing it to self-destruct (apoptosis). In case everything looks normal the drug does not get released.

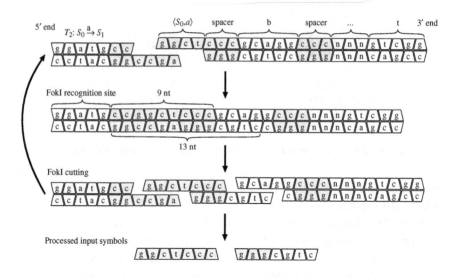

Figure 2.6. Software computation of the input sequence *ab*...

Critics question what is special about this method. In truth, the novelty of their study does not lie in the method of recognizing cancer; they do rely on existing diagnostic methodology. What is different about their approach is its potential for performing diagnostics and therapy within tissue itself. Traditional diagnostic methods require tissue extraction, isolation of the molecular marker in question, and its comparison to the normal tissue. The end goal for the medical computer in the future is to be administered as a drug, and distributed throughout the body by the blood stream to detect specific disease markers autonomously and independently in every cell. In this manner, a single cancer cell could be detected and destroyed before the tumor develops.

The most interesting achievement to date in DNA computing is the programmable and autonomous in vivo computer using E. coli by Nakagawa et al. [2005]. They implemented finite state automata based on their previous four base codon framework by employing the protein synthesis mechanism of E. coli. The best student paper award went to this paper. This is the first in vivo implementation of a Turing machine paving the way to doctor-in-a-cell/programmable DNA computer.

2.3 Synthetic biology

Synthetic biology is a nascent field that is leveraging natural structures as ways of building things on the molecular scale. This is different from molecular computing (or DNA computing) which is trying to solve NP-complete mathematical problems with molecules, or compute with DNA. The first synthetic Biology 1.0 conference was held at MIT on June 10–12, 2004. The purpose of Synthetic Biology 1.0 was to ignite the formation of a tight-knit, cooperative community of researchers that can tackle the scope and complexity of synthetic biology's challenges and consider potential misguided applications of future biological technology. 'Synthetic biology' is the blanket term for a multidisciplinary attempt to identify a class of standard operational components that can be assembled into functioning molecular machines. Central to that effort is the ability to isolate discrete biomolecular mechanisms and define standard interfaces for them so that they can be assembled in much the same way as electronic circuits. This confluence of computer science and biology is so remarkable that this new movement rises to the level of moon shot initiatives: to reverse-engineer life itself. Nanotechnologists are finding that naturally occurring biological functions can be redirected to tasks such as building molecular circuits.

Pharmaceutical companies are finding that naturally occurring metabolic pathways in bacteria that produce useful drugs, are not as efficient as re-engineered pathways. The approach taken by synthetic biology is that engineered organisms represent the quickest route for a true nanotechnology that could manufacture materials and systems on a molecular scale. There is a new research promise by DNA computing and synthetic biology in various application areas. Traditional genetic engineering has focused at adding or deleting genes that confer specific traits. Now, synthetic biologists are trying to assemble an entire genome from scratch and have predictable interactions with existing and new chemically synthesized genomes. They use only the genes necessary for an organism's survival and those that are of functional value for the specific task. The complexity of such engineering is quite high. Applications of synthetic biology span a vast variety of application domains, such as: energy production (photosynthesis, hydrogen production), biochemical synthesis, e.g. production of materials such as natural polymers, engineering materials, bioactive substances, communication systems, sensing processes, e.g. biological sentinels, and intelligent therapeutic agents.

The idea introduced Weiss et al. at MIT was to assemble genes into networks designed to direct cells to perform almost any task programmers can conceive [Weiss et al. 2003]. To achieve the engineering goal of designing complex circuits, they started with a genetic component library and a biocircuit design methodology for assembling these components into compound circuits. The main challenge in biocircuit design lies in selecting well-matched genetic components that, when coupled, reliably produce the desired behavior. In this process they apply rational design and directed evolution to optimize genetic circuit behavior. In the rational design approach, they used simulation tools to guide circuit design, to select the appropriate components, and to genetically modify existing components until the desired behavior is achieved. Directed evolution is the process that directs cells to mutate their own DNA until they find gene network configurations that exhibit the desired system characteristics (analogous to selective breeding of animals).

Here, instead of electrical signals representing streams of binary ones and zeros, the chemical concentrations of specific DNA-binding proteins and inducer molecules act as the input and output signals of the genetic logic gates. Within the cell, these molecules interact with other proteins, bind specific DNA sites, and ultimately regulate the expression of other proteins. An advantage offered by synthetic biology is the ability to study cellular regulation and behavior using novel regulatory networks. The hope is that the future applications of synthetic systems will also extend to the fields of medicine and biotechnology [McDaniel et al., 2005]. There are engineered E. coli populations programmed to exhibit spatial patterns with various shapes (e.g. polka dot pattern). This type of programmed behaviour can possibly extend to physical structures (e.g. tissue) useful in making new biomaterials and tissue engineering.

2.4 Conclusion and future directions

DNA computing has made huge strides since Adleman's experiment in 1994. The field has made quite a lot of progress in theoretical aspects, experimental computing with biomolecules, as well as practical aspects in dealing with nano-sized robots and microfluidics. With the current advancements, there is a great promise that DNA computing can have impact on diagnostic and therapeutic methods in the future. With all the ongoing work there are more research questions that answers. However, future directions can be along the following questions:

- What are the potential experimental designs that can lead to discovery of gene regulation mechanisms? How can we use the various Web resources and available computational genomics tools to hypothesize different experimental designs in order to discover essential computable elements for molecular computing outside and inside the cell?

- Does the environment within the cell provide a mechanism that would enable computations that are (1) predictable, (2) safe, and (3) controllable? When computations are performed in vitro, we have a highly controlled environment. When computations are done in vivo, the input can be any of the products of the signaling pathways of the cell. All these byproducts can interfere with the computable procedure that we want the cell to carry out, either in a positive or in a negative way.

- How can we build up complete understanding of the elements that would perform molecular computations in a failsafe mode? What are all the mechanisms by which one can stop a DNA computing process at desirable points in time?

- How can we extend any of the current imaging modalities for reading the results? Various types of fluorescent protein (green, red, yellow) are used today to show that a certain reaction has happened. In this manner, gene activity can be visualized and 'seen' under ordinary light microscope. However, this approach is not practical for many tissues in the human body. Is there an equivalent of fluorescent protein that 'lights up' under a different imaging modality?

- If certain genes can be expressed only under specific conditions (presence/absence of repressor/promoter binding proteins), then, can we combine this knowledge with one of the diagnostic imaging modalities in order to get more specific readout from specific tissue cells which are in the 'middle' of their activities?

- Can we use any of the imaging modalities to monitor the progress of the diagnostic and therapeutic computations?

- Can any of the molecular imaging modalities be used for controlling the biomolecular computations in-vivo?

In the future, many of these research questions will open up a plethora of new engaging topics. Both DNA computing and synthetic biology are nascent areas and have much to answer and evolve from the current systems to true understanding of real dynamic systems.

References

Adleman, L. [1994]. Molecular computation of solutions to combinatorial problems. *Science*, 266:1021–1024.

Benenson, Y., T. Paz-Elitzur, R. Adar, E. Keinan, Z. Livneh, and E. Shapiro [2001]. Programmable computing machine made of biomolecules. *Nature*, 414:430–434.

Benenson, Y., R. Adar, T. Paz-Elizur, Z. Livneh, and E. Shapiro [2003]. DNA molecule provides a computing machine with both data and fuel. *Proc. Natl. Acad. Sci. USA*, 100(5):2191–2196.

Benenson, Y., B. Gil, U. Ben-Dor, R. Adar, and E. Shapiro [2004]. An autonomous molecular computer for logical control of gene expression. *Nature*, 429:423–429.

Brauer, W., H. Ehrig, J. Karhumäki, and A.K. Salomaa (eds.) [2002]. *Formal and Natural Computing, Essays Dedicated to Grzegorz Rozenberg, Lecture Notes in Computer Science*, 2300.

Brown, T.A. [2002]. *Genomes*. 2nd edition, John Wiley & Sons.

Ekins, R., and F.W. Chu [1999]. Microarrays: their origins and applications. *Trends in Biotechnology*, 17:217–218.

Garey, M.R., and D.S. Johnson [1979]. *Computers and Intractability: A Guide to the Theory of NP-Completeness*. W.H. Freeman and Company.

Head, T., Gh. Paun, and D. Pixton [1997]. Language theory and molecular genetics: Generative mechanisms suggested by DNA recombination. *Handbook of Formal Languages*, G. Rozenberg and A. Salomaa (eds.), Vol. 2, Chapter 7. Springer-Verlag, pages 295–360.

Head, T., M. Yamamura, and S. Gal [1999]. Aqueous computing: Writing on molecules. *Proc. Congress on Evolutionary Computation*, IEEE Service Center, Piscataway, NJ, pages 1006–1010.

Head, T., G. Rozenberg, R. Bladergroen, C.K.D. Breek, P.H.M. Lomerese, and H. Spaink [2000]. Computing with DNA by operating on plasmids. *Bio Systems*, 57:87–93.

Head, T., X. Chen, M.J. Nichols, M. Yamamura, and S. Gal [2002a]. Aqueous solutions of algorithmic problems: Emphasizing knights on a 3X3. *DNA Computing – 7th International Workshop on DNA-Based Computers, Lecture Notes in Computer Science*, 2340:191–202.

Head, T., X. Chen, M. Yamamura, and S. Gal [2002b]. Aqueous computing: A survey with an invitation to participate. *J. Computer Science and Technology*, 17:672–681.

Lipton, R.J. [1995]. DNA solution of computational problems. *Science*, 268:542–545.

McDaniel. R., and R. Weiss [2005]. Advances in synthetic biology: On the path from prototypes to applications. *Current Opinion in Biotechnology*, 16(4):476–483.

Nakagawa, H., K. Sakamoto, and Y. Sakakibara [2005]. Development of an in vivo computer based on Escherichia coli. *11th International Meeting on DNA Computing*, London, Ontario, Canada.

Paŭn, G. [2000]. Computing with membranes. *Journal of Computer and System Sciences*, 61(1):108–143.

Regev, A., and E. Shapiro [2002]. Cellular abstractions: Cells as computation. *Nature*, 419:343.

Sa-Ardyen, P., N. Jonoska, and N.C. Seeman [2003]. Self-assembling DNA graphs. *DNA-Based Computers VIII, Lecture Notes in Computer Science*, 2568:1–9.

Watson, J.D., and A. Berry [2003]. *DNA: The Secret of Life*. Random House.

Weiss, R., S. Basu, S. Hooshangi, S., A. Kalmbach, D. Karig, R. Mehreja, and I. Netravali [2003]. Genetic circuit building blocks for cellular computation, communications, and signal processing. *Natural Computing*, 2(1):47–84.

Chapter 3

BIO-INSPIRED DATA MANAGEMENT

Martin L. Kersten and Arno P.J.M. Siebes

Abstract   The pervasive penetration of database technology may suggest that we have reached the end of the database research era. The contrary is true. Emerging technology, in hardware, software, and connectivity, brings a wealth of opportunities to push technology to a new level of maturity. Furthermore, ground breaking results are obtained in Quantum- and DNA-computing using nature as inspiration for its computational models. This chapter provides a vision on a new brand of database architectures, i.e. an Organic Database System where a large collection of connected, autonomous data cells implement a semantic meaningful store/recall information system. It explores the analogy of a biological complex to charter the contours of this research vision. A concrete computational model is defined and illustrated by examples as a step into this direction. [1]

Keywords   Embedded databases, DBMS, evolving software systems, ubiquitous data management.

3.1   Introduction

The innovation thrust of current database research comes from attempts to deploy its technology in non-trivial application areas. Enhancements proposed to the core functionality are primarily triggered by the specific needs encountered, e.g. multi-media, data mining, and sensor systems.

This chapter presents a vision on an alternative track for database architecture research. One grown over a decade as a temporary escape from our contract-research work, which dictated most of our agenda. As such, the vision presented is by no means complete, nor explored in all its depths, let alone be implemented in a eye-catching demonstrator. It is, however, a significant step forward from the first version, presented at the RIDE 97 workshop on research issues in databases and published by Kersten [1997], master student projects

---

[1]The work is carried out in Bsik/BRICKS project *Databases for personalised ubiquitous intelligent devices.*

Wim F.J. Verhaegh et al. (Eds.), Intelligent Algorithms in Ambient and Biomedical Computing, 37-56.
© 2006 *Springer. Printed in the Netherlands.*

to construct a prototype since, presentations given at VLDB [Kersten et al., 2003] and ICDE [Kersten, 2003], and interaction with people from Philips Nat. Lab. to isolate tangible intermediate versions as described by Fontijn & Boncz [2004a; 2004b]. All are minor steps on a long road ahead.

The premises is that database technology has contributed significantly to society over several decades, *but* it is also time to challenge its key assumptions. A few issues considered to be dogmatic and a bottleneck for progress are:

- A database is larger then the main memory of the computer on which the DBMS runs and a great deal of effort should be devoted to efficient management of crossing the chasms between disk and memory.

- A DBMS should adhere to a standard datamodel, whether it be relational, an object-relational, xml-based, and leave functional and deductive models as a playground for theoretical research.

- A DBMS provides quick response to any (unrealistically complex) query, optimizing resource usage wherever possible without concern on the effect of concurrent users.

- A DBMS should support concurrent access by multiple users at the smallest granularity level (record) and reconcile the different perspectives on the database contents transparently.

- A DBMS provides a transaction models based on the ACID (atomicity, consistency, isolation, and durability) principles, or a semantically enriched version, regardless its primary domain of application.

This list is by no means complete, but merely indicates the delineation of research activities in the database community. A prototype DBMS ignoring these points is not taken seriously in the research realm. Albeit, in each of the assumptions reality is threatening, e.g. the web challenges the rigid data models, the transaction models, and replication management.

The research agenda derived at the Asilomar workshop[2] rightfully acknowledge that in the near future all but the largest relational tables will be memory resident, calling for a complete overhaul of the current data structures, algorithms and system architecture. This observation is re-iterated in the context of personalized and organic databases at the Lowell workshop on the database research agenda in 2003.[3]

A grand challenge for the database community is: *The information utility*: make it easy for everyone to store, organize, access and analyse the majority of

---

[2]http://www.acm.org/sigmod/record/issues/9812/asilomar.html
[3]http://research.microsoft.com/ gray/lowell

human information online. The key question then boils down to "Is the current architectural conception of database technology a sufficient basis to meet this challenge?" Our preliminary answer to this question is negative. The threats to the database dogmas are evidence of their failure. Instead we need a more unorthodox approach to break our historical bonds. A broader perspective on computer science research may be of help here.

Recent major fundamental advances in computer science have found their inspiration in nature. DNA computing uses biochemistry to implement massive parallel computation and the engineering of DNA computing device seems tractable [Landweber & Baum, 1996; Rubin & Wood, 1997; Paŭn et al., 1998]. Quantum computing uses the physics of light to design a new computational model. Hurdles to be taken here include non-destructive observation of the result of a computation. Theoretically both quantum- and DNA-computation have been shown to crack hard problems in cryptography.

Theoretical models of computation based in the bio-analogy have received quite some attention in recent years. For example, the research school around *membrane computing* has studied a rich class of variations of the traditional Turing model using a direct analogy of cells in biology. It has been shown that such models can capture inherent massive parallelism and still provide insight [Paŭn, 2001; Paŭn, 2002].

Further back into the history of computing, we find the notion of self-reproducing automata [Von Neumann, 1966]. A study severally hindered by the state-of-the-art in computer architecture, but nevertheless an intriguing concept. A more mundane use of nature as a stimulus for novel programming paradigms have let to such broad fields as evolutionary computing and neural nets. A large community deploys these concepts to realize new kinds of applications, e.g. adaptive, intelligent, and even socially acceptable agents [Weiss, 1999; Hunds & Singh, 1998] .

With these examples in mind, it is worth considering how nature can inspire us in the design of a new database management paradigm. The remainder of this chapter charters the contours of such an Organic Database System, i.e. a large collection of connected, autonomous data cells that implement a semantic meaningful store/recall information system.

In a nutshell, the architecture is centered around the concept of a *data cell*, characterised by three components: an interface, a cell body and a nucleus. The cell interface is a semi-permeable membrane taking two forms; RECEPTORS, where objects in the cell's environment may enter the cell; and EMITTERS, which enable objects in the cell's interior to migrate to the outer world. The cell's body is a memory structure for the tree-structured objects received. It is a persistent store organised by object entrance/creation time. The nucleus consists of genetic code strings interpreted under triggering events, i.e. objects stored in the cell's body.

As such, a data cell models a physical container capable of managing a small database, limited in capacity, without a fixed foothold, and equipped with behavioral knowledge, described by RECALL-FORGET-KEEP genes, which replace the role of procedural methods. The cells live in a resource rich environment, which enable them to migrate or to clone as soon as the physical boundaries are met.

The Internet is assumed as the underlying communication network, where cells are addressable with a URN and live in a hierarchically organized namespace. To survive in this dynamic world, a cell may decide to seed a copy of its state. It is resurrected upon request to recover from a physical disaster.

Querying the organic database is a non-atomic process. A query is mapped to a membrane modification that allows answers to pass to the communication interchange, where they can be picked up by the issuer. Since cells may be temporary dormant or inaccessible, the issuer should be prepared to wait for all cells to respond or be satisfied with partial answers. The net effect is that querying becomes probabilistic, much like searching the Web. One never knows for sure if all information has been obtained.

The remainder of this chapter is organized as follows. Section 3.2 introduces the data cell, its internal architecture, and a notation for reasoning. Section 3.3 introduces the modalities of communication. Section 3.4 addresses the life cycle of an individual cell, its sensors, its cloning, and seeds to survive disasters. Section 3.5 presents a small application with mockup traces to illustrate the projected behavior. Section 3.6 boils down to a summary, and raises some fundamental research issues to be considered next.

3.2 Data cell overview

In this section we introduce the notion of a data cell, the basic building block of an Organic Database System. We take an inward exploration, starting with the cell's membrane, followed by its memory structure, and to finish with the behavior described by its nucleus.

3.2.1 The data cell and its membrane

A data cell is a physically bounded resource to store and recall persistent information. Physically boundedness is interpreted as anything from a simple smart-card in a mobile phone, up to large multi-processor SMP machine. In our search for a new architecture we favor the former, because it challenges us to go for minimal and razor-blade components. An SMP context merely leads to challenging engineering issues related to scale.

The data cell and its membrane are defined as follows:

Definition 3.1. *A data cell type $D = \langle M, N, T \rangle$ named D, consists of a set of receptor- and emitter- membranes $M = \{M_i\}$, a set of genetic code strings in the nucleus $N = \{N_j\}$, and $T = \{T_k\}$ a collection of data types.*

Definition 3.2. *The instances of a data cell type D are denoted by the type Cid, i.e. globally unique, life-time tags.*

Definition 3.3. *Object structures are defined by a context-free grammar $G = \langle N, V, P, D \rangle$ with non-terminals N, terminal variables V, their productions P.*

Definition 3.4. *A* RECEPTOR *membrane is defined by the structure: Name* RECEPTOR *W* WHERE *$P(B)$ where Name is an optional membrane tag, W a derivation tree for a term in $L(G)$ starting at D, and B the binding table for variables in W.*

Definition 3.5. *An* EMITTER *membrane is defined by the structure: Name* EMITTER *W* WHERE *$P(B)$ where Name is an optional membrane tag, W a derivation tree for a term in $L(G)$, and B the binding table for variables in W.*

The membrane definitions are based on our conjuncture that most objects for information exchange can be described formally, and exhaustively, with a context-free grammar.[4] This grammar provides a structured name space to access and to reason about the components. The degrees of freedom lie in the production rules, i.e. the type constructors, and the lexical tokens, i.e. the data types. This relationship between structure and values is factored out in the binding table B. A parse tree for an object then contains bound variables as leafs.

The last component of a membrane is a predicate over the object components, represented by the (dynamically typed) variables V. The predicate is safe when all its variables are bound. Otherwise the predicate fails. The predicate language relies on operators defined for the data types T. For all practical purposes considered in this chapter, we assume T to include the standard set of basic types available in the programming environment. Furthermore, T includes N, the grammar non-terminals and Cid, the cell identities.

Definition 3.6. *A membrane M_i for data cell D accepts (emits) objects from (to) its environment if it satisfies both the structure implied by the grammar G, the values T, and the predicate $P(B)$ holds.*

The parser derived from the RECEPTOR membrane grammar looks for external object structures tagged by the cell type name, i.e. the grammar's start

---

[4]Notational convention: identifiers starting with a lower-case character act as cell names, their components, and as object structure tags. Those starting with an upper-case character are used as variables.

symbol. The object structures are taken from the environment and stored in the cell's body. The EMITTER looks for qualifying object structures in the cell body and emit them to the environment.

Receptors and emitters map to autonomous threads. They may inspect objects concurrently in their environment, but only one may finalise the transaction (passage of a cell's membrane). Objects rejected by all RECEPTORS are left to the responsibility of the cell's environment. Objects failing the EMITTER membrane remain in the cell body.

A full fledged implementation of the data cell may exploit standard notational conventions, such as XML structures. The sole requirement for the object description language is that a message can be mapped onto the grammar and binding table of the membrane, i.e. the parse is unambiguous irrespective the lexical convention. A concrete syntax for cell identities *Cid* could be an URN.

Example. As a running example we consider a toy database of colored marbles. The data cell defined below looks in its environment for red marbles. All encountered are catched and stored in the cell's body as marble("red"). The first emitter may pick up the marbles again and thrown them back into the environment. The second emitter looks for marbles changed (by magic) to those with a primary and secondary color. They are sent to cells interested in multi-coloured marbles. The last part illustrates initialization with a few marbles.

```
CELL marble;
    RECEPTOR K WHERE K = "red";
    EMITTER marble(K) ;
    EMITTER marble(Primary,Secondary)
        WHERE Secondary = "orange";
    marble("orange");
    marble("green");
    marble("yellow","orange");
END marble;
```

3.2.2   Structure unification terms

The organic database system exploits the equivalent of Watson-Crick complementary feature provided 'for free' by the nature. This feature stipulates the programming power in DNA-strands, where bases opposite each other are complementary. During the construction of the double helix strands, genes are *unified* with the nucleotides to find matches.

The equivalent notion exploited here is to unify object terms in predicates with the parse trees of the object structures received. Unification is supported by the operator ':', i.e. the term $X : Y$ succeeds if the operands can be unified. The terms considered are classified into (un)ordered- and (un)tagged-object terms.

- *ordered* $(X_0, \cdots, X_j)$, which exhausts a single object component list structure.

- *unordered* $\{X_i, \cdots, X_j\}$, where all elements mentioned denote path expressions binding different components in a single object structure.

- *prefixed*, $\mathtt{cntxt}(X_0, \cdots, X_j)$ and $\mathtt{cntxt}\{X_i, \cdots, X_j\}$ are called prefixed object terms, all components mentioned belong to a single object structure reachable through the path expression $\mathtt{cntxt}$.

Example. Consider the term $\mathtt{marble}(\mathtt{primary}(P), \mathtt{secondary}(S))$, which unifies with any variable Z. Subsequently Z can also be unified with $\mathtt{marble}(X,Y)$ and $M(X,Y)$ where M binds with $\mathtt{marble}$. The unordered unification $Z : \{\mathtt{marble}.\mathtt{primary}(X)\}$ and $Z : \{\mathtt{primary}(X)\}$ hold, because the paths exist in the structure referenced by Z. The prefixed terms $Z : marble(primary, secondary)$ and $Z : marble\{primary, secondary\}$ hold, while $Z : \{\mathtt{marble}.P, \mathtt{marble}.Z, \mathtt{marble}.\mathtt{secondary}\}$ fails, because the three arguments can not be bound to different object components. Finally, we permit unification with a type name to denote membership, e.g. "red": $\mathtt{string}$ also holds. With this notational convention, predicates over the hierarchical object base becomes condense and easy to interpret.

3.2.3 The cell's body

The cell's body is a persistent memory structure, where objects passing the membrane are kept. Its organization affects the subsequent computational model, both internally and externally. One extreme is to consider memory as a set of tree structured objects, freely floating within the cell's body. The effect is that all sequential behavior calls for 'sorting', or the cell behavior becomes purely probabilistic. Given that nature also processes cell DNA strands in sequential fashion, we choose for a time-organized sequence.

The memory sequence comes with two maintenance operations: KEEP and FORGET. A KEEP V operation adds the object V to the end of this sequence in an atomary step, while FORGET V 'zaps' the (bound) object from the memory sequence, leaving no traces behind. Information in the memory sequence can be located with a RECALL operation followed by an ordered list of object terms. Its semantics is to traverses the memory sequence in reverse direction, i.e. it unifies terms to the latest objects entered. Moreover, no two terms in the recall list unify to the same object (component).

Example. The table below illustrates a memory sequence. The right part illustrates the successive term unifications that result from the RECALL marble(X), marble(Y,Z). Note, the two red marbles are distinct objects, although their structure and value are identical.

marble	Memory sequence	$X, (Y,Z)$
0	marble("red")	0, 1
1	marble("red","orange")	2, 1
2	marble("yellow")	3, 1
3	marble("red")	

3.2.4 The cell's nucleus

The nucleus of a cell contains a set of chromosomes, gene strands, that define its behavior. A gene strand consists of RECALL – FORGET-KEEP statements. Each gene inspects and changes the memory sequence under all possible variable bindings.

Definition 3.7. *A gene G is described by the structure:*

G RECALL L WHERE $P(L)$
    FORGET Fl WHERE $P(L \cup Fl)$
    KEEP Sl WHERE $P(L \cup Fl \cup Sl)$;

where G is an optional gene tag, L, Fl, and Sl term lists to locate objects, and $P(\alpha)$ a clause over the variables in the term lists indicated.

The interpretation of a gene is that each completed binding leads to an atomary action against the memory sequence. Some objects bound are prepared for removal, and new object structures are prepared to inclusion. These changes take immediate effect for each term binding encountered.

Definition 3.8. *A chromosome is a structure $G = $ RECALL $Cl\{G_0; \cdots; G_k\}$ where Cl is a term list and G_i is either a gene or a chromosome sequence.*

The scope of variables introduced in the chromosome RECALL list is defined by the corresponding gene sequence. For simplicity we assume no redefinition of variables.

Definition 3.9. *A chromosome G is independently activated for each bindings of its memory recall list Cl.*

The chromosome describes a hierarchical sequence of behavioral actions. The RECALL is an implicit loop through memory and the qualified update statements are guarded commands. Conceptually, each time an object appears in the cell's body it will arbitrarily activate a chromosome interested in the object. All valid bindings are explored before the chromosome ceases activity. Since binding works its way back into object history, and KEEPs are always at the head of the sequence, this process will eventually terminate. A limited set of additional functions controls the life-cycle of a cell. This includes EN-ABLE/DISABLE of cell components, WAKEUP peer cells, going to HIBERNATE, RUNning linked in routines, and CLONEing itself.

Example. The intended behavior of the cell nucleus is illustrated using our marble toy database. Each time the membrane stores an object marble("red"), the nucleus is inspected for a qualifying chromosome, i.e one whose first element in the RECALL term list unifies with the object. Once detected, a process thread interprets the chromosome, consuming the object and creating a new object for emission later on.[5]

```
CELL marble;
    RECEPTOR K WHERE K = "red";
    EMITTER marble(K) WHERE K ≠ "red" ;
NUCLEUS
    RECALL marble(Msg)
        FORGET IT
        KEEP marble("green");
    END marble;
```

The probabilistic behavior of chromosome selection and their inter-relationships are illustrated below. In the next fragment one chromosome arbitrarily transforms a red marble into either green or orange. With each color change we also remove any trace of the red marble received. Note that the probabilistic behavior envisioned may also lead to emission of red marbles, before they are inspected by any of the chromosomes. They may then end-up in cells capable to react.

```
CELL marble;
    RECEPTOR K WHERE K = "red";
    EMITTER marble(K);
NUCLEUS
    RECALL marble("red")
        FORGET IT
        KEEP marble("green");
    RECALL marble("red")
        FORGET IT
```

---

[5]The keyword IT stands for all objects bound in the memory recall list.

```
        KEEP marble("orange");
    END marble;
```

The second fragment replicates a red marble into both red and green by being bound with each chromosome once. Moreover, it accumulates the red marbles in its memory, because they are never forgotten.

```
    CELL marble;
        RECEPTOR K WHERE K = "red";
        EMITTER marble(K) WHERE K ≠ "red";
    NUCLEUS
        RECALL marble("red")
            KEEP marble("green");
        RECALL marble("red")
            KEEP marble("orange");
    END marble;
```

To get rid of the red marbles too, we have to encode state information in the data cell. A possible solution using tagged intermediate results is shown below. The tag is attached to each object indicated by the KEEP to indicate the chromosome responsible for its creation.

```
    CELL marble;
        RECEPTOR K WHERE K = "red";
        EMITTER marble(K);
    NUCLEUS
        RECALL marble("red")
            KEEP m1("green");
        RECALL marble("red")
            KEEP m2("orange");
        RECALL m1(A), m2(B), marble("red")
            FORGET IT;
            KEEP marble(A), marble(B);
    END marble;
```

3.3 The communication infrastructure

Data cells in isolation are of limited use. A communication infrastructure gives the Organic database system access to its sensory components and circumvents the physical boundaries imposed by hardware. This section describes the analogue of biological communication schemes in the context of our Organic Database System.

3.3.1 Artery system

Nature has found an efficient solution for communication in the form of arteries, where the transport medium need not worry about the message content. It merely passes objects around and leaves it to the autonomous cells attached to the artery to filter out objects of interest. In practice, the artery system can be seen as an unfolded hierarchical communication scheme. The top contains a 'pump' to move information down the hierarchy to the smallest components, whereafter the flow is reversed and aggregated upward for the next cycle.

We use this hierarchical scheme for the organic database as the backbone communication infrastructure. It consists of artery segments, which are effectively containers for a limited number of message objects floating through the system. Furthermore, each segment shares a membrane with (a limited number of) data cells, looking for messages of interest passing by. These cells get access to the messages in a probabilistic manner. Furthermore, the segment may be linked with sensors to the outside world, giving it eyes and ears to communicate with the user.

The artery system metaphor provides a natural communication scheme, but also possesses some dangers. First, the artery system may become polluted with messages of no interest to any cell. Second, the probabilistic flow does not guarantee that a message will reach a destination cell in acceptable time. Although this reflects real-life on the Internet, it may be unacceptable in a confined application environment. The solution to consider then is to introduce many cells on the artery system, such that the probability steeply increases. Alternatively, a multi-level artery system can be designed through which messages quickly reach their intended destination. For example, nature often uses a nerve system to sent simple information around quickly. This includes intermediate control centers to handle local issues and shortcuts.

Example. Pollution of an artery segment with unwanted messages can be controlled by tagging them with an age component. A single cell removes them as waste when they get too old. The artery cell below is charged with this functionality.

```
CELL message
    RECEPTOR (M, age(C));
    EMITTER message(X,Y);
NUCLEUS
    RECALL (M, age(C))
        WHERE C <= 1000
        FORGET IT
        KEEP message(M, age(N)) WHERE N is C+1;
    RECALL (M, age(C))
```

```
        WHERE C > 1000
        FORGET IT;
    END message;
```

3.3.2   Neurons

Every data cell carries an unique identifier. Knowing this identifier permits direct addressing of a target cell using a dedicated transport scheme. It merely has to be constructed. Nature's realisation for this can be found in neurons. It is a one-to-one communication channel, orthogonal to the artery system. They have to be 'learned' and they involve much less overhead in terms of communication and analysis.

A neuron can only fire when the target cell is alive, leaving an object at the target cell's membrane for direct inclusion. In this sense, neurons communication can be seen as a kind of synchronous communication. This makes them part of the processing thread(s) of the nucleus, where they block progress until the message is delivered.

Example. The fragment below illustrates how a marble cell handles a query of a client. The client issues the request getAll(SELF) to pass its identity to the marble cell. It expects copies of the objects to arrive at its membrane in return. The marble catches the request with the first chromosome, and activates the neuron stream of answers. It also illustrates a complex chromosome with sequential behavior.

```
CELL marble
    RECEPTOR getAll(Mid);
NUCLEUS
        RECALL getAll(Msg) {
        FORGET IT;
        RECALL client(Mid), Object
            NEURON Mid(Object);
        }
    END marble;
```

3.3.3   Membrane sharing

The third communication scheme between cells is based on sharing a MEMBRANE definition and being alive in the same environment. In nature it occurs directly after a cell split, before both have evolved by taking their autonomous role in the environment.

Temporarilly sharing the object collection with peers provide a powerful construct to built data-distributed applications. The objects satisfying the membrane freely move between the cells.[6] It is based on common definitions and mutual trust. As time progress, the cell may shut down this feature and regulate all access through its membrane.

Example. The fragment below illustrates two instances of a marble cells sharing the term membrane. This makes the marbles stored directly accessible to the other cell. When cell `marble#1` finds an orange marble, it will remove any green marble in cell `marble#0` and `marble#1`. Conversely, `marble#0` looks up any red marble in both bodies and transform it to green. As such, the cell are functionally specialized.

```
CELL marble#0;
    RECEPTOR K WHERE K ≠ "orange";
    MEMBRANE marble(X);
    marble("green"); marble("red");
NUCLEUS
    RECALL marble("red") KEEP marble("green");
END marble;

CELL marble#1;
    RECEPTOR K WHERE K = "orange";
    MEMBRANE marble(X);
NUCLEUS
    RECALL marble("orange"), marble(Z)
        WHERE K = "green"
        FORGET marble(K);
END marble;
```

3.4 The life cycle

The textual definitions given for the data cells are their 'seed' state. They can be resurrected from this state by an external entity, which is typically a organic database system kernel implementation or a cell using a WAKEUP call. Once active, it can CLONE itself, and return to HIBERNATE state as part of its nucleus behavior. These issues are described below.

3.4.1 Cloning a data cell

A data cell may CLONE itself to form a data cell tissue, a collection of cells with identical behavior. This process is triggered by a nucleus action and consists of two phases. In the first phase, all activity is stopped as if the cell M

---

[6]Subject to a proper semantics of the memory sequence.

goes into hibernation state. Then a textual 'seed' copy *C* is created, which contains half of the memory sequence, a MEMBRANE definition, and the object parent(*M*). This phase ends with forgetting the objects moved to the clone and placing the object child(*C*) and the MEMBRANE definition in the body of *M*. Following, in phase two, cell *M* becomes receptive to external requests again when the object child is forgotten. Cell *C* follows the normal awakening sequence, where it will react to the *parent* object before it accepts any further request.

A major difference between cloning and the creation of a cell is that a clone is connected to its heritage via membrane sharing. This sharing allows the cell tissue to act as one cell as far as data storing and retrieving is concerned. To an outsider it is immaterial which cell in a tissue acts upon his request as long as it is acted upon.

At the moment a cell is cloned -or created- the artery system has to be adjusted as well. The identity of the new cell should be announced to this communication channel.

3.4.2 Hibernation and wakeup

Hibernation is a multi-step procedure. First, the cell's receptors are deactivated, the chromosomes are instructed to stop as soon as possible in a recoverable state, the emitters finish sending all qualifying objects. They also stop when the environment does not accept the objects emitted anymore. Finally, the cell status is saved to disk.

Example.   The marble cells goes into hibernation after receiving a "blue" marble.

```
CELL marble
      RECEPTOR marble("blue");
NUCLEUS
      RECALL marble("blue")
            FORGET IT
            HIBERNATE;
END marble;
```

A dormant cell can be awakened by any cell using a WAKEUP call passing the cell's identity. This typically takes place in an artery segment, triggered by the cell name in a message header. If an artery segment runs out of resources, it may decide in a probabilistic manner what cell to ask for hibernation or to migrate a cell under its control to another segment.

An awakened cell starts with RECEPTOR and EMITTER elements in passive mode first. They should be activated by a chromosome. The triggering event is existence of the object 'resurrected' in the cell's body. This unifies with the corresponding initialization chromosome. The default -shown below- looks up all (still passive) membrane structures and activates each.

```
NUCLEUS RECALL resurrected {
    RECALL E:EMITTER {ACTIVATE E;}
    RECALL R:RECEPTOR {ACTIVATE R;}
    FORGET resurrected;
    }
```

3.5 Application challenges

In this section we illustrate the Organic Database System using a distributed phone book, one whose data cells may indeed live in our digital organizer, our PC, and mobile phone concurrently. As such it is close to what one would expect from a store/recall information system. We start with a sensory interface, the eyes and ears of the system. Following we give a concrete definition of the phone book, one that will not (!) immediately work, but which highlights the issues to be dealt with. Finally, we indicate routes to implement an organic database system.

3.5.1 Sensors

The data cells introduced so far were blind actors. They communicate amongst one-another using the biological inspired schemes. However, at least one cell should provide a bridge to the real-world, where we observe and control the behavior of an organic database.

This calls for the equivalent of sensors, the eyes and ears of the system. Since the functionality of sensors are tightly coupled with the environment where they operate, it has to rely on linked-in libraries. The minimal set to be considered for a first implementation are a direct link to the stdio library and XML for web-based interaction.

A sensor cell has an event loop triggered both by the external interface and the messages from the cells. The latter are screened for type correctness. Subsequently, they may be picked up by a chromosome to be executed. This essentially makes a sensor cell a wrapper around a user-supplied interface library.

Example. The ascii sensor below assumes an io-library, which interacts with the user through an text-based interface.

```
CELL ascii USE stdio;
    RECEPTOR print(Msg:string);
    RECEPTOR printf(Format:string,Msg:string);
NUCLEUS
    RECALL Action RUN stdio.Action;
END ascii;
```

3.5.2 A phone book

```
CELL phone;
    RECEPTOR person(name(N),tel(I))
        WHERE N:string AND I:integer;
    RECEPTOR lookup(name(N))
        WHERE N:string;
    RECEPTOR delete(N));
    EMITTER answer(Msg);
NUCLEUS
    RECALL lookup(N) {
        RECALL person(name(Nme),T)
            WHERE Nme == N {
                FORGET lookup(N);
                KEEP answer(N,T));}
    }
    RECALL delete(N) {
        RECALL person(name(Nme),T)
            WHERE Nme == N FORGET IT;
        FORGET delete(N);
    }
    RECALL count(P:person) > 10 CLONE;
END phone;
```

Figure 3.1. The phone book.

Figure 3.1 illustrates the starting position of the phone cell. This definition is no more complex than a class definition in an object-oriented paradigm, or an SQL-3 table definition. Each time an object passes the membrane it is added to the persistent store, as is to be expected from a database system. Using the textual interface we might add some persons.

```
> phone.person(name("Smith"),tel(808717));
> phone.person(name("Jones"),tel(828503));
> phone.person("Jones",tel(808717));
```

The structure for Jones does not match the receptor, leaving it in the artery. A waste recovery cell (see Section 3.3.1) can be used to get rid of these messages. Alternatively, we could accept a broader class of person structures and emit an error message where appropriate. The necessary additions become:

```
RECEPTOR person;
EMITTER error(X);
RECALL P:person
    WHERE NOT(P:person(name(N),tel(T)))
        KEEP error(P);
```

Lookup of Smith's telephone number is straight forward requested by:

```
> phone.lookup("Smith");
```

However, the user does not know when the answer will arrive, because, due to cloning, the actual cell containing this information may be temporarily inaccessible. This is generally the case with interrogation of an organic structure. Getting an answer to the question "Is Smith not part of the phone book", can only be answered from the contextual knowledge that all cells are active and have dealt with this question.

The cell also contains a chromosome to clone itself when it accumulates too many telephone entries. Its controlling predicate is an aggregate over the memory sequence. However, such constructs require extreme care, because membrane-based sharing implies that we always count **all** elements in the data tissue. This leads to a cascade of clones after receiving 10 persons. A way out of this dilemma is to consider non-sharing clones or quantified bindings, i.e.

```
RECALL count(ALL LOCAL P:person) > 10 CLONE;
```

Being able to inject a receptor, emitter, or chromosome into a cell modifies its behavior. This is particularly useful if we want to extract information, i.e. query its content. A 'virus' cell penetration of the phone book might be a road to explore. It may take the following form:

```
CELL virus;
    EMITTER steal(X);
NUCLEUS
    RECALL resurrected
        ACTIVATE steal;
    RECALL Y KEEP steal(Y);
END virus;
```

Once we are able to bind this virus with an object passing the phone book membrane (possibly as a Trojan horse), it awakes from its hibernated state. Both symbiotic and harmful viruses are easy to design.

3.5.3 Implementation strategies

The organic database system outlined does not require a start from scratch. Many ingredients for its realization are readily available.

The history for the cell architecture can be traced back as far as the Von Neumann cellular automata, a dream where computers conquer free space to grow and solve intricate problems [Von Neumann, 1966]. The abundance of literature on cellular automata in the 60s provide further hints to formalize parts of the data cell semantics [Wolfram, 1982]. The work on associative memories [Oskarahan, 1986; Su, 1988] in the 70s can be regarded as preliminary steps to improve the cell's memory.

The declarative model underlying the interrogation of the cell's memory sequence is a natural extension to SQL- and logic-based systems. Modern prototype database engines to be considered are Lore [McHugh et al., 1997] and MonetDB [Boncz et al., 1996a; Boncz et al., 1996b; Boncz et al., 1998]. They provide a lean implementation to start from. Furthermore, Java-beans technology may be a pivot in realisation of cells with a small footprint.

Likewise, the problems posed by cloning find their analogy in distributed belief systems. Recent developments in agent technology, especially the libraries being developed for agent-based applications on the web, may provide the seeds to quickly build a prototype organic database system [Weiss, 1999].

The data sharing that results from the membrane replication, may use techniques from distributed computation models, such as explored in Linda [Gelertner, 1985]. The temporal aspects of the memory sequence can be borrowed from [Snodgrass et al., 1994].

3.6   Conclusion

Much of the research activities in the database area take the underlying datamodel, query language, and transaction features as a fact of life. The consequence is that penetration of database technology into non-administrative application domains is a slow, engineering-rich and tedious process. A quantum leap in technology is required instead.

The vision developed in this chapter provides an innovative computational model, data model, and architecture for database processing. We consider this chapter a success, if the reader has raised questions about the limitations of the organic database approach, thought of refinements, or envisioned a concrete realization. Moreover, the biological metaphor may have to be extended into other fruitful directions or being corrected as a result of our limited knowledge on the biological mechanics.

The research road ahead is thus marked with many fundamental questions calling for in-depth studies. Amongst these, the most pregnant are to describe cloning as reflexive behavior, to use symbiotic behavior, and to better understand the implications for large-scale application development. At an architectural level, embedding data cells in hardware ranging from smart-cards up to super-computers with their wildly differing communication infrastructure stresses the need for strong interface definitions.

Fortunately, the AmbientDB project undertaken at the cross-section of industrial research and curiosity-driven research, may provide the setting to take slow, but consciuous steps into the realisation of our grand vision.

References

Boncz, P., W. Quak, and M. Kersten [1996]. Monet and its geographical extensions. *Proceedings EDBT'96, Lecture Notes in Computer Science*, 1057: 147–166.

Boncz, P., F. Kwakkel, M.L. Kersten [1996]. High performance support for O-O traversal in Monet. *proceedings BNCOD'96*, Edingburgh, UK.

Boncz, P., A.N. Wilschut, and M.L. Kersten [1998]. Flattening an object algebra to provide formance. *Proceedings of the IEEE Intl. Conf. on Data Engineering*, Orlando, FL.

Fontijn, W., and P. Boncz [2004a]. AmbientDB: P2P data management middleware for ambient intelligence. *Workshop on Middleware Support for Pervasive Computing (PerCom)*.

Fontijn, W., and P. Boncz [2004b]. *The Data Management Challenge of Ambient Intelligence*. Philips Report PR-TN-2004/0089.

Gelertner, D. [1985]. Generative communication in Linda. *ACM Trans. Prog. Lang. and Systems*, 7(1): 80–112.

Hunhs, M.N., and M.P. Singh [1998]. *Readings in Agents*. Morgan-Kaufmann, ISBN 1-55860-495-2.

Apers, P.M.G., C.A. van den Berg, J. Flokstra, P.W.P.J. Grefen, M.L. Kersten, and A.N. Wilschut [1992]. PRISMA/DB: A parallel, main-memory relational DBMS. *IEEE KDE, special issue on Main-Memory DBMS*, 4(6): 541–554.

Landweber, L., and E. Baum [1996]. *DNA Based Computers II*. DIMACS Vol 44, ISBN 0-8218-0765-0.

McHugh, J., S. Abiteboul, R. Goldman, D. Quass, and J. Widom [1997]. Lore: A database management system for semistructured data. *SIGMOD Record*, 26(3): 54–66.

Kersten, M.L. [1997]. A cellular database system for the 21th century. *Proceedings ARTDB'97*, Springer-Verlag.

Kersten, M.L. [2003]. Databases for ambient intelligence. *Proceedings ICDE*, Bangalore (India), page 795.

Kersten, M.L., G. Weikum, M. Franklin, D. Keim, and A. Buchmann [2003]. A database striptease; or how to manage your personal data. *Proceedings VLDB 2004*, Berlin (Germany), pages 1043–1044.

Oskarahan, E. [1986]. *Database Machines and Database Management*. Prentice Hall, ISBN 0-13-196031-8.

Paǔn, G., G. Rozenberg, and A. Salomaa [1998]. *DNA Computing. New Computing Paradigms*. Springer-Verlag, ISBN 3-540-64196.

Paǔn, G. [2001]. *Computing with Cells and Atoms* Tayler & Francis (London), ISBN 0-7484-0899-1.

Paǔn, G. [2002]. *Membrane Computing*. Springer-Verlag, ISBN 3-540-42601-4.

Rubin, H., and D. Wood [1997]. *DNA Based Computers III*. DIMACS Vol 48, ISBN 0-8218-0842-7.

Su, S.Y.W. [1988]. *Database Computers, Principles, Architectures & Techniques*. McGraw Hill, ISBN 0071003290.

Snodgrass, R.T., et al. [1994]. *The Tsql2 Temporal Query Language*. Kluwer International Series in Engineering and Computer Science; No. 330.

von Neumann, J. [1966]. The general and logical theory of automata. In *J. von Neumann. "Collected Works"* (ed. A.H. Taub), Vol. 5, page 288; J. von Neumann, "Theory of Self-Reproducing Automata", (ed. A.W. Burks), Univ. of Illinois Press(1966); ed. A.W. Burks, "Essays on Cellular Automata", Univ. of Illinois Press (1970).

Weiss, G. [1999]. Multiagent Systems, A Modern Approach to Distributed Artifical Intelligence. MIT Press, ISBN 0-262-23203-0.

Wolfram, S. [1982]. http://www.wolfram.com/s.wolfram/articles/82-cellular/index.html

Chapter 4

AN INTRODUCTION TO
MACHINE CONSCIOUSNESS

Kees van Zon

Abstract   Neuro-scientific research suggests that consciousness plays an important role in human cognition, and may even be required for human-level intelligence. Machine consciousness is an emerging 'technology of mind' that aims at achieving consciousness in man-made systems. It is in an early stage, and much is yet to be discovered and proven. The hope is that machine consciousness will add to the existing suite of artificial intelligence techniques and enhance the performance and capabilities of autonomous agents. In this chapter, we briefly review the concept of biological consciousness, then introduce machine consciousness, present some practical approaches, address its relevance, and discuss applications.

Keywords   Machine consciousness, artificial consciousness, autonomous agents, robotics.

4.1   Introduction

In an age in which electronics and data networks are to be found practically everywhere, one might wonder if technology is close to reaching a saturation point. There are good reasons however to believe that quite the opposite is the case — that technology is merely getting started. Since technology supports the development of more advanced technology, it will tend to grow exponentially or even faster (e.g. [Moravec, 1998; Kurzweil, 1999]. As a result, the world may find itself in a complexity spiral for as long as technology translates into money.

With electronics and networks invading every aspect of our lives, industries will find a continuing challenge in shielding consumers from needless complexity while providing added value. Consumers expect products that are simple to use and that work in intuitive, natural interaction modalities. Advanced technologies that support modern lifestyles in this manner are commonly believed to require an increasing degree of smartness. Besides artificial

Wim F.J. Verhaegh et al. (Eds.), Intelligent Algorithms in Ambient and Biomedical Computing, 57-70.
© 2006 *Springer. Printed in the Netherlands.*

intelligence (AI) in support of consumers, it will also be required for solving more and more complex problems in other areas, e.g. physics, bio-chemistry, and healthcare.

When searching for new AI techniques, we often tend to look at nature. Well-known examples of bio-inspired techniques are *evolutionary* computing, *genetic* algorithms, and *neural* networks, but to name a few. Many improvements in AI are actually based on progress in the cognitive sciences. An important recent finding of neuroscience, for example, is the understanding that human decision making is not predominantly driven by rational thinking but by emotional states. Cued by dozens of neurotransmitters and neuromodulators, decisions are only retro-rationalized by inner speech after a decision has actually been taken [Damasio, 1999]. Such findings have led to the field of *emotive* computing.

While many aspects of the human mind and brain have been considered by AI, the role of *consciousness* has been rather overlooked until recently. Consciousness is intimately related to intuitive, natural interaction. It has been shown to play a role in functions that range from speculation, planning, and decision making to inner speech, social skills, and empathy, among many others. Owing to rapid progress in the neurosciences, consciousness has become one of humankind's last scientific challenges. This challenge is being tackled by a multi-disciplinary effort of chemists, physicists, psychologists, philosophers, neurobiologists, cognicists, computer scientists, and many others who have joined into a new scientific endeavor.

Recent findings suggest that the limbic system not only plays a major role in emotional processes but also is an important element of consciousness. Even extended damage to the neocortex can leave a patient conscious, while localized damage to the limbic system or the thalamic system usually impairs consciousness anywhere from mildly to severely [Damasio, 1999]. This reveals that in the human brain, consciousness and emotions are anatomically linked. If man-made systems must have the kinds of functions that involve consciousness in humans, it is therefore reasonable to assume they may have to possess — or at least mimic — consciousness themselves. Such ideas have lead to a new AI discipline called *machine consciousness* (MC). Also known as *artificial consciousness*, MC is an emerging technology that aims at achieving consciousness in man-made systems, with the expectation that this will enable a wide variety of new products and applications. The current state of affairs is far from being merely theoretical; concrete system architectures have been proposed and are being tried out. In this chapter we introduce MC, present some practical examples, address its relevance, and discuss applications. For reference, we start with a brief review of biological consciousness.

4.2 Biological consciousness

As human beings we are all intimately familiar with consciousness; most generally, it is equated with *experience, awareness, subjectivity, sentience.* For many people, the word 'consciousness' has strong metaphysical or even religious connotations, and for a long time consciousness was a scientific non-starter. Driven by, amongst others things, dramatic advances in neurosciences and brain imaging (e.g. [Changeux, 1995; Freeman, 1999]), it has, over the past two decades, come to be recognized as a fascinating research topic, giving rise to an explosion of academic research and several scientific societies with conferences and peer-reviewed journals[1,2].

Despite progress on many fronts however, consciousness remains very hard to define or explain. The lack of a clear definition confuses discussions, for instance when consciousness gets mixed up with other mental capacities such as mind, intelligence, and cognition — for which there are no generally accepted definitions either. Another complicating aspect is that consciousness (like many other mental phenomena) is inherently a first-person phenomenon, and is therefore not accessible for direct observation or study[3,4]. And while mental capacities like intelligence and reasoning are rather well understood and can even be modeled, the very essence of consciousness remains unclear[5]. [Holland, 2004] sums up the situation: "we can't define it; we don't know how it arises; and it's peculiar — and that's what makes it interesting."

In order to have somewhat of a working definition, we propose the following: *mental phenomena* are all non-physical, high-level phenomena that go on in the human brain[6]; they include perception, proprioception; drives, emotions, feelings; memory, thinking, inner speech, reasoning, knowledge, intelligence, qualia; egoic structures; and developmental lines. We define the complete set of these phenomena as the *mind*. As such, *intelligence* and *cognition* are

---

[1]Including the *Association for Scientific Study of Consciousness* (http://www.assc.caltech.edu/, with an annual conference and two journals); the *Journal of Consciousness Studies* (http://www.imprint.co.uk/jcs.html); and the *Towards a Science of Consciousness* conference (http://consciousness.arizona.edu/conference/tucson2004/) that is held every two years.

[2]For a comprehensive textbook on consciousness, see [Blackmore, 2004]. [Velmans, 2000] is also recommended.

[3]Giving rise to the 'Zombie Hypothesis', which asserts that if there were people among us with normal behavior but no subjectivity, we would be unable to tell them from normal people. First mentioned in [Kirk, 1974].

[4]Many academic efforts are underway that try to identify the 'neural correlates of consciousness'. The hope is that observing such correlates will reveal what goes on in consciousness; but it has been pointed out that a correlate of a thing is not that thing itself.

[5]Giving rise to a spectrum of philosophies about the nature of consciousness, ranging from materialism (everything is matter, consciousness is an illusion) through dualism (matter and consciousness are fundamentally different aspects of reality), to idealism (everything is consciousness, matter is an illusion).

[6]Physical phenomena being e.g. blood flow and neuron activity; an analogy is information flow vs. electron flow in a computer.

subsets of the mind. *Consciousness*, as a mental phenomenon, is also a subset of he mind; but it is one that stands out, because *all other* mental phenomena can be partially conscious, partially unconscious.

An aspect of biological consciousness that is important for our purpose is that it is not an on-off thing, but comes in degrees instead. There is increasing evidence that animals have minds like we do, be they simpler to some degree [Hauser, 2000]; and acceptance of the idea that at least higher-order animals have consciousness is growing. Our own human consciousness also exhibits growth, both through the evolution of our species (phylogeny) and through the course of our individual lifetimes (ontogeny)[7]. The former is exposed when we compare for instance the human rights paradigm of primeval societies with that of modern democracies; the latter becomes apparent by comparing one's awareness of self and the world as a child to the awareness one has as an adult.

A final point is that biological consciousness makes physical matter aware of itself, a fact that is rather hard to explain. Many indeed agree that, as [Chalmers, 1995] famously puts it, "The really hard problem of consciousness is the problem of experience. ... In this central sense of 'consciousness', an organism is conscious if there is something it is like to be that organism, and a mental state is conscious if there is something it is like to be in that state. Sometimes terms such as 'phenomenal consciousness' and 'qualia' are also used here, but I find it more natural to speak of 'conscious experience' or simply 'experience.'" The question we have to ask now is if it is, even in principle, possible to make artificial systems that can experience themselves and the world like we do. How would one do that? And why?

4.3 Machine consciousness

[Chalmers, 1995] contrasts the hard problem of consciousness with a number of related phenomena, for instance, "the reportability of mental states; the ability of a system to access its own internal states; the focus of attention." He considers these phenomena as the *easy* problems of consciousness because they are "straight-forwardly vulnerable to explanation in terms of computational or neural mechanisms." This is a reference to theories that consider the human brain to be a computational device, and state that the computations it performs could be implemented by silicon hardware just as well as by our brain's neural wetware.

Among such theories is the *Computational Theory of Mind* (CTM), which was "first expressed by the mathematician Alan Turing, the computer scientists Alan Newell, Herbert Simon, and Marvin Minsky, and the philosophers

---

[7]While controversial, these two strands of evolution are sometimes said to show parallels, in the sense that our individual growth resembles a miniature version of the growth of our species. This effect is summarized in the maxim "ontogeny recapitulates phylogeny" [Haeckel, 1899].

Hillary Putnam and Jerry Fodor" [Pinker, 1997]. Underlying CTM is "what is known as the physicalist assumption: Mind is what brain does, or something very like it in relevant ways" [Franklin, 1995]. In this view, the human brain is the physical substrate from which our mind emerges. CTM claims that alternative substrates can implement the basic computations of our brain, and thereby that human minds can emerge from artificial substrates. [Chalmers, 1993] puts it as follows: "there is a certain class of computations such that any system implementing that computation is cognitive" and, more specifically, "a model that is computationally equivalent to a mind will itself be a mind." Such artificial minds solve the 'easy' problems of consciousness; but can they come to have experiences? MC researchers think that they can indeed.

One instance of CTM is *Global Workspace Theory* (GWT), proposed in [Baars, 1988; Baars, 1997]. GWT is based on the concept of a human working memory — the "inner domain in which we can rehearse telephone numbers to ourselves or, more interestingly, in which we carry on the narratives of our lives. It is usually thought to include *inner speech* and *visual imagery*." The contents of this small, short-term memory closely correspond to what we are conscious of, and get broadcast to the vast multitude of unconscious cognitive brain processes which require entrance into the workspace to be globally accessible. These unconscious processes, operating in parallel with limited communication between them, can form coalitions whose participants cooperate to achieve certain goals. Individual as well as allied processes compete for access to the global workspace, striving to disseminate their messages to all other processes in an effort to obtain more cohorts and thereby increase the likelihood of achieving their goals. While [Baars, 1997] concedes that the global workspace "is closely related to conscious experience, though not identical to it," GWT can successfully explain several characteristics of consciousness, such as its role in handling novel situations, its limited capacity, and its sequential nature. Moreover, it lends itself well to computational modeling.

GWT has been put into practice; an example instantiation is IDA (Intelligent Distribution Agent), developed at the Conscious Software Research Group at the University of Memphis. In line with [Baars, 1988; Baars, 1997], IDA's mental father Stan Franklin observes in [Franklin, 2003] that "consciousness has many functions. It helps us deal with novel or problematic situations for which we have no automatized response. It makes us aware of potentially dangerous situations. It alerts us to opportunities presented by the environment. It allows us to perform tasks that require knowledge of location, shape, size or other features of objects in our environments." An autonomous agent, which is "a system situated in, and part of, an environment, which senses that environments and acts on it, over time, in pursuit of its own agenda," can thus be said to possess functional consciousness "if its architecture and mechanisms allow it a number of these and, perhaps, other functions." With this in mind,

Franklin defines a functionally conscious autonomous agent as an agent that implements global workspace theory, emphasizing, that "it's functional consciousness that's being claimed, not phenomenal consciousness." IDA is a software implementation of GWT, and, by Franklin's own definition, is therefore functionally conscious. IDA's task is to negotiate new assignments for sailors in the US Navy after they end a tour of duty, by matching each individual's skills and preferences with the Navy's needs. IDA interacts with Navy databases and communicates with the sailors via natural language email dialog "while adhering to some ninety [Navy] policies." The IDA computational model was developed during 1996–2001 at the Conscious Software Research Group at the University of Memphis. It "consists of approximately a quarter-million lines of [Java] code, and almost completely consumes the resources of a 2001 high-end workstation." It heavily relies on codelets, which are "special purpose, relatively independent, mini-agent[s] typically implemented as a small piece of code running as a separate thread." Figure 4.1 shows IDA's architecture, which arose from a top-down approach in which high-level cognitive functions are explicitly modeled. [Franklin, 1995] and [Franklin, 2003] explain the various functions and their underlying concepts.

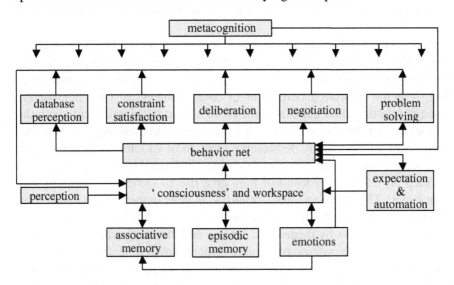

Figure 4.1.   IDA architecture (source: [Franklin, 2003]).

Whereas GWT is a top-down approach since it starts from a high-level model of consciousness, an extreme bottom-up CTM approach is found in [Cotterill, 2003]. This paper presents "a project aimed at searching for the neural correlates of consciousness through computer simulation. The underlying model is based on the known circuitry of the mammalian nervous system, the neuronal groups of which are approximated as binary composite units. The simulated nervous system includes just two senses — hearing and touch — and it drives a set of muscles that serve vocalisation, feeding and bladder control. These functions were chosen because of their relevance to the earliest stages of human life, and the simulation has been given the name CyberChild." The computer-bound CyberChild is born with just a few basic reflexes, and has to learn to ease discomfort (hunger caused by an empty stomach, pain caused by a full diaper) by getting the operator's attention, for instance to obtain a new bottle of milk or to have its diaper changed. Starting with random muscle movements, it also needs to learn from scratch how to bring the milk bottle to its mouth in order to feed itself. The model will be made more precise and complete over time, counting on Moore's Law to maintain real-time operation. The underlying assumption is that "consciousness will one day emerge from the blinking lights that are the simulation's graphical representations of the neural units." While Cotterill concedes that "it takes a lot of faith in the reductionist canon" that this will happen, he believes that "it may be possible to infer the presence of consciousness in the simulation ... from the monitoring of its ability to ontogenetically acquire novel reflexes." He suggests that "this ability is the crucial evolutionary advantage of possessing consciousness." [Cotterill, 2004] reports that there is no evidence of consciousness in CyberChild as of yet.

An incremental bottom-up approach somewhat resembling Cotterill's is suggested by [Holland and Goodman, 2003]. Rather than modeling the nervous system however, they "emphasize a single mechanism — internal modeling — as the possible underpinning of consciousness. ... Their approach is rooted in robotics; their claim is that a robot able to deal intelligently with the complexities of the real world will have to engage in planning, and that this requirement will inevitably demand the creation of an internal model not just of the world, but of many aspects of the embodied agent itself. They speculate that such an internal agent-model may give rise to some consciousness-like phenomena. Their strategy, like Cotterill's, is developmental, but rather than allowing an entity to modify and extend its own capabilities, they propose to re-engineer the robot by themselves, adding and changing whatever is necessary to deal with the progressively more difficult environmental contingencies to which they intend to expose it. ... [their] starting point is a robot that they claim is definitely not conscious; from there, as they remark, 'The only way is up.'"

A very different bottom-up approach to MC is proposed by [Haikonen, 2003]. His starting point is that classical rule-based computing is inadequate: "the brain is definitely not a computer. Thinking is not an execution of programmed strings of commands. The brain is not a numerical calculator either. We do not think by numbers." Conventional artificial neural network approaches are also of no avail, as they can be implemented on computers.

Rather than trying to achieve mind and consciousness by identifying and implementing their underlying computational rules, Haikonen proposes "a special cognitive architecture to reproduce the *processes* of perception, inner imagery, inner speech, pain, pleasure, emotions and the cognitive functions behind these. This machine would produce higher-level functions by the power of the elementary processing units, the artificial neurons, without algorithms or programs"[8]. Haikonen believes that a machine based on this architecture, which is shown in Figure 4.2, can develop consciousness, which he sees as "a style and way of operation, characterized by distributed signal representation, perception process, cross-modality reporting and availability for retrospection. There is no need for special 'conscious neurons', conscious matter or a special seat of consciousness. *There is no discrete machine supervisor self, the supervision is distributed in the machine.*"

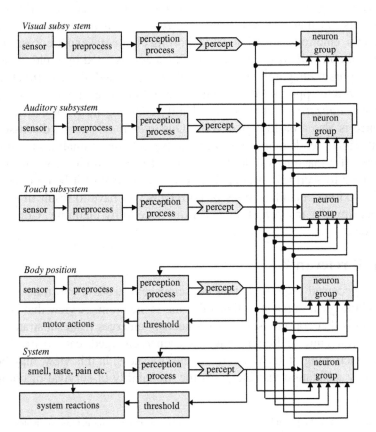

Figure 4.2.   IDA architecture (source: [Haikonen, 2003]).

---

[8]Emphasis added; refer to [Haikonen, 2003] for details.

Haikonen is not alone in this process view of consciousness, or in the view that MC will spontaneously emerge in autonomous agents that have a suitable neuro-inspired architecture of complexity; these are shared by many, as witnessed for instance in [Freeman, 1999] and [Cotterill, 2003]. Interestingly, Haikonen claims that the artificial neurons that underlie his architecture "cannot be implemented by existing digital microcircuits," apparently because they are hybrid analog/digital circuit that require dedicated chips[9].

Along the way, all MC researchers will face the same daunting problem: how to know if the contraption they just made is consciousness? This important problem is being approached in various ways. [Alexander and Dunmall, 2003] are trying to construct an *axiomatic framework* to test for consciousness, taking the view that "it is only through an assessment and analysis of the mechanisms of the organism that we can conclude whether such mechanisms can support consciousness or not within that organism." [Damasio, 1999] points out that consciousness has *observable manifestations*, such as wakefulness, emotions, attention, and reporting of internal experiences. Others try to find the *neural correlates* of consciousness, assisted by rapid advances in real-time brain imaging[10]; the hope is that once found, those correlates can be generalized to apply to MC. [DeGroot, 2004] proposes the use of powerful *computer visualization* techniques to identify consciousness. Of considerable importance is also a basic human trait. Our *intentional stance*, discussed by [Dennett, 1987], tends to make us attribute typical human qualities to animals or objects. If a machine would behave as if it were conscious, we would be very tempted to ascribe consciousness to it irrespective of it being conscious or not[11]. Axioms, manifestations or correlates will have to result in reliable identifiers in order to prevent anyone from deceitfully claiming 'Consciousness Inside'.

4.4 Is it relevant?

A well-accepted premise is that consciousness needs a substrate to arise from. In biological organisms, that substrate is the brain. A view that is gaining ground is that consciousness spontaneously emerges from substrates with suitable architecture and complexity, and, once emerged, can positively impact the functioning of such systems. Consciousness, in this view, may be an evolutionary adaptation that increases the probability of survival of complex

---

[9]In principle it ought to be possible to implement the synapses on a general purpose computer. Such a sequential implementation would be rather inefficient however for a system containing millions or billions of neurons all operating in parallel; dedicated hardware would indeed be a better choice for implementing such a system.

[10]Especially PET scans and fMRI.

[11]In fact, this is all we can do even with our fellow humans, as there is no way to know for sure whether other humans are conscious because of the first-person nature of consciousness.

organisms in a complex environment (e.g. [Humphrey, 1983]). If that is correct, it should be possible to identify the benefits of being conscious. [Baars, 1997], which states that "consciousness is a supremely functional adaptation," suggests a variety of functions in which consciousness plays a role: prioritization of alternatives, problem solving, decision making, brain processes recruiting, action control, error detection, planning, learning, adaptation, context creation, and access to information. Based on extensive neurological research, [Damasio, 1999] even considers consciousness among the life regulation devices of an organism that possesses it[12]. Consciousness allows emotions and feelings to be known to the organism; this awareness gives it a concern for its own survival and thereby increases its survival drive. Crucially, Damasio argues that consciousness comes into play before high reason, *suggesting that human-level cognitive abilities require the presence of consciousness.*

Our discussion of IDA showed that attempts to instill functions that require consciousness into an autonomous agent by designing around a consciousness mechanism can lead to systems that adequately handle complex human tasks. An alternative view is that an agent whose architecture mimics the cognitive processes of the human brain with sufficient detail will spontaneously develop consciousness; Haikonen's work falls in this school of thought. It has not been proven that high-levels of cognitive functioning cannot be achieved without consciousness being involved. An important consideration however is that high-level cognitive processes can be partially conscious, partially unconscious (cf. section 4.2) which means that the ontological nature of consciousness differs from that of other mental phenomena, including intelligence, knowledge, learning etc. Consciousness, whether biological or artificial, therefore clearly adds something unique.

An important point is that the pursuit of MC has got people thinking about architectures for autonomous systems in new ways, creating the potential for a variety of new products and applications. [Haikonen, 2003] puts it bluntly: "New technology will arise. Cheap common sense in a chip will be in demand. Those who master the new technology of artificial cognition and consciousness will reap magnificent profits in growing markets. There will be new applications and products, ones never seen before. Some of them will be trivial, others quite unexpected." His role as Principal Scientist at Nokia Research gives this statement weight. Several industrial activities can actually be identified today, which is relevant given that MC is in an early stage and most of the efforts go on in academia.

---

[12]These devices include *basic life regulation* (metabolism, reflexes, drives), *emotions* (stereotyped responses that are physical and public), *feelings* (representations of emotions that are mental and private), and *high reason* (planned responses).

While Haikonen may well be right, the drive for MC is, in the author's opinion, not *only* of a commercial nature. Consciousness is perhaps the ultimate scientific challenge[13], and if MC is fundamentally possible, mankind's insatiable curiosity and drive for innovation make it inevitable that it *will* some day be achieved. And whether that quest succeeds or not, it is likely to result in a variety of spin-offs with commercial value. To stick with the metaphor — it is safe to assume that there will be plenty of crusaders on the lookout for MC.

MC undeniably has a philosophical aspect; but engineers have by no means lagged in the field, as witnessed by the efforts mentioned above. Artificial intelligence and expert systems evolve at a fast pace, supported by the increasing understanding of cognitive processes and the independent field of information processing and theory, which all come together in fields such as Artificial Life (e.g., [Steels and Brooks (eds.), 1994]). Context-, system- and self-awareness may prove to be the right paradigm for advanced systems to deal with the fuzzy complexity of our world. This is for instance reflected in European Commission's calls for IT proposals such as Bio-I^3, which focuses on three topics: *reverse-engineering brains*; *growth and plasticity*, and *self-aware control systems*[14]. European Integrated Projects in the making as part of this framework will cover topics such as *awareness engines* and *self-aware robots*. Self-awareness is very close to consciousness, and it is therefore no surprise that MC is explicitly on the Bio-I^3 agenda[15].

In terms of ultimate relevance, machine consciousness will become significant to the extreme if the predictions of futurologists like Hans Moravec and Ray Kurzweil come true. Comparing estimates of the compute power of the human brain (in the order of 10^{16} operations per second) with the growth of computer power as predicted by Moore's Law, [Moravec, 1998] predicts that "computers suitable for humanlike robots will appear in the 2020s." [Kurzweil, 1999] likewise predicts that a "$1,000 computing device (in 1999 dollars) is [in 2019] approximately equal to the computational ability of the human brain." He goes on to extrapolate all the way up to 2099, and foresees that not only will machines be conscious by then, but humans will have evolved into conscious machines...

4.5 Applications

Current MC thinking is mainly concerned with the plausibility of the concept and the feasibility of proposed mechanisms. In comparison, not that much is being said about the application of conscious technology. If we assume that

---

[13]Even Nobel-prize winners like G.M. Edelman and F. Crick have embarked on the study of consciousness.
[14]Bio-I^3 stands for Bio-Inspired Intelligent Information Systems; see http://www.cordis.lu/ist/fet/bioit.htm.
[15]See for example http://fp6.cordis.lu/ist/fet/proposal_details.cfm?ID_EVENT=49.

MC technology will some day be mature and available though, where and how will we apply it? We will speculate on an example application, but first attempt to sketch some general characteristics of MC application areas and of the conscious machiness operating in it.

To identify MC application areas, we propose to look for environments in which consciousness is considered to be beneficial. These would predominantly be complex environments that are subject to frequent, unpredictable changes. For MC technology to benefit an artificial autonomous agent operating in this type of environment, the agent would be required to fulfill one or more non-trivial tasks in which it can only succeed by properly reacting to environmental changes. The high degree of adaptability this would require from the agent ought, as we saw above, to benefit from consciousness. Within the boundaries of its mission, the agent would need a large degree of autonomy, as the capability to influence both itself and its environment would allow it the 'learn and live' type of survival that played a role in the rise of biological consciousness. Responsibility for its own survival would in fact have to be among the agent's primary drives, in the sense that substandard performance would lead to its deterioration and possible termination. The agent should moreover be capable of at least simple emotions and feelings, derived from more elementary pain and pleasure mechanisms. Its goals would then be defined by the avoidance of certain events (pain) and the pursuit of others (pleasure).

With this in mind, we now briefly speculate, as promised, on an actual application. As our complex environment, we choose a modern home in a technologically advanced society of the not-too-distant future[16]. This home consists of the common framework of walls, doors, windows etcetera, which define its living spaces. These spaces interact with each other and with the environment outside, for instance through human activity, movement of objects, various electronic communication channels, and the effects of the weather. Integrated into the framework is a large set of sensors and actuators. The sensors provide the status of such properties as light, temperature, and humidity, they accept user commands, they identify objects and/or persons along with their positions and activities, they measure the usage of electronic and other resources, etc. Likewise, the actuators provide for instance heating, cooling, and lighting, and include a variety of appliances, infotainment sources, and user interfaces. When these sensors and actuators are tied together via MC technology, *the entire home can be regarded as a conscious agent*. Appropriate 'pleasure drives' for this agent would be maintaining the integrity of the house and the safety, health and comfort of all occupants, as well as receiving compliments and system upgrades from the occupants. Likewise, appropriate

---

[16]For instance, some Western society ten years from now.

'survival drives' could be the avoidance of damage, abuse, and neglect of the home framework, preventing reprimands, and steering clear of any system downgrade, replacement, or discard. The normal lives of the occupants, including their communications with the outside world, along with the weather with its, well, weathering effects on the home, make for a complex and dynamic environment; and the tasks of keeping a family happy and a house in good shape are far from trivial, as most of us probably know. If we consider the increased complexity and the fast pace of today's life and extrapolate, the idea that people will try to turn their passive shelters into proactive support entities is not that inconceivable. A relatively simple MC concept like this can clearly spur a lot of technical and philosophical debate[17], but may also make clear that a key application as well as challenge for MC technology will be to make things better and simpler for us humans, not harder and more complicated.

4.6 Conclusion

Neuro-scientific research suggests that consciousness plays an important role in human cognition, and may even be required for human-level intelligence. Machine consciousness is an emerging 'technology of mind' that aims at achieving consciousness in man-made systems. It is in an early stage, and much is yet to be discovered and proven. The hope is that MC will add to the existing suite of AI techniques and enhance the performance and capabilities of autonomous agents. First-generation architectures exist and were proven to be meaningful. If MC succeeds and lives up to expectations, it can be an enabling technology for new products and applications, suggesting that there is commercial potential. In fact, there is already some industrial activity despite the early stage the field is in. Products and applications that come to possess MC will be self-aware; they will in a certain sense be alive, posing ethical questions never encountered before. All in all, good reasons exits for keeping an eye on the development of MC technology.

Acknowledgments

The author would like to thank Michel Decré of Philips Research for valuable contributions.

References

Alexander, Igor, and Barry Dunmall [2003]. Axioms and tests for the presence of minimal consciousness in agents. In *Machine Consciousness*, Owen Holland (ed.). Imprint Academic, Exeter, UK.

---

[17]Consider for instance that if some of the sensors or actuators of a conscious home were conscious in and of themselves, the home would turn into some sort of artificial society...

Baars, Bernard [1988]. *A Cognitive Theory of Consciousness.* Cambridge University Press, Cambridge, MA.

Baars, Bernard [1997]. *In the Theater of Consciousness.* Oxford University Press, New York, NY.

Blackmore, Susan [2004]. *Consciousness: An Introduction.* Oxford University Press, New York, NY.

Chalmers, David [1993]. A computational foundation for the study of cognition. *Papers on AI and Computation,* http://www.u.arizona.edu/ chalmers/ai-papers.html.

Chalmers, David [1995]. Facing up to the problem of consciousness. *Journal of Consciousness Studies,* 2(3):200–219.

Changeux, Jean-Pierre [1995]. *Neuronal Man.* Princeton University Press, Princeton, NJ, reprint 1997.

Cotterill, Rodney [2003]. Cyberchild: A simulation test-bed for consciousness studies. In *Machine Consciousness,* Owen Holland (ed.). Imprint Academic, Exeter, UK.

Cotterill, Rodney [2004]. Neuronal dynamics and cyberchild's behaviour. Presented at *Workshop on Models for Machine Consciousness* Antwerp, Belgium.

Damasio, Antonio [1999]. *The Feeling of what Happens – Body and Emotion in the Making of Consciousness.* Harcourt Inc., New York, NY.

Dennett, Daniel [1987]. *The Intentional Stance.* MIT Press, Cambridge, MA.

DeGroot, Doug [2004]. Visualizing high-level cognitive processes: A proposed architectural approach. Presented at *Workshop on Models for Machine Consciousness,* Antwerp, Belgium.

Franklin, Stan [1995]. *Artificial Minds.* MIT Press, Boston, MA.

Franklin, Stan [2003]. IDA: A conscious artefact? In *Machine Consciousness,* Owen Holland (ed.). Imprint Academic, Exeter, UK.

Freeman, Walter [1999]. *How Brains make up their Minds.* Phoenix, London, UK.

Haeckel, Ernst [1899]. *Riddle of the Universe at the Close of the Nineteenth Century,* cited at http://www.ucmp.berkeley.edu/history/haeckel.html.

Haikonen, Pentti [2003]. *The Cognitive Approach to Conscious Machines.* Imprint Academic, Exeter, UK.

Hauser, Marc [2000]. *Wild Minds: What Animals Really Think.* Henry Holt Publishers, New York, NY.

Holland, Owen, and Rod Goodman [2003]. Robots with internal models: A route to machine consciousness? In *Machine Consciousness,* Owen Holland (ed.). Imprint Academic, Exeter, UK.

Holland, Owen [2004]. *Towards a Technology of Consciousness* Presented at ASSC8, Tucson, AZ.

Humphrey, Nicolas [1983]. *Consciousness Regained: Chapters in the Development of Mind.* Oxford University Press, New York, NY.

Kirk, Robert [1974]. Sentience and behaviour. *Mind,* 83:60–61.

Kurzweil, Ray [1999]. *The Age of Spiritual Machines.* Penguin Group, New York, NY.

Moravec, Hans [1998]. When will computer hardware match the human brain? In *Journal of Evolution and Technology* 1, http://www.transhumanist.com/volume1/moravec.htm.

Pinker, Stephen [1997]. *How the Mind Works.* W.W. Norton & Company, New York, NY.

Steels, Luc and Brooks, Rodney (eds.) [1994]. *The 'Artificial Life' Route to 'Artificial Intelligence'. Building Situated Embodied Agents.* Lawrence Erlbaum Associates, New Haven, CT.

Velmans, Max [2000]. *Understanding Consciousness.* Routledge, Philadelphia, PA.

Part II

LIFESTYLE

Chapter 5

OPTIMAL SELECTION OF TV SHOWS
FOR WATCHING AND RECORDING

Wim F.J. Verhaegh

Abstract   In this chapter we address a problem from the area of personalized electronic
program guides (EPGs), concerning the selection of a number of TV shows for
watching and recording, given a number of available tuners, such that the total
value of the selected shows is maximized. Furthermore, the shows selected for
watching are to be scheduled in a given time interval. We give a mathematical
model for this problem, and show that it is NP-hard. Next, we present a dynamic
programming approach that solves the problem to optimality. Furthermore, we
present a few options to reduce the run time, albeit at the cost of losing the
guarantee of finding an optimal solution. Finally, we perform some experiments
on actual EPG data.

Keywords   Electronic program guide, recommender system, channel, TV show, dynamic
programming.

5.1   Introduction

With the ever increasing number of available television channels, either
through terrestrial connections, satellite, or cable, the problem for the user to
select TV shows is becoming too big to handle manually. With the advent of
digital television, the number of channels and hence the number of options to
choose from are becoming even larger. As a result, users will lose the overview,
and it is very likely that a user will miss TV shows that he would find very
interesting.

Because of the large number of TV shows available each week, printed program guides are no longer a viable option, as they would simply become too
big and heavy to handle. Electronic program guides (EPGs) form a solution to
this problem, by presenting the available TV shows for a number of channels
onto the TV screen. Unfortunately, the number of channels and the length of
the time interval of the portion shown on the screen is very limited, because of

73

Wim F.J. Verhaegh et al. (Eds.), Intelligent Algorithms in Ambient and Biomedical Computing, 73-87.
© 2006 *Springer. Printed in the Netherlands.*

the poor resolution of a TV screen for this kind of information. However, even if the resolution were high enough, simply showing all available TV content would overwhelm the average user.

In order to alleviate the problem of what TV shows to select, EPGs offer the option to search for keywords, in this way reducing the number of TV shows to a tractable number. Another way to weed out many uninteresting TV shows is by the use of recommender systems [Gutta et al., 2000; Smyth & Cotter, 2000]. These systems maintain a preference profile of the user, indicating what he likes and dislikes, and use that to predict to what extent the user will like the newly offered TV shows. Then, the top-scoring shows can be highlighted in an EPG, or the user can be given a list of top-scoring shows. A problem with these solutions, however, is that they do not take into account whether shows overlap in time or not, so it is still left up to the user to compile a nice program of TV shows to watch during e.g. an evening. In other words, whereas filters and recommenders come up with a list of *individual* shows, we would like to offer the user a *sequence* of shows to watch. This problem, which has hardly been addressed in the literature, is the topic of this chapter.

When considering the problem of determining an optimal TV experience for a user for a certain time interval, there are numerous aspects to be taken into account. On the one hand we are currently compiling a list of aspects that play a role and are assessing their relevance, while on the other hand we have started investigating the problem area in a bottom-up fashion, concentrating on a few aspects at a time. By starting with a simplified problem setting and investigating how it can be solved, we gain first insights into the problem area. Building on this, future research will focus on extending the problem setting to include more aspects, and on investigating how the developed algorithms can be elaborated to solve more complex settings. Furthermore, in this chapter we concentrate on a mathematical formulation and solution technique of the discussed problem, and discard the user interaction aspects for the moment.

The scope of this chapter covers the problem of what shows to select for watching or recording with a given number of available tuners. We assume here that the number of tuners is limited, whereas recording multiple shows at the same time is no issue, using today's hard-disk recorders. Furthermore, we assume that we are given a preference value for each show, indicating how much the user likes it. This value may be given explicitly by the user, but more realistically it is given by a recommender tool. Finally, we assume for the moment that the total value of a number of shows is given by the sum of the values of the individual shows.

In our problem setting we consider the option to time-shift shows that have been selected for watching, thereby also introducing a scheduling aspect next to the selection aspect. Including this makes the problem formally hard to solve optimally, but we nevertheless present a dynamic programming approach that does so in a reasonable time for problem instances of practical size.

The remainder of this chapter is organized as follows. In Section 5.2 we present a formal definition of the problem that we consider, of which we show in Section 5.3 that it is formally hard to solve. In Section 5.4 we show that the TV shows can be considered in a particular order for scheduling, which we use in Section 5.5 to derive a dynamic programming approach. A few ways to reduce the run time are presented in Section 5.6. Finally, we perform some experiments on actual EPG data, in Section 5.7.

5.2 Problem definition

Informally, the problem we consider consists of two aspects. First, we have to select the shows we want to receive (which will be recorded). Secondly, from this first selection we have to decide which shows we want to watch during a given interval (e.g. that evening) and at what time. We assume a certain value per show to be given for receiving it, and an additional value is given if it is also selected for watching. For instance, a sports program gets a certain value for being received (and recorded), and an additional value if it can be watched during the evening of broadcast.

More formally, the problem we consider is the following.

Definition 5.1 (Time-shifted show selection problem (TSSP)). *Given is a set S of TV shows, where each show s is broadcast from begin time b_s to end time e_s, and a number m of tuners. Furthermore, for each show a value v_s^r is given for receiving it, and an additional value v_s^w for watching it. Finally, a begin time b and an end time e are given, indicating the interval $[b, e)$ during which the user wants to watch shows.*

The question is to determine a subset $S^r \subseteq S$ of shows to be received (and recorded), and a subset $S^w \subseteq S^r$ of shows that are going to be watched, such that at all times at most m shows have to be received, i.e., for all times x we have

$$|\{s \in S^r \mid b_s \leq x < e_s\}| \leq m,$$

such that the set S^w of shows selected for watching can be scheduled in the time interval $[b, e)$, and such that the total value of this solution, given by

$$\sum_{s \in S^r} v_s^r + \sum_{s \in S^w} v_s^w,$$

is maximized.

The constraint that the set S^w selected for watching can be scheduled in the time interval $[b, e)$ means that we also have to determine for each of the shows s in this set a time w_s at which it is started to be watched, such that for all

$s \in S^{\mathrm{w}}$,

$$w_s \geq b_s$$
$$w_s \geq b$$
$$w_s + e_s - b_s \leq e,$$

and such that for all $s, t \in S^{\mathrm{w}}$, $s \neq t$,

$$w_s \geq w_t + e_t - b_t \quad \lor \quad w_t \geq w_s + e_s - b_s.$$

Note that scheduling the shows selected for watching may introduce gaps in the schedule. Such a gap will, for instance, be introduced if it is better to wait some time for a show to come that has a very high value.

The above problem is typically solved for a limited time interval, e.g. an evening. The shows in S^{w} can be watched that evening, while the shows in $S^{\mathrm{r}} \setminus S^{\mathrm{w}}$ can be watched at another occasion. For example, they can be used if shows in S^{w} turn out to be disappointing, or they can be used on other days when there are no good shows broadcast. Although a following instance of the problem, e.g., the next day, may hence have a set of shows already recorded, which may be selected for watching, we discard that for the time being.

5.3   Computational complexity

First we show that TSSP is formally hard to solve. To this end, we define a decision variant TSSP-D in which the question is whether a solution exists with a value of at least a certain bound V.

Theorem 5.1. *TSSP-D is NP-complete.*
Proof. First, we note that we can verify a solution to TSSP-D, given by the sets S^{r} and S^{w} and the times w_s for all $s \in S^{\mathrm{w}}$, straightforwardly in polynomial time. Hence TSSP-D is in NP.

Next, we give a reduction from Subset Sum [Garey & Johnson, 1979]. In Subset Sum we are given a multiset $X = \{x_1, x_2, \ldots, x_k\}$ of numbers and an integer B, and the question is whether a subset $X' \subseteq X$ exists with $\sum_{x \in X'} x = B$.

Subset Sum can be reduced to TSSP-D by means of the following transformation. For each number $x \in X$, we create a show s with $b_s = 0$, $e_s = x$, $v_s^{\mathrm{r}} = 0$, and $v_s^{\mathrm{w}} = x$. Furthermore, we set $b = 0$, $e = B$, $m = k$, and $V = B$. Then, the Subset Sum instance has a solution if and only if the constructed TSSP-D instance has a solution. This can be seen by using the straightforward relation that we include a number x in X' if and only if we select the corresponding show for watching (and reception). $\qquad \square$

The above theorem basically states that selecting a set of shows for watching that completely fills the interval $[b, e)$ is hard.

Even though TSSP is NP-hard, we are going to solve it to optimality at first by means of dynamic programming. Before doing so, we address the actual scheduling of the shows selected for watching.

5.4 Scheduling shows for watching

The actual scheduling of the shows for watching, once they have been selected, is not a hard problem, as the next theorem states.

Theorem 5.2. *Without loss of feasibility, a set S^w of shows selected for watching can be scheduled in a time interval $[b, e)$ in order of increasing begin time.*
Proof. Suppose we have a feasible schedule, given by the start times w_s for all $s \in S^w$, and suppose we have two shows s, t, with s being scheduled directly before t, but with t being available before s, i.e., $b_t < b_s$. Then, swapping s and t also gives a feasible schedule, which can be seen as follows.

Consider a new schedule w' which is given by the same start times for all shows except for shows s and t, which are given by

$$
\begin{aligned}
w'_t &= w_s \\
w'_s &= w'_t + e_t - b_t.
\end{aligned}
$$

See Figure 5.1 for a visualization of this swap. This swap only affects shows

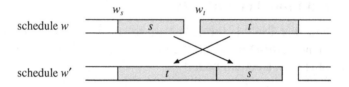

Figure 5.1. Swapping directly-succeeding shows s and t.

s and t and not the other shows, so we only have to check s and t for meeting the constraints. To this end, we derive

$$
\begin{aligned}
w'_t &= w_s \geq b \\
w'_t &= w_s \geq b_s > b_t \\
w'_t + e_t - b_t &\leq w_t + e_t - b_t \leq e \\
w'_s &\geq w_s \geq b \\
w'_s &\geq w_s \geq b_s \\
w'_s + e_s - b_s &= w_s + e_t - b_t + e_s - b_s \leq w_t + e_t - b_t \leq e.
\end{aligned}
$$

□

So, scheduling the shows for watching can be done in order of increasing begin time. Furthermore, assigning tuners to the shows to be received corresponds to

coloring an interval graph, which can also be done in order of increasing begin time by means of the so-called left-edge coloring algorithm [Korst, 1992].
This allows us to apply a dynamic programming approach [Papadimitrious &
Steiglitz, 1982], where we iteratively consider the shows in order of increasing
begin time, and determine for each show whether to skip, receive, or receive
& watch it. For the remainder, we hence assume the shows to be numbered
$s = 1, 2, \ldots, n$, with $b_s \leq b_{s+1}$ for all $s = 1, \ldots, n - 1$.

5.5   A dynamic programming approach

As mentioned, we are going to apply a dynamic programming approach in
which the shows are considered in order of increasing begin time. The state
space that we choose should be such that when considering a certain show s,
we have all the relevant information of previous decisions in order to determine
the effect of skipping, receiving, or receiving & watching it (possibly time-
shifted).

Therefore we define a state by giving for each of the m tuners the time from
which onwards it is (again) available for receiving a next show, plus for the user
the time from which onwards he can watch another show. Given this information, we can determine whether we can skip, receive, or receive & watch a next
show s under consideration, and what the effect is of these choices. Without
loss of generality, we can keep the times for the tuners sorted in non-decreasing
order, as all tuners are identical.

In addition to the above parameters to keep track of the state, we also maintain for each state the value of the already selected shows for reception and
watching. Furthermore, we store for each state the sets S^r and S^w of shows
selected for reception and watching, respectively, so afterwards we can simply
read out what the solution is.

More formally, a state is given by an $(m+4)$-tuple $(a_1, \ldots, a_m; a; v; S^r, S^w)$,
where $a_1, \ldots, a_m$ indicate the times at which the tuners are again available, with
$a_i \leq a_{i+1}$ for all $i = 1, \ldots, m - 1$, a indicates the time at which the user is again
ready to watch the next show, and v is the total value of shows selected up to
now.

The begin state for dynamic programming is given by $a_i = b_1$ for all $i = 1, \ldots, m$, i.e., the tuners are available from the first beginning of any show,
$a = b$, i.e., the user can start watching from time b onwards, a value $v = 0$,
and sets $S^r = S^w = \emptyset$. The set of possible states is hence initialized by $P_0 = \{(b_1, \ldots, b_1; b; 0; \emptyset; \emptyset)\}$.

Next, given the set of possible states P_{s-1} at the beginning of iteration
$s = 1, \ldots, n$ of the dynamic programming approach, we can calculate the set
of possible states P_s at the end of this iteration as follows. For each state

$(a_1,\ldots,a_m;a;v;S^{\mathrm{r}},S^{\mathrm{w}}) \in P_{s-1}$, we can do (at most) three possible actions for the show s under consideration.

- The first action is to simply skip show s. This does not change the state at all, so we get an identical state $(a_1,\ldots,a_m;a;v;S^{\mathrm{r}},S^{\mathrm{w}})$ in P_s.

- The second action is to receive show s, which can be done if and only if there is a tuner available at time b_s. As the tuner availability times are sorted, this can be checked by checking whether $a_1 \leq b_s$ holds. If so, we can use any available tuner for receiving the show without loss of optimality, as all shows $t = s+1,\ldots,n$ to be considered in later iterations have a begin time $b_t \geq b_s$. So, we choose the first tuner for this, and it will again be available at the end time e_s. As a result, we get a new state $(e_s,a_2,\ldots,a_m;a;v+v_s^{\mathrm{r}};S^{\mathrm{r}} \cup \{s\},S^{\mathrm{w}})$ in P_s, where the availability times $e_s,a_2,\ldots,a_m$ should next be sorted.

- Thirdly, we can decide to select show s for reception & watching (possibly time-shifted), which is possible if and only if there is a tuner available at time b_s, i.e., $a_1 \leq b_s$, and there is still time to watch show s. For the latter, the user is again available from time a onwards, and the show is available from time b_s onwards, so watching this show can start at time $w_s = \max\{a,b_s\}$. Note that if $a > b_s$, the show will be time-shifted. Watching shows should be finished by time e, so $\max\{a,b_s\}+e_s-b_s \leq e$ has to hold. If these two conditions are met, show s can be selected for watching. Again, the first tuner is used for receiving the show, so it is available again at time e_s. Furthermore, the show is scheduled for watching at time $w_s = \max\{a,b_s\}$, so the user can continue watching from time $w_s + e_s - b_s$ onwards. So, this gives a new state $(e_s,a_2,\ldots,a_m;w_s+e_s-b_s;v+v_s^{\mathrm{r}}+v_s^{\mathrm{w}};S^{\mathrm{r}} \cup \{s\},S^{\mathrm{w}} \cup \{s\})$ in P_s, where again the availability times $e_s,a_2,\ldots,a_m$ should be sorted.

Next, the set P_s of states may be pruned, as some states may dominate other states. Before doing so, however, we can update the states in P_s as follows. Because the next show begins at time b_{s+1} (where we may define $b_{n+1} = \infty$) and the following ones do not begin earlier, a tuner availability time $a_i < b_{s+1}$ might as well be replaced by an availability time $a_i = b_{s+1}$. Similarly, the time a at which the user is again available for watching may as well be replaced by b_{s+1} if $b_{s+1} > a$. Note that in the latter case the watching schedule of the corresponding state will have a gap.

After the updating step, we remove dominated states. A state $\sigma = (a_1,\ldots,a_m;a;v;S^{\mathrm{r}},S^{\mathrm{w}}) \in P_s$ is called to dominate another state $\sigma' =$

$(a'_1, \ldots, a'_m; a'; v'; S'^r, S'^w) \in P_s$ if and only if the next three conditions hold:

$$a_i \leq a'_i \quad \text{for all } i = 1, \ldots, m$$
$$a \leq a'$$
$$v \geq v'.$$

So, the availability of the tuners and the user in state σ are no later than in state σ', whereas the value of σ is at least as good as the value of σ'. In this situation, σ' can be removed from P_s without loss of optimality. This results in a smaller set of states, and hence in a shorter run time of the algorithm.

To remove all dominated states, we would have to do a check for each pair of states, which takes a number of steps that is quadratic in the number of states. As this will be too time consuming in practice, we perform a less complete check by first sorting the states in lexicographical order, and next comparing each pair of successive states. This results in a good trade-off between the run time required for domination checks and the run time saved by reducing the state space. Pseudo code for the resulting dynamic programming approach is given in Figure 5.2.

5.6 Run time improvements

In this section we discuss two approaches to reduce the run time, at the cost of losing the guarantee of finding an optimal solution.

5.6.1 Reducing the state space

Key in the dynamic programming approach presented in Section 5.5 is that the state space can be reduced after each step by removing dominated states, and that in this way the state space can be kept reasonably small. An important factor in this is the time granularity of the problem instance at hand. If the time granularity is relatively large, compared to the total time span, then the number of possible availability times is quite small. For instance, if all shows begin and end at multiples of half an hour, and the earliest time point is 18:00 and the latest one is 23:00, then each availability time (of each tuner and of the user) can assume only 11 values. If we have two tuners, then there are 66 possible pairs (a_1, a_2) of tuner availability times, so the state space contains at most $66 \cdot 11 = 726$ states, regardless of the number of shows. If, however, the time granularity is one minute, then the number of possible availability times is much larger, and hence the state space may become very large.

A way to prevent the state space from becoming very large, is by artificially making the time granularity larger. For instance, all times may be rounded to multiples of five minutes.

For the availability time a of the user, we can stay on the safe side by only rounding upwards. Then, the amount of shows selected for watching with

```
{initialization}
P₀ = {(b₁,...,b₁;b;0;∅;∅)};
for s = 1 to n do
begin
    for σ = (a₁,...,aₘ;a;v;Sʳ,Sʷ) ∈ Pₛ₋₁ do
    begin
        {first action: skip s}
        copy σ into Pₛ;
        {second action: receive s}
        if a₁ ≤ bₛ
        then add (sort(eₛ,a₂,...,aₘ);a;v + vₛʳ;Sʳ∪{s},Sʷ) to Pₛ;
        {third action: receive & watch s}
        if a₁ ≤ bₛ ∧ max{a,bₛ} + eₛ − bₛ ≤ e
        then add (sort(eₛ,a₂,...,aₘ);max{a,bₛ} + eₛ − bₛ;v + vₛʳ + vₛʷ;Sʳ∪{s},Sʷ∪{s}) to Pₛ;
    end;
    {update states}
    for σ = (a₁,...,aₘ;a;v;Sʳ,Sʷ) ∈ Pₛ do
    begin
        for i = 1 to m do
        begin
            aᵢ = max{aᵢ,bₛ₊₁};
        end;
        a = max{a,bₛ₊₁};
    end;
    sort the states σ₁,...,σₚ in Pₛ in lexicographically increasing order
    {remove dominated states}
    for j = 1 to p − 1 do
    begin
        if σⱼ dominates σⱼ₊₁
        then remove σⱼ₊₁ from Pₛ;
    end;
end;
{end result: the (single) state in Pₙ}
```

Figure 5.2.   Pseudo code for the dynamic programming approach for TSSP.

rounded times can certainly be watched if times are not rounded. In other words, the found solution is guaranteed to be feasible, but we may lose optimality. Rounding the availability time to the nearest (not necessarily higher) multiple of the time granularity may also be done, but then the end time e may be exceeded by the eventual solution.

For the availability times a_i of the tuners $i = 1,\ldots,m$, we can also stay on the safe side by only rounding upwards. A drawback of this approach, however, is that two shows that directly succeed each other on a channel, may not be selected together. For instance, if the first show is selected, and the corresponding tuner is available again from the end time of this show, then by rounding up this availability time we may conclude that it is not possible to also receive the next show with this same tuner. If the begin and end times of the shows are such that there is a gap between the end of one show and the beginning of the next one on the same channel (for instance because of commercials in between), then this effect of rounding need not occur. Furthermore, if it is not an issue that the beginning of a show is missed, one may resolve the problem caused by rounding times upwards by simply increasing all begin times by a certain amount.

5.6.2 Filtering shows

A second way to reduce the run time of the dynamic programming approach, is to prune the set of shows before solving the instance. To this end, we set a lower bound p_{min} on the preference p_s of a show, and we simply remove all shows with $p_s < p_{min}$. In Section 5.7 we show the effect of this step. Key in this pre-filtering step is that a sufficient number of uninteresting shows are removed, but still a sufficient number of interesting shows is kept to choose from.

5.7 Experiments

The dynamic programming approach has been implemented in C++. For a first experiment, we ran the algorithm on an instance of Dutch EPG data of June 27, 2004, as shown in Figure 5.3. We selected all shows that started between 18:00h and 00:00h, giving 191 shows in total. The preference values for reception in this example were randomly drawn between 1 and 20. Furthermore, when a show is selected for watching, its value gets doubled (i.e. $v_s^w = v_s^r$). The begin and end time for watching were $b = 18:00h$ and $e = 23:00h$, and we used four tuners. The run time for this instance was about 2 minutes and 54 seconds on a Linux server with two 2.4 GHz Intel Xeon processors, with a maximum number of states reached in any iteration of 176,571. The shows selected for reception and watching are also shown in Table 5.1. Note that the time interval for watching has been filled completely, and that the shows selected for watching have an almost maximal preference value.

Reducing the number of tuners to only two, drastically reduces the run time to about 2.7 seconds, with a maximum of 6144 states in any iteration. The resulting shows selected for reception and watching are shown in Table 5.2.

5.7.1 Varying the time granularity

Next, we are going to increase the granularity of the user availability time a from 1 to 30 minutes to determine the effect on the number of states, the run time, and the total value. For the effect on the number of states, see Figure 5.4.

The effect of the granularity on the run time of the dynamic programming algorithm is shown in Figure 5.5. The effect of the granularity on the total value of the solution is less than 1%. It turned out, for instance, that for a granularity of 30 minutes the optimal solution contained shows selected for watching that all last a multiple of 30 minutes. For these shows, the effect of rounding the user availability time is void.

5.7.2   Filtering shows

Next, we are going to filter shows on their preference density value, and see what the effect is on the run time and total value. To this end, we applied a threshold $1, \ldots, 20$ on the preference density value of the shows, and solved

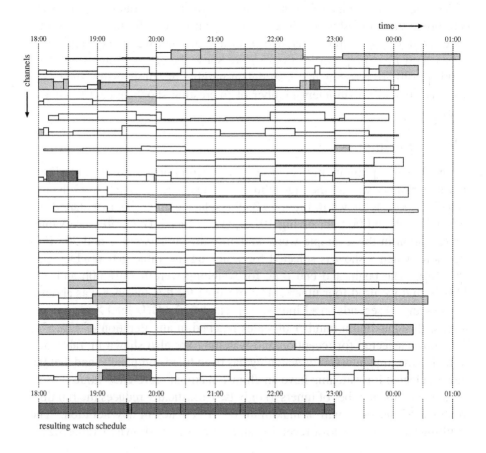

Figure 5.3.   A practical example of TV shows and the resulting schedule. The dark grey boxes indicate the shows that have been selected for watching, for which the schedule is given at the bottom, and the light grey boxes indicate the shows that have been selected for recording. The height of a box indicates the preference value of the corresponding show. The number of tuners is four in this example.

Table 5.1. The list of shows selected for reception and watching. For shows to be watched, the time indicates when they are scheduled for watching, and between brackets the corresponding time shift is given. The value is given by the weight (2 for watching, 1 for reception only) times the duration in minutes times the preference value.

time (shift)	channel	show	value
receive & watch:			
18:00–19:00	TMF	Babetrap	$2 \cdot 60 \cdot 20$
19:00–19:31 (0:52)	ARD	Sportschau: EK voetbal	$2 \cdot 31 \cdot 20$
19:31–19:32 (0:52)	ARD	Ein gutes Los für alle	$2 \cdot 1 \cdot 20$
19:32–19:35 (0:32)	Nederland 3	Nederland 3	$2 \cdot 3 \cdot 18$
19:35–20:25 (0:30)	Nickelodeon	Saved by the bell	$2 \cdot 50 \cdot 20$
20:25–21:25 (0:25)	TMF	Interactive charts	$2 \cdot 60 \cdot 20$
21:25–22:50 (0:50)	Nederland 3	Tegenlicht	$2 \cdot 85 \cdot 20$
22:50–23:00 (0:15)	Nederland 3	NOS-Journaal	$2 \cdot 10 \cdot 19$
receive only:			
18:00–18:05	KETNET/Canvas	Hopla	$1 \cdot 5 \cdot 12$
18:00–18:15	Nederland 3	Sesamstraat	$1 \cdot 15 \cdot 20$
18:00–18:55	SBS 6	Klussen & wonen	$1 \cdot 55 \cdot 19$
18:05–18:08	ARD	Der 7er Sinn	$1 \cdot 3 \cdot 4$
18:15–18:25	Nederland 3	Melatten	$1 \cdot 10 \cdot 15$
18:25–18:30	Nederland 3	Buikzingen	$1 \cdot 5 \cdot 20$
18:30–19:00	RTL 5	Stapel op auto's	$1 \cdot 30 \cdot 16$
18:40–19:05	Nickelodeon	Kuifje	$1 \cdot 25 \cdot 15$
18:55–20:30	Veronica	The Dukes of Hazzard	$1 \cdot 95 \cdot 19$
19:00–19:30	Yorin	De modepolitie	$1 \cdot 30 \cdot 19$
19:03–19:33	Nederland 3	Gewe(e)st	$1 \cdot 30 \cdot 14$
19:30–20:00	RTL 4	RTL Nieuws	$1 \cdot 30 \cdot 16$
19:33–20:35	Nederland 3	R.A.M. Compilatie	$1 \cdot 62 \cdot 19$
19:58–20:00	ARD	Heute Abend im Ersten	$1 \cdot 2 \cdot 13$
20:00–20:15	WDR Fernsehen	Tagesschau	$1 \cdot 15 \cdot 15$
20:15–20:45	Nederland 1	De vakantierechter	$1 \cdot 30 \cdot 18$
20:30–22:20	NET 5	Outrageous fortune	$1 \cdot 110 \cdot 17$
20:45–22:28	Nederland 1	KRO Detectives: Second sight	$1 \cdot 103 \cdot 20$
21:00–22:00	Discovery Channel	Xtreme martial arts	$1 \cdot 60 \cdot 20$
22:00–23:00	Discovery Channel	Xtreme martial arts	$1 \cdot 60 \cdot 19$
22:00–23:00	National Geographic	Innovation: Spy catchers	$1 \cdot 60 \cdot 14$
22:25–22:35	Nederland 3	D66	$1 \cdot 10 \cdot 18$
22:30–00:35	Veronica	The Emerald forest	$1 \cdot 125 \cdot 16$
22:45–23:40	Yorin	The bachelor	$1 \cdot 55 \cdot 17$
23:00–23:15	BBC 1	Nieuws en weerbericht	$1 \cdot 15 \cdot 11$
23:08–01:07	Nederland 1	KRO Filmtheater: Musicbox	$1 \cdot 119 \cdot 11$
23:15–00:20	SBS 6	Reportage: Oog in oog met ...	$1 \cdot 65 \cdot 20$
23:45–00:25	Nederland 2	De nieuwe pest	$1 \cdot 40 \cdot 18$

Table 5.2.   The list of shows selected for reception and watching in case of two tuners.

time (shift)	channel	show	value
receive & watch:			
18:00–18:55	SBS 6	Klussen & wonen	$2 \cdot 55 \cdot 19$
18:55–19:55 (0:55)	TMF	Babetrap	$2 \cdot 60 \cdot 20$
19:55–20:45 (0:50)	Nickelodeon	Saved by the bell	$2 \cdot 50 \cdot 20$
20:45–20:47 (0:47)	ARD	Heute Abend im Ersten	$2 \cdot 2 \cdot 13$
20:47–21:17 (0:32)	Nederland 1	De vakantierechter	$2 \cdot 30 \cdot 18$
21:17–23:00 (0:32)	Nederland 1	KRO Detectives: Second sight	$2 \cdot 103 \cdot 20$
receive only:			
18:55–20:30	Veronica	The Dukes of Hazzard	$1 \cdot 95 \cdot 19$
19:00–19:03	Nederland 3	Nederland 3	$1 \cdot 3 \cdot 18$
20:00–20:15	WDR Fernsehen	Tagesschau	$1 \cdot 15 \cdot 15$
20:35–22:00	Nederland 3	Tegenlicht	$1 \cdot 85 \cdot 20$
22:00–23:00	Discovery Channel	Xtreme martial arts	$1 \cdot 60 \cdot 19$
22:30–00:35	Veronica	The Emerald forest	$1 \cdot 125 \cdot 16$
23:00–23:15	BBC 1	Nieuws en weerbericht	$1 \cdot 15 \cdot 11$
23:15–00:20	SBS 6	Reportage: Oog in oog met ...	$1 \cdot 65 \cdot 20$

the instance with the shows having this value or higher. The results are shown in Figure 5.6. As we can see, we can save quite some run time by applying a threshold before the quality of the end solution drops.

5.8   Conclusion

We have presented the problem of how to optimally select a number of TV shows for reception and watching, and showed that the problem is NP-hard. Next, we have presented a dynamic programming approach that finds optimal solutions, and works in a reasonable run time in practice. We also presented and tested a few ways to reduce the run time of the approach, thereby finding approximate solutions.

Future work concerns the extension of the presented problem to include additional relevant problem aspects, to refine the objective function by including dependencies between shows, and the extension of the solution approach. Furthermore, when the techniques are to be used in real applications, actual values for the threshold, the time granularity, the time horizon, etcetera, have to be made.

Next, the second stage of TV evening programming has to be addressed, which concerns the compilation of sequences of shows from the set of selected and stored shows, and eventually shows also have to be chosen for deletion, in order to make room for new ones.

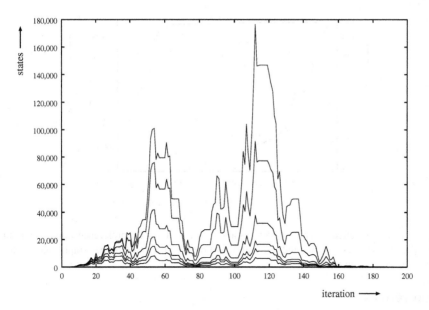

Figure 5.4. The number of states as a function of the iteration number $1, \ldots, 191$, for a granularity of 1 (top line), 2, 5, 10, 15, and 30 (bottom line) minutes, for the example with four tuners.

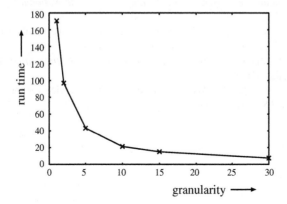

Figure 5.5. The run time of the dynamic programming approach (in seconds) as a function of the granularity (in minutes), for the example with four tuners.

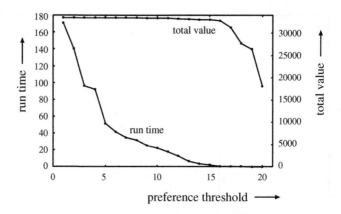

Figure 5.6. The effect of applying a threshold of $1, \ldots, 20$ for the preference value on the run time (in seconds, lower graph and left scale), and on the total value (upper graph, right scale), for the example with four tuners.

References

Garey, M.R., and D.S. Johnson [1979]. *Computers and Intractability: A Guide to the Theory of NP-Completeness.* W.H. Freeman and Company, New York.

Gutta, S., K. Kurapati, K.P. Lee, J. Martino, J. Milanski, D. Schaffer, and J. Zimmerman [2000]. TV content recommender system. In *Proceedings of the 17th National Conference on AI,* pages 1121–1122.

Korst, J. [1992]. *Periodic Multiprocessor Scheduling.* Ph.D. thesis, Eindhoven University of Technology.

Papadimitriou, C.H., and K. Steiglitz [1982]. *Combinatorial Optimization: Algorithms and Complexity.* Prentice-Hall.

Smyth, B., and P. Cotter [2000]. A personalized TV listing service for the digital TV age. *Knowledge-Based Systems,* 13:53–59.

Chapter 6

MOVIE-IN-A-MINUTE: AUTOMATICALLY GENERATED VIDEO PREVIEWS

Mauro Barbieri, Nevenka Dimitrova and Lalitha Agnihotri

Abstract   *Movie-in-a-minute* is a summarization method that enables quick browsing and access to hundreds of hours of stored video programs. A *movie-in-a-minute* is a short video sequence composed of automatically selected portions of the original video that aims at conveying key aspects of a program and its story in an efficient but entertaining way. In this chapter we discuss an approach to generating *movie-in-a-minute* summaries using film grammar rules to guide the selection of video segments that are indexed using automatically computed signal-level features.

Keywords   Automatic content analysis, video content analysis, multimedia summarization, video browsing, personal video recorders.

6.1   Introduction

Summarization has become a highly necessary tool in browsing and searching home video collections and produced video archives, saving users' time and offering great control and overview. Various types of summarization methods have been offered in the literature: visual table of contents, skimming, and multimedia summaries [Barbieri et al., 2003]. Also, various domains have been explored such as structured video summarization for news, music videos, and sports. On the other hand, the Holy Grail remains to be narrative video summarization, which includes methods for summarizing narrative content such as movies, documentaries and home videos. In this chapter we present *Movie-In-A-MInute (Miami)*, also known as 'video preview' or 'video thumbnail'. *Miami* is a short video sequence dynamically composed of selected portions from the original video. It aims at conveying key aspects of a program and its story with an array of important images and segments. *Miami* videos help users selecting programs (instead of zapping channels), deleting, downloading or simply recalling watched programs. These features are

89

Wim F.J. Verhaegh et al. (Eds.), Intelligent Algorithms in Ambient and Biomedical Computing, 89-101.
© 2006 *Springer. Printed in the Netherlands.*

indispensable for large video archives, such as personal video recorders and home network systems [Paulussen et al., 2003].

While video summaries aim at conveying all the information of the original content in shorter and more efficient versions, *Miami* videos aim only at giving users clues for selecting programs. Therefore a *Miami* video does not need to be comprehensive, or to include all highlights. Video trailers are somewhat similar to *Miami* videos although they purely aim at teasing consumers, attracting their attention and creating expectations that can or cannot be met by consuming the real content.

In this chapter we will first discuss related work to video summarization in Section 6.2. We will detail the requirements for *Miami* video in Section 6.3. In Section 6.4 we will introduce a formal model within an optimization framework that translates the requirements into constraints. Implementation and results of this model will be given in Section 6.5. Section 6.6 will introduce the need for personalization of video summaries, and Section 6.7 will conclude the chapter.

6.2   Related work

In the recent literature various video summarization methods have been introduced: *video skim, highlights,* and *multimedia summaries.*

Video skim is a temporally condensed form of the video stream that preferably preserves the most important information. It is a set of short video sequences composed of automatically selected portions of the original video. A method for generating visual skims based on scene analysis and using the grammar of film language is presented by [Sundaram et al., 2002]. [Ma et al., 2002] proposed an attention model that includes visual, audio, and text modalities for video summarization. With respect to these summarization methods, *Miami* videos are not meant to convey all the information of the original content but aim at including only key aspects of a program to allow users to quickly see what it is about and make a selection.

Video highlights is a form of summary that aims at including the most important events in the video. Various methods have been introduced for extracting highlights from specific genres of sports programs: goals in soccer video, hits in tennis video or pitching in baseball [Zhong et al., 2001], important events in car racing video [Petkovic et al., 2002], and others.

Multimedia video summary is a collection of audio, visual, and text segments that preserve the essence and the structure of the underlying video (e.g. pictorial summary, story boards, surface summary). A multimedia video summary of audio-video presentations is presented in [He et al., 1999]. The summarization system uses slide-transitions in video, pitch in audio and user interactions with presentations in order to generate a multimedia summary.

6.3 Requirements

The automatic creation of a *Miami* video can be formulated as the problem of selecting the best set of segments of a given duration of the original program that satisfies a certain list of requirements.

Two different approaches can be followed for the design of an algorithm that generates *Miami* videos: the *machine learning* approach and the *knowledge-explicit* approach. In the machine learning approach a statistical classifier is trained with positive and negative examples, selected by humans, with the aim of generalizing the common underlying properties of the examples in such a way that the classifier learns to distinguish between 'good' previews and 'bad' previews. This approach has the advantages of being generic and potentially reusable for all types of video and video previews. However, in practice the main problem is the amount of proper positive and negative examples required to train a classifier that can achieve a proper level of generalization. A simpler problem that could be tackled using machine learning is deciding for each segment of the original program whether or not it is suitable for being included in a preview. Although appealing, this method neglects to consider that a good preview is not simply a preview formed by including 'good' segments. The meaning conveyed by a video lies largely in the relationships and the temporal order of the segments of which it is composed.

In the *knowledge-explicit* approach, the designer embeds in the algorithm the knowledge on how to make a video preview in the form of requirements and constraints that drive the search for the best subset of the original program and the composition of the video preview. Machine learning can then still be used to fine-tune the parameters of the model in an objective and systematic way. Based on a study of cinematic production rules we have developed our own knowledge-explicit approach.

To allow fast and convenient content selection, a video preview should meet more than thirty requirements that have been collected by analyzing related literature on video summarization [Ma et al., 2002; Pfeiffer et al., 1996] and film production [Mascelli, 1965; Zettl, 2001] and interviewing a restricted number of 'expert' users. The requirements can be grouped into seven categories: *duration, continuity, priority, uniqueness, exclusion, structural* and *temporal order*:

- *Duration* requirements deal with the durations of the preview and of its sub-parts. Each segment chosen for the preview has a minimum time required for comprehension depending on its type, complexity, and generically speaking, amount of information it conveys.

- *Continuity* requirements necessitate that a video preview should be as continuous as possible; users will not appreciate a preview with many abrupt 'jumps'.

- *Priority* requirements indicate which content should be included in the preview to convey as much information on the program as possible in the shortest amount of time. Examples are: including sequences with close-ups of the main actors as well as action segments and dialogues giving clues on the story line.

- *Uniqueness* requirements aim at maximizing the efficiency of the preview by minimizing redundancy.

- *Exclusion* requirements indicate which content should not be included in the preview. For example a preview of a recorded broadcast program should not include any commercial advertisement. Additionally, in order not to spoil the plot of the program and allow users to later view the content in its entirety, the video preview should not disclose the end of the story.

- *Structural* requirements dictate rules that pay attention to the structural properties of video. For example, in order to provide a good overview, a video preview should cover uniformly the entire program and mimic its original mood and tempo.

- *Temporal* order requirements concern the temporal order of the sequences included in the preview. In this category, users have indicated conflicting requirements. Keeping the original order certainly helps users to understand the story line given the few clues provided by the preview. On the other hand, changing the order prevents revealing too much of the story line in case users want to later view the entire content. The choice of which requirement to follow can be left to the final user of the system.

The requirements are formalized in computable constraints in order to be used to select a subset of segments from the original program that is admissible for being a video preview. At the same time, to create a good preview it is necessary to compare admissible sets of segments to select 'the best' set. The comparison is based on a function that numerically estimates the value of a preview, the *importance score*. Given a computational model of the constraints and an importance score function to maximize, the problem of automatic generation of video previews is a constrained optimization problem and its solution can be found with known methods (e.g. constraint logic programming, local search techniques [Aarts & Lenstra, 1997]).

6.4   Formal model

In this section, the previously listed requirements are mapped to a part of an objective function or to a constraint that the *Miami* video should fulfill.

In formalizing the problem of automatic preview generation, the original program can be seen as a finite sequence of successive video segments, with synchronized audio and subtitle $V = v_1, \cdots, v_M$ where v_i is the i-th video segment in the original program. M depends on the original program duration and on the actual segmentation. The desired video preview can be represented as a finite sequence of successive positions that can be taken by any video segment belonging to the original program: $S = s_1, \cdots, s_N$ where s_j is the j-th position in the preview. N depends on the duration of the preview that is fixed by the user to a certain amount D: $\sum_{j=1}^{N} duration(s_j) \leq D$. The duration of each video segment should not be shorter than a certain minimum amount, to be understandable out of its original context and, at the same time, it should not be longer than a certain maximum value in order not to give away too many details of the story. This can be formalized by requiring: $d_{min} \leq duration(s_j) \leq d_{max}$.

The objective function whose absolute maximum denotes the best preview, has the following structure:

$$eval(S) = \alpha \cdot \pi(S) - \beta \cdot \rho(S) + \gamma \cdot \eta(S) + \delta \cdot \omega(S) . \qquad (6.1)$$

In (6.1) the priority requirements are taken into consideration in $\pi(S)$ that is defined as:

$$\pi(S) = \sum_{j=1}^{N} \pi(s_j) \qquad (6.2)$$

where $\pi(s_j)$ is the *priority score* of segment s_j and it is defined as follows:

$$\pi(s_j) = \mathbf{w} \cdot \mathbf{A}(s_j) \qquad (6.3)$$

in which $\mathbf{w}$ is a vector of weighting factors and $\mathbf{A}(s_j)$ is a column vector of attributes associated to segment s_j in the range $[0, 1]$. These attributes are computed by applying several low- and mid-level content analysis algorithms such as: computation of contrast, audio loudness, detection of action [Peker et al., 2001], faces [Abdel-Mottaleb & Elgammal, 1999], dialogues [Sundaram & Chang, 2000], music/speech/noise/silence [McKinney & Breebaart, 2003], and camera motion [Tan et al., 2000]. The relative importance of the various attributes can be linearly tuned using the weighting factors $\mathbf{w}$.

The term $\rho(S)$ in equation (6.1) estimates the degree of redundancy of the preview that in our case has a negative sign, which means we promote uniqueness and penalize redundancy. It is defined as a linear combination of visual and textual redundancy:

$$\rho(S) = \beta_1 \cdot \sum_{i=1}^{N-1} \sum_{j=i+1}^{N} \sigma_v(s_i, s_j) + \beta_2 \cdot \sum_{i=1}^{N-1} \sum_{j=i+1}^{N} \sigma_t(s_i, s_j) \qquad (6.4)$$

where $\sigma_v(s_i, s_j)$ represents the visual similarity of segments s_i and s_j and is computed based on automatically extracted visual features. Textual redundancy is measured by extracting keywords in the closed captions or in the speech transcript, $K(s_i)$, and by counting the number of times they are repeated in the preview segments:

$$\sigma_t(s_i, s_j) = |K(s_i) \cap K(s_j)| . \tag{6.5}$$

The continuity requirements can be taken into account by considering the shots as elementary segments constituting the program. The shot boundaries can be found by performing shot cut detection. Additionally, sentences should be included entirely and not be abruptly cut while subtitles should be displayed for a sufficient amount of time to be read. If we represent the synchronized audio stream A as $A = a_1, \cdots, a_Q$ (a_j being the j-th audio segment and Q the number of audio segments), the synchronized subtitles C as $C = c_1, \cdots, c_R$ (c_k being the k-th subtitle and R the total number of subtitles), and we indicate with b_s, e_s, and $\triangle_s$ respectively the start time, the end time and the time-span of the video, audio or subtitle segment s, the continuity requirement of including complete audio segments can be formalized with the following constraints (the same applies to subtitles):

$$\begin{aligned} b_a \in \triangle_s \quad \forall s \in S, \forall a : e_a \in \triangle_s \\ e_a \in \triangle_s \quad \forall s \in S, \forall a : b_a \in \triangle_s \end{aligned} \tag{6.6}$$

The requirement of not including commercial blocks can be easily fulfilled by removing the segments indicated by our commercial block detector [Schaffer et al., 2002]. Additionally, we take into consideration the requirement of not disclosing the end of the program by discarding a certain percentage[1] of segments at the end (e.g. 10%).

The structural requirement of uniform coverage of the whole program can be fulfilled by considering a segmentation of the program into L different scenes (U_j, $j = 1, \cdots, L$) and maximizing $\eta(S)$ in equation (6.1) that is the product of the relative durations of the selected segments belonging to each scene:

$$\eta(S) = \sqrt[L]{\prod_{j=1}^{L} \frac{\sum_{s \in S, s \in U_j} duration(s)}{duration(U_j)}} . \tag{6.7}$$

Scene boundaries are computed using a time-constrained clustering procedure similar to the one described in [Boreczky et al., 2000].

Temporal order requirements different from the original order are taken into consideration in the term $\omega(S)$ of equation (6.1). For example, to generate a preview having all the action segments at the end, $\omega(S)$ is defined as follows:

$$\omega(S) = \sum_{i=1}^{N} i \cdot action(s_i) \tag{6.8}$$

---

[1] A statistically sound percentage can be found by identifying for a large set of programs at which point the end is disclosed.

where $action(s_i)$ indicates whether segment s_i is classified as action segment. The original order constraint is implicitly solved during optimization by keeping the video segments chronologically ordered.

Each factor in equation (6.1) has a different scale and the weights α, β, γ, and δ are used as normalization factors for the different contributions. They can also allow personalizing the generation of the preview (as discussed in Section 6.6) by changing the relative impact of the different types of requirements on the value of the objective function.

6.5 Implementation and results

The generation of a *Miami* video can be divided into four main steps (see Figure 6.1):

1. audio and video feature extraction,

2. audio and video segmentation and classification,

3. segment selection, and

4. preview composition.

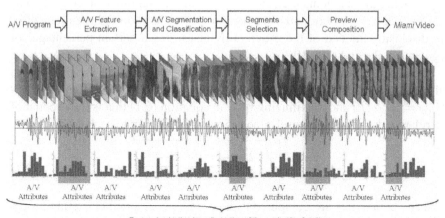

Figure 6.1. Steps for the generation of a *Miami* video

6.5.1 Audio and video feature extraction

Various algorithms are applied to the audio and video signals to extract the features required by the next steps and for computing the *priority score* according to equation (6.3) after normalization over the entire video. Video features include low-level attributes such as contrast, color distribution, motion activity and mid-level attributes such as face location and size, and camera motion. Audio features include RMS value, spectral centroid, bandwidth, zero-crossing rate, and MFCC (see [McKinney & Breebaart, 2003] for a detailed description).

6.5.2 Audio and video segmentation and classification

This step can be divided into five sub-steps:

a. *Shot segmentation and clustering*: a standard shot cut detection algorithm [Lienhart, 1999] is applied to divide the video stream into continuous shots. Time-constrained clustering [Boreczky et al., 2000] is then applied to group together visually consistent shots that are not far apart in time.

b. *Audio classification*: the synchronized audio stream is classified into coherent audio classes such as silence, speech, music, noise, etc. Changes in the audio class indicate the audio segment boundaries used to verify constraints (6.6).

c. *Micro-segmentation*: segments exceeding the maximum duration after the shot segmentation are further divided into sub-segments with durations bigger than d_{min} and with boundaries possibly aligned with content-based clues such as: a change in the audio class, appearance or disappearance of a detected face, a change in camera motion or object motion. The micro-segmentation step can be easily formalized as an integer linear programming problem and solved with standard methods (e.g. simplex method).

d. *Segment compensation*: successive segments violating constraints (6.6) are merged until the continuity requirement is fulfilled without violating the maximum segment duration d_{max}. When this is not possible, the segmentation induced by the audio classifier is used as primary instead of the shot-based one.

e. *Pre-filtering*: commercial detection [Schaffer et al., 2002] is performed over the entire video and the detected commercials are discarded from the set of segments available for the generation of the *Miami* video. In order not to disclose the end of the program, an extra 10% of segments is removed from the end.

6.5.3 Segment selection

The segment selection step consists of searching the best set of segments that maximize the objective function (6.1) in the space of all possible previews.

The space of all possible previews can be explored using a local search method such as simulated annealing or a genetic algorithm [Aarts & Lenstra, 1997] because at this point each requirement or constraint has either been solved in the previous steps or it is mapped to a priority or penalty score term in the objective function (6.1). However, considering that the actual usage of the *Miami* video is to provide a rough overview of the content of a program, the goal of finding the absolute maximum of (6.1) can be relaxed to finding a good approximation, a preview with a reasonably high value of $eval(S)$.

We have implemented a heuristic search strategy that iteratively improves an initial set of selected segments. The starting set is constructed by selecting for each scene the segment with the highest priority score $\pi(s_j)$ that generates the minimum redundancy $\rho(S)$. At every iteration, the first segments of each scene that improve the objective function are added to the set. The algorithm stops after a certain fixed number of iterations or if $eval(S)$ cannot be significantly improved. The solution is not optimal but usually good enough for the typical *Miami* video usage.

6.5.4 Preview composition

The last step consists of the actual composition of the preview by fusing the selected segments into one continuous audiovisual stream. Abrupt audio and video transitions between segments are smoothed using fading and dissolve effects.

6.5.5 Prototype implementation

A prototype of *Miami* video has been implemented in C++ (MPEG-2 decoding and content analysis algorithms) and Java™(local search, segment selection and preview composition) for the generation of previews of recorded broadcast programs in MPEG-2 video format. The generation of a *Miami* video on a state-of-the-art personal computer requires no longer than the actual program duration. Most of the CPU time is used for video decoding and content analysis algorithms; the segment selection step requires only a fraction of the total running time.

In preliminary tests the system has been manually tuned and tested with a large set of narrative programs such as feature films and documentaries. The typical duration of a *Miami* video for a two-hours-long feature film is usually set to 60 or 90 seconds.

6.5.6 Results

The first reaction of most of the users to the seeing the generated previews was always very positive. However evaluation of the results has always been a difficult task for video summarization. Just as there are many ways to describe an event or a scene, users can produce many video previews that they consider acceptable. Objective evaluation and benchmarking of different algorithms are still open challenges.

To judge whether the *Miami* video algorithm fulfils actual users' requirements, whether we should consider other requirements and, ultimately, if a *Miami* video provides a good overview of a program, we performed a user study involving ten subjects, male and female in various age categories. None of the participants were in any way involved in the development of the *Miami* video.

We conducted guided interviews organized in three parts. The first part was aimed at getting an impression of how much of the story line is comprehensible, the second part contained questions related to requirements and the third part consisted of a benchmark of against a preview generated by uniform sub-sampling.

In the first part participants had to write down a description of the story line of four movies after seeing only the corresponding *Miami* videos 60 or 120 seconds long. Some of the users had seen some of the four movies at least once in the near past. However only half of the participants who did see the movie and one third of the participants who did not see the movie could give a correct description of the story. Overall, 23% of all participants gave a wrong description. These results indicate that it is difficult to grasp the story line of a movie from a 60 or 120 seconds long *Miami* video. However presenting the ambiance of a movie is just as important. To this respect, most of the users indicated that *Miami* video is a useful tool.

In the second part, participants were shown examples of *Miami* videos and were asked various questions related to each of the seven categories of requirements presented in Section 6.3. The results indicate that the set of requirements considered by the *Miami* algorithm is relevant and complete. Generally speaking, participants were moderately positive about the degree of fulfillment of the requirements. In particular, segment duration and speech continuity were not perceived as satisfactory in many cases. Fulfillment of these requirements can be improved by using a more accurate and robust audio classifier and video segmentation algorithm.

In the third part of the interview subjects were shown two versions of a video preview (for five movies of various genres) and were asked to choose which one they preferred and why. The two versions were a *Miami* video and a preview generated by uniformly sub-sampling the program while preserving shot boundaries. The tests indicate that *Miami* video is only slightly more appreciated than uniform sub-sampling. Moreover users found it very difficult to choose between the two previews.

This could be related to the fact that some requirements (e.g. continuity) were not fully met. Users might have perceived the *Miami* videos as randomly composed as the sub-sampled versions (although this type of randomness is different from the randomness introduced by uniform sub-sampling). To verify this hypothesis, a new user test should be performed using *Miami* videos 'manually repaired' to fully meet the users' requirements.

6.6   Need for personalization

As content availability continues to grow, personalized summaries become important. For example a movie mainly containing action scenes could also have a poignant love story embedded in it. Persons who particularly like love stories might like previews highlighting these love story elements. Users will require summaries to be personalized so that they can choose the movie they like to watch and not miss out on a movie because the preview did not include sections that might appeal to them more.

Any of the requirements of duration, continuity, priority, uniqueness, exclusion, structural, and temporal order, that were presented in Section 6.3, can be subject to personalization. For example, a user might desire to see more of the introduction segments, which will then affect the structural requirements. The priority requirements can also be based on a user profile: for a person who prefers 'dark', 'silent' scenes, we should include those as opposed to 'bright', 'dialog' scenes.

So far the user preferences on summarization have not been fully explored by the research community. An exploration panel of experts and users [Agnihotri et al., 2003] on issues of multimedia summarization indicated in general that summaries should be personalized. For what concerns previews for the purpose of making a selection among a large collection of available content, users indicated in another study that previews should include scenes that might shocking. In this way they can more easily decide not to watch the entire content.

As with any personalization, the problem is twofold: to have an extensive good profile that reflects the user's needs and to have an accurate model for performing the computational matching of the user profile to the video analysis features. The challenge here is to ask the 'right' questions in order to generate this user profile. One approach is to pose this as a problem of learning from examples where users would be shown many previews and would need to select the one that appeals to them the most. Once the system is trained to the type of previews that a user likes, the different weights that were presented in Section 6.4 can be worked out in order to generate personalized previews. However a question arises whether changing the weights is sufficient to influence the segment selection step (see Section 6.5.3) in order to generate a preview that is really perceived as personalized.

6.7 Conclusions

Producing movie trailers in the production world is an art in itself. On the other hand, previews, or as we call them *movie-in-a-minute* (*Miami*) videos are not available for all the different types of narrative programs and home videos. Moreover, people with various tastes would like to see personalized previews. In this chapter we introduced a knowledge-based computational framework for generating *Miami* videos that includes: audio and video feature extraction, audio and video segmentation and classification, segment selection, and preview composition.

The framework has been implemented and manually tuned for narrative type of content such as feature films and documentaries. We conducted an initial user study that confirms that the requirements considered by the *Miami* video algorithm are relevant and complete. Users find it rather difficult to understand the story line of a movie if they see only the *Miami* video. However they also indicate that *Miami* video gives a good representation of the ambience of a program. The ambience or mood of a program seems to be an important characteristic for selecting among large collections at least in the case of home use. The study also revealed that some requirements are not yet completely fulfilled. More precise and robust content analysis algorithms could help achieving a higher level of user satisfaction.

Future work will include a formal method to fine-tune the model parameters, a more accurate benchmark of the quality of the *Miami* video algorithm with respect to other methods of generating video previews and another user study aiming at assessing the usefulness of *Miami* video in selecting programs in large video archives.

Acknowledgments

Albertine Visser the department of Psychology of the University of Utrecht The Netherlands conducted the user study described in Section 6.5.6 during an internship at Philips Research under the supervision of the first author and in collaboration with Jettie Hoonhout and Jan Engel from Philips Research, The Netherlands.

References

Aarts, E.H.L., and J.K. Lenstra [1997]. *Local Search in Combinatorial Optimization*. John Wiley & Sons, Chichester, England.

Abdel-Mottaleb, M., and A. Elgammal [1999]. Face detection in complex environments from color images. In *Proceedings of International Conference on Image Processing, ICIP*.

Agnihotri, L., N. Dimitrova, J.R. Kender, and J. Zimmerman [2003]. Study on requirement specifications for personalized multimedia summarization. In *Proceedings of the IEEE International Conference on Multimedia and Expo, ICME*, Baltimore, USA.

Barbieri, M., L. Agnihotri, and N. Dimitrova [2003]. Video summarization: Methods and landscape. In *Proceedings of SPIE International Conference on Internet Multimedia Management Systems IV, ITCom*, Orlando, USA.

Boreczky, J., A. Girgensohn, G. Golovchinsky, and S. Uchihashi [2000]. An interactive comic book presentation for exploring video. In *Proceedings of the ACM Conference on Computer-Human Interaction, CHI*, Vol. 2, No. 1, Den Haag, The Netherlands.

He, L., E. Sanocki, A. Gupta, and J. Grudin [1999]. Auto-summarization of audio-video presentations. In *Proceedings of the ACM Multimedia Conference*, Orlando, USA.

Lienhart, R. [1999]. Comparison of automatic shot boundary detection algorithms. In *Proceedings of Storage and Retrieval for Image and Video Databases VII*, Vol. 3656, USA.

Ma, Y.-F., L. Lu, H.-J. Zhang, and M. Li [2002]. A user attention model for video summarization. In *Proceedings of the ACM Multimedia Conference*, Juan Les Pin, France.

Mascelli, J.V. [1965]. *The Five C's of Cinematography – Motion Pictures Filming Techniques*. Silman-James Press, Los Angeles, CA, USA.

McKinney, M., and J. Breebaart [2003]. Features for audio and music classification. In *Proceedings of the Fourth International Symposium On Music Information Retrieval, ISMIR*, Washington DC, USA.

Paulussen, I., M. Barbieri, and G. Mekenkamp [2003]. The SPATION project: Embedding content analysis in consumer electronics networks. In *Proceedings of the Third International Workshop on Content-Based Multimedia Indexing, CBMI*, Rennes, France.

Peker, K.A., A. Divakaran, and T.V. Papathomas [2001]. Automatic measurement of intensity of motion activity of video segments. In *Proceedings of SPIE Storage and Retrieval for Media Databases Conference*, Vol. 4315.

Petkovic, M., V. Mihajlovic, and W. Jonker [2002]. Multi-modal extraction of highlights from TV formula 1 programs. In *Proceedings of the IEEE International Conference on Multimedia and Expo, ICME*, Lausanne, Switzerland.

Pfeiffer, S., R. Lienhart, S. Fischer, and W. Effelsberg [1996]. Abstracting digital movies automatically. *Journal of Visual Communication and Image Representation*, 7(4).

Schaffer, D., L. Agnihotri, N. Dimitrova, T. McGee, and S. Jeannin [2002]. Improving digital video commercial detectors with genetic algorithms. In *Proceedings of the Genetic and Evolutionary Computation Conference*, New York, USA.

Sundaram, H., and S.-F. Chang [2000]. Determining computable scenes in films and their structures using audio-visual memory models. In *Proceedings of ACM Multimedia*, Marina del Rey, USA.

Sundaram, H., L. Xie, and S.-F. Chang [2002]. A utility framework for the automatic generation of audio-visual skims. In *Proceedings of the ACM Multimedia Conference*, Juan Les Pin, France.

Tan, Y.-P., D. Saur, S. Kulkarni, and P. Ramadge [2000]. Rapid estimation of camera motion from compressed video with application to video annotation. *IEEE Transactions on Circuits and Systems for Video Technology*, 10(1).

Zettl, H. [2001]. *Sight Sound Motion – Applied Media Aesthetics*. Third Edition, Wadsworth Publishing Co., Belmont, USA.

Zhong, D., R. Kumar, and S.-F. Chang [2001]. Demonstrations: Real-time personalized sports video filtering and summarization. In *Proceedings of the ACM Multimedia Conference*, Ottawa, Canada.

Chapter 7

FEATURES FOR AUDIO CLASSIFICATION: PERCUSSIVENESS OF SOUNDS

Janto Skowronek and Martin McKinney

Abstract   Automatic classification of audio and music provides a core technology for various applications, e.g. database management tools. State of the art algorithms use various low-level signal parameters for this type of classification but recent developments focus on higher-level features that are more tangible to users. In an experimental study we investigated the *percussiveness* of sounds as one possible higher-level feature. We implemented an algorithm to describe the percussiveness of individual sounds and use this information for further classification of music. We were able to discriminate between and detect several music genres based only on these percussiveness features.

Keywords   Music information retrieval, timbre, percussiveness, automatic classification.

7.1   Introduction

Current Internet, broadcast and storage technologies enable users to access large amounts of multimedia content. For many applications and products (e.g. portable devices), users require simple and automatic tools to access, filter, process and store these amounts of data. Automatic classification is such a tool and typically consists of two stages: (1) feature extraction and (2) statistical classification.

We have investigated various low-level features such as zero-crossing rates as well as simple auditory models for different classification tasks [McKinney & Breebaart, 2003]. Our current focus is to extend these bottom-up approaches with some top-down-knowledge. We are investigating higher-level features, which are for instance based on knowledge from musicology like musical key or tempo. The idea is to use them for improving classification performance and providing features that are more tangible for users. The work presented here focuses on those higher-level features that are related to the timbre and in particular to the percussiveness of an audio signal.

Wim F.J. Verhaegh et al. (Eds.), Intelligent Algorithms in Ambient and Biomedical Computing, 103-118.
© 2006 *Springer. Printed in the Netherlands.*

7.1.1   Percussiveness of sounds

In general the term timbre is used for those characteristics of a sound that help humans to distinguish it from another sound, even if both sounds have the same loudness and the same pitch. We know from timbre research that it is rather difficult to describe timbre with one single parameter. Various signal characteristics have been found to contribute to the perception of timbre, including spectral characteristics, such as the number and distribution of spectral components, and temporal parameters, such as the attack time.

Automatic classification of sounds based on such timbre parameters has been attempted in several studies (e.g. [Jensen, 1999]) although such studies deal with the identification of single instrument sounds. There have been a few attempts (e.g. [Gouyon & Herrera, 2001]) at automatic identification of a single instrument in polyphonic music, however the problem remains unsolved. A major problem is that in polyphonic music we are confronted with overlapping instruments and quite often with synthesized or manipulated sounds, which introduce new and often time-variant timbres. The known timbre features typically refer to single monophonic instruments and are not easily applied to such sounds. This means that there is a need for more general features, which describe the timbre of arbitrary sound textures without determining the source type (e.g. music instrument) itself.

One of these general features that we are investigating is the percussiveness of a sound. "Percussiveness" does not necessarily mean a sound coming from a percussion instrument. It is rather a term used for short sounds having a sharp attack-decay characteristic. In order to get a better idea of what the term percussiveness means, consider the following four examples.

- Flute-sound: you would classify it as a typical non-percussive sound.

- Snare-drum: you would classify it as a typical percussive sound.

- Violin played sustained (the string is excited with a bow): you would classify it as a typical non-percussive sound.

- The same violin, but this time played pizzicato (the string is plucked with the fingers): you would agree that the sound is now more percussive than the latter one. The reason is that the sound has now a different time characteristic, whereas the underlying spectral structure (e.g. regularity of spectral components) is almost the same.

The examples show that we define percussiveness as a generalized description of the signal's temporal envelope.

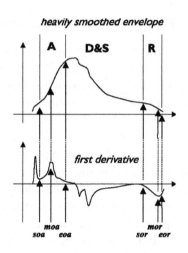

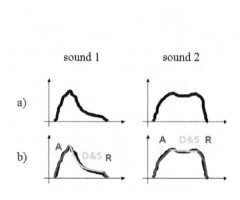

Figure 7.1. The A-D&S-R phases of two sounds.

Figure 7.2. Detection of A-D&S-R phases using the envelope and its first derivative.

7.1.2 Study on percussiveness

In order to investigate the concept of percussiveness as a feature for classifying audio and music, we performed an experimental study that comprised three major tasks: first find a parametric description of percussiveness and implement a corresponding feature extraction algorithm (Section 7.2). Then test, whether the extracted features, in fact, do describe the percussiveness of sounds (Section 7.3.1). Finally extend the algorithm to polyphonic music and test if a classification of different music genres is possible (Section 7.3.2).

7.2 Feature extraction algorithm

Because percussiveness refers to the envelope of a sound, the feature extraction algorithm should compute a parametric description of that envelope. Our approach applies first a three-phase approximation (see Figure 7.1) of the signal's envelope using parameters from synthesizer technology: Attack (A), Decay & Sustain (D&S) and Release (R). Secondly, several features are computed relating to the time durations, level differences and curve shape of these phases.

7.2.1   A-D&S-R approximation

The first step of determining the desired A-D&S-R approximation is to determine the phases' start and end points. Our approach is similar to that proposed by Jensen [1999], which consists of a three stage process. First we compute a heavily smoothed envelope and determine the desired start and end points. Secondly we adjust these points step by step using less and less smoothed versions of the envelope until the unsmoothed version is reached.

Jensen's procedure of detecting the time instances from the heavily smoothed envelope has been developed for single harmonic components of instrument sounds. He computed the first derivative of the smoothed envelope and used different derivative thresholds in order to find good candidates for the desired start and end points. But to make it work properly for the broadband signals in which we were interested, we extended Jensen's approach by using combinations of thresholds for the first derivative and for the envelope itself (see Figure 7.2):

1. The algorithm searches for the steepest point (derivative criterion) having a reasonable value (envelope criterion) and claims this as the middle of attack phase (*moa*). Starting from this *moa* point, the algorithm goes backward until certain derivative and envelope criteria are fulfilled and defines this point as start of attack phase (*soa*). Then starting from *moa* again, the algorithm goes forward and uses another derivative and envelope criterion for finding the end of attack phase (*eoa*)[1].

2. The algorithm looks for the start and end points of the release phase (*sor*, *eor*) in a similar way, this time starting with the identification of the middle of release (*mor*) and using negative derivative criteria.

3. Finally the Decay/Sustain phase is defined as the period beginning at the end of attack (*eoa*) and ending at the start of release (*sor*).

This gives start end and points of the three phases for the smoothed envelope.

In the second stage – the adjustment of the found time instances to the unsmoothed case – we used an iterative procedure. Step by step a less smoothed version of the envelope is computed and the time instances (*soa*, *eoa*, *sor*, *eor*) are adjusted using a certain time and level criterion: The new candidate must not be too far away from the former time instance and its new envelope value not too far from the former envelope value[1].

Once we found the above mentioned start and end points, we can apply the three-phase approximation of the signal envelope. Since we are interested in

---

[1] The used criteria were chosen empirically by testing the algorithm with about 40 different instrument sounds including piano, violins played sustained or pizzicato, flutes played vibrato or non-vibrato, different drums etc.

an efficient parametric description of the envelope, we applied for each phase a curve shape approximation proposed by Jensen [1999]: $Curve(x) = v_0 + (v_1 - v_0)(1 - (1-x)^n)^{1/n}$. The boundary conditions v_0 and v_1 are the envelope values for the start and end points of the phase. The variable x is the time normalized between zero and one ($t = start \rightarrow x = 0$, $t = end \rightarrow x = 1$). The scalar parameter n determines the curve form: If n is equal to 1, then the curve form is linear; if n is smaller than 1, then the curve form has an exponential characteristic; and if n is greater than 1, then the curve form is logarithmic. The optimal curve form parameter n_{opt} is found by minimizing the least-square error between the resulting curve form and the envelope.

In summary the algorithm provides a three-phase parametric description of the envelope with 11 parameters: 4 time instances (*soa*, *eoa*, *sor*, *eor*), 4 level values (*env(soa)*, *env(eoa)*, *env(sor)*, *env(eor)*) and 3 curve shape parameters (one for each phase: n_A, $n_{D\&S}$, n_R).

7.2.2 Multi-band analysis

The above described A-D&S-R approximation has been designed for calculating a parametric envelope description in one band over the whole frequency range. But in our study we used this method for a multi-band analysis as well.

For that purpose the algorithm filters the signals with a filter bank (24 ERB-rate scaled [Moore & Glasberg, 1996] approximately rectangular band-passes, linear-phase FIR filters) and computes the A-D&S-R parameters for each filter output separately.

7.2.3 Adaptation to polyphonic music

Preliminary tests with about 40 sounds showed that the algorithm works properly for single instrument sounds. In order to apply it to polyphonic music we chose an approach consisting of two steps.

1. Slice a continuous music or audio stream into (broadband) pieces starting at occurring onsets and ending at the subsequent onsets. For the onset detection we used a method implemented by Schrader [Schrader, 2003].

2. Apply the A-D&S-R approximation (broadband and multi-band analysis) and compute the features for estimating percussiveness for each audio piece.

7.2.4 Feature sets

From the derived A-D&S-R parameters, we computed an extensive list of features which were likely to be useful for our classification tasks. We organized the features into nine general groups.

- Group 1 (broadband computation): low level features per A-D&S-R phase: Time duration of phase, level difference between start and end point of phase, steepness (level difference over time duration) of phase.

- Group 2 (broadband computation): curve form description per A-D&S-R phase.

    - Curve form parameter n of phase.
    - An additional parameter that describes the error between approximation curve and real signal envelope. It is a parameter based on the autocorrelation function (ACF) of that error function. The parameter is the height of the first peak besides the zero-lag point of the ACF. It describes the 'strength' of the periodicity of the error function and is therefore called 'error regularity'.

- Group 3 (multi band computation): features that describe the asynchrony of start and end points of the phases per band. We defined asynchrony as the deviation of the time instances *soa*, *eoa*, *sor* and *eor* in one band from their mean value across all bands. Two scalar features are then computed: The mean and the variance of the asynchrony values across bands.

- Group 4 (multi band computation): mean values across bands of group 1 features, which were computed per band beforehand.

- Group 5 (multi band computation): mean values across bands of group 2 features, which were computed per band beforehand.

- Group 6 (multi band computation): variance values across bands of group 1 features, which were computed per band beforehand.

- Group 7 (multi band computation): variance values across bands of group 2 features, which were computed per band beforehand.

- Group 8 (multi band computation): features that describe the *shape* of group 1 feature values across all bands.
  Shape means the distribution of the single-band feature values when they are plotted as a function of bands. The shape is described by two parameters similar to the curve form parameter and the error regularity parameter mentioned above:

    - One parameter that describes the approximation of the shape using a linear curve. The parameter is the gradient m of the linear approximation.
    - One parameter describing the regularity of the error between shape and linear approximation. Similar to the error regularity parameter

it is based on an ACF of the error between the linear approximation and the real shape.

- Group 9 (multi band computation): Shape parameters for group 2 features.

7.3 Experiments

We investigated the performance of the implemented algorithm in two experiments that comprised a number of different classification tasks. We used a framework providing a classification algorithm based on quadratic discriminant analysis (QDA) [Duda & Hart, 1973] and a feature ranking procedure (see below). In the first experiment we tested whether the extracted features, in fact, describe the percussiveness of sounds, while in the second experiment we tested their ability to classify different music genres.

7.3.1 Experiment 1: Percussiveness of single instrument sounds

Experiment set up. We used single instrument sounds as test and training data, because for these sounds, the envelope approximation algorithm worked best. This minimized the probability that misclassifications occurred due to wrong detections of the A-D&S-R phases. Thus the classification results were as independent as possible from the feature extraction algorithm, meaning that the results were mainly affected by the type of features themselves.

The database comprised 722 single instrument sounds (classical instruments, acoustical and synthesizer drums) that we grouped into three classes: (1) percussive and non-harmonic, (2) percussive and harmonic, (3) non-percussive. The labels were assigned per instrument and play style. For instance all sounds coming from a cello that is played pizzicato were assigned to class 2, all sounds coming from a cello that is played sustained were assigned to class 3.

Goals and method. Some of the questions that we investigated with this experiment were: What is a good number of features? What is the value of the more complex features? What is the best feature set?

Since these questions are meant in terms of classification performance, we performed several classification runs, which consisted of four steps:

1. Define the conditions of the classification run:

    (a) The feature set that we want to investigate.

    (b) The number of features that the available ranking method shall select.

2. Run the feature ranking procedure:

   (a) Take the complete feature set.

   (b) Eliminate one feature and estimate the error probability ε of the remaining set, based on the so called Bhattacharyya distances [Papoulis, 1991].

   (c) Repeat this for all other features.

   (d) Take that feature, whose elimination from the feature set yielded the lowest error, as the least important feature (last ranking place).

   (e) Repeat steps (b) to (d) for the remaining feature set.

   (f) Continue this procedure until all features are ranked.

3. Estimate the classification performance using a 70/30-fold method with 25 bootstrap repetitions:

   (a) Pick randomly 30% of the feature vectors as test data and 70% as training data.

   (b) Estimate the classes of the test data using the QDA method and compare them with their real labels.

   (c) Store the number of correct and incorrect classifications per class in a confusion matrix.

   (d) Put the test data back to the training data ("Bootstrap") and repeat this procedure 25 times.

   (e) Compute the mean value and the standard deviation over the resulting 25 confusion matrices leading to a confusion matrix containing the average classification rates and their deviation intervals.

4. Compute the mean classification performance by averaging the values in the main diagonal (correct classifications) of the mean confusion matrix.

Investigation A: Dependency of performance on the number of features.
Here we checked how the classification accuracy depends on the number of features by computing the classification performance for the different feature sets using the 3, 6 and 9 best ranked features, as well as all features from each set.

We see in Table 7.1 that the classification performance increases in most cases with an increasing number of used features. The improvement between using the best three and using the best six features is in most cases significant. A further addition of features has a slightly lower impact on the classification performance, indicating a saturation effect. Therefore we had to find a good compromise between classification performance and classification effort in terms of number of used features: we chose to use nine features.

Table 7.1. Classification performance depending on the number of used features. For a detailed description of the feature groups see Section 7.2.4.

no.	feature set	number of ranked features	classification performance	no.	feature set	number of ranked features	classification performance
1	group 1	3	$69 \pm 7\%$	6	group 6	3	$74 \pm 7\%$
		6	$76 \pm 8\%$			6	$78 \pm 8\%$
		9	$87 \pm 8\%$			9	$80 \pm 9\%$
2	group 2	3	$74 \pm 8\%$	7	group 7	3	$69 \pm 14\%$
		6	$82 \pm 8\%$			6	$82 \pm 8\%$
3	group 3	3	$62 \pm 8\%$	8	group 8	3	$70 \pm 7\%$
		6	$71 \pm 8\%$			6	$76 \pm 8\%$
		8	$72 \pm 8\%$			9	$79 \pm 9\%$
4	group 4	3	$76 \pm 8\%$			18	$84 \pm 9\%$
		6	$84 \pm 9\%$	9	group 9	3	$61 \pm 7\%$
		9	$90 \pm 9\%$			6	$69 \pm 7\%$
5	group 5	3	$87 \pm 8\%$			9	$72 \pm 7\%$
		6	$86 \pm 7\%$			12	$72 \pm 7\%$

Investigation B: Classification performance dependent on feature complexity. If we have a closer look at the nine general feature groups, we see that each group requires a different computational effort. Based on the computation time, we assigned each feature set into one of three groups (A, B, C) of computational efficiency (A being most efficient, C being least efficient). Since the multi-band features are a kind of secondary features based on primary features computed per band, we distinguish between the feature complexity for the primary (*per-band*) and secondary (*across-bands*) features. Note that most of the different features sets are computed by a systematic step-by-step increase of computational complexity. Only group 3 does not fit into that system and is therefore omitted in the following discussion. Table 7.2 gives an overview about the remaining feature groups, their major computational requirements and their assigned levels for both the primary and secondary features. Especially for future applications it is interesting to see which complexity level is actually needed in order to achieve a good performance. In order to be fair, we decided to use the same number of features per set. For

Table 7.2.   Complexity levels of the different feature sets.

feature set	primary (per-band) features		secondary (across-bands) features	
	requirements	level	requirements	level
group 1	three-phase approximation + some basic calculations	A	–	–
group 2	three-phase approximation + optimization method for finding the best curve form parameter	B	–	–
group 4	three-phase approximation + some basic calculations	A	mean-value operations	A
group 5	three-phase approximation + optimization method for finding the best curve form parameter	B	mean-value operations	A
group 6	three-phase approximation + some basic calculations	A	variance-value operations	B
group 7	three-phase approximation + optimization method for finding the best curve form parameter	B	variance-value operations	B
group 8	three-phase approximation + some basic calculations	A	optimization method for finding the best shape parameter over bands	C
group 9	three-phase approximation + optimization method for finding the best curve form parameter	B	optimization method for finding the best shape parameter over bands	C

that reason we chose the six (the size of the smallest feature set) best ranked features per set.

First we considered the different complexity levels of the primary (*per-band*) features for each level of secondary (*across-bands*) features separately. That means that we compared the performance results of group 1 with group 2, group 4 with group 5, and so on. We see in Table 7.3 that using the curve shape parameters (primary level B) has a slightly positive influence on the results, except for the last case, where the performance even decreases significantly. Regarding this case, we saw in the data that group 9 showed a bad performance in general.

In a second comparison we considered the different levels of secondary features for both primary feature levels separately (e.g. group 4 vs. group 6 vs.

Table 7.3. Classification performance depending on the primary level of feature complexity.

no.	feature set	primary level	classification performance
1	group 1	A	$76 \pm 8\%$
	group 2	B	$82 \pm 8\%$
2	group 4	A	$84 \pm 9\%$
	group 5	B	$86 \pm 7\%$
3	group 6	A	$78 \pm 8\%$
	group 7	B	$82 \pm 8\%$
4	group 8	A	$76 \pm 8\%$
	group 9	B	$69 \pm 7\%$

Table 7.4. Classification performance depending on the secondary level of feature complexity.

no.	feature set	secondary level	classification performance
1	group 4	A	$84 \pm 9\%$
	group 6	B	$78 \pm 8\%$
	group 8	C	$76 \pm 8\%$
2	group 5	A	$86 \pm 7\%$
	group 7	B	$82 \pm 8\%$
	group 9	C	$69 \pm 7\%$

group 8). Table 7.4 shows that the mean values (level A) as secondary *across-bands* features led to the best performance, followed by the variances (level B) and the shape parameters (level C).

An open issue is whether it is beneficial to use the group 3 features (synchrony of envelopes). Because they are not based on the single-band features, we had to examine them differently. We compared the classification performances between group 3 and the other feature sets. With a performance of $72 \pm 8\%$, the group 3 features belong to the lowest performing feature sets (compare with results in Table 7.3). As a consequence we skipped these group 3 features.

Investigation C: Best feature set. Following the discussions in Investigation A, we were interested in the best nine features. Considering Investigation B we decided to keep both levels of primary features (low-level parameters & curve shape parameters). Regarding the secondary features, we chose only the mean values due to their best performance among the multi-band features. This pre-selection led us finally to the following feature groups: 1, 2, 4 and 5.

We tested various combinations of the preselected groups: each group alone, both single-band groups together, both multi-band groups together, all four groups together. The classification accuracies in Table 7.5 are relatively high (about 80 to 90%). In addition we see a slight tendency that with about 90% the combinations No. 3 (group 4 alone) and No. 6 (4 & 5 together) achieved the best results.

Table 7.5. Classification performance for different combinations of the feature sets.

no.	feature set	classification performance
1	group 1	$87 \pm 8\%$
2	group 2	$82 \pm 8\%$
3	group 4	$90 \pm 9\%$
4	group 5	$86 \pm 7\%$
5	groups 1 & 2	$85 \pm 8\%$
6	groups 4 & 5	$89 \pm 8\%$
7	groups 1, 2, 4 & 5	$83 \pm 8\%$

In summary we can state that we are able to predict the percussiveness of sounds with the extracted features.

7.3.2   Experiment 2: Percussiveness for classifying polyphonic music

In this second experiment we checked, whether the concept of percussiveness is useful for music genre classification. Our approach for that purpose is a three stage process. First we extracted single sounds from a 6-seconds long audio stream and estimate their percussiveness using an adequately trained classifier (first classifier). Secondly we computed secondary features (e.g. statistics) over these percussiveness estimations per audio file. And thirdly we used those secondary features for the final genre classification of the audio file (second classifier).

The idea behind this approach is relatively straight forward: we assume that different music genres contain different degrees of percussive sounds, e.g. dance music should contain a relative greater percentage of percussive sounds than classical music.

Percussiveness predictor. The task of the first classifier is to predict the percussiveness of the extracted sounds. Based on the results in Section 7.3.1 we decided to implement seven predictors consisting of the different combinations of the selected feature sets (Table 7.5), in order to figure out which predictor performs best. The training material of this first classifier consisted of sounds automatically extracted by the algorithm from 455 audio files. With the help of a user interface, a human annotator assigned the sounds into one of three categories: percussive, between percussive and non-percussive, non-percussive.

Secondary features. The secondary features used for the second classifier (music genre classification) were computed per audio file. First we took all percussiveness predictions from the first classifier belonging to one audio file. Then we computed per audio file how often (in %) the sounds are assigned to the different classes of percussiveness, e.g. 50 % percussive, 30 % between percussive and non-percussive, 20 % non-percussive. Finally we used these percentages as secondary features.

Experiment set up. As test and training data for the final classification (second classifier) we used the above mentioned audio database, but according to the experiments in [McKinney & Breebaart, 2003], we took only those files from that database (188 in total) that belong to one of seven music genres: Jazz, Folk, Electronica, R&B, Rock, Reggae and Vocal music. We used the same performance evaluation method as described in Section 7.3.1: QDA and 70/30-fold bootstrapping procedure. We computed the performances for three different tasks: classify all seven genres at once, detect each genre out of the other six, discriminate between two music genres.

Results. Table 7.6 shows the average classification performance for the seven predictors for the 'all at once' experiment. First we see that the differences between the used percussiveness predictors are lower than the performance variability within the single predictors. That means there is no dominatant percussiveness predictor that outperforms the others. Secondly we see that the results are in general quite poor. Only average performance values far below 50 % are achieved.

The poor performance shows that the approach used here does not allow classification of all seven music genres at once. However, the algorithm might be able to detect certain classes out of the remaining ones. In order to check this, we assigned one music genre to class 1 and all others to class 2 and we rerun the classification. This we repeated for all music genres. Since there was no clear best performing percussiveness predictor, we performed these tests for all seven percussiveness predictors. Table 7.7 shows the detection performance for the seven music genres of those predictors that achieved the highest accuracy. First we see that for detecting different classes, different predictors allowed the best detection. Secondly the detection of classes is moderate (70%). The results suggest that this algorithm might be useful for tasks in which a certain type of music is to be detected.

In a last test we investigated the algorithm's ability to discriminate between two music genres. For that purpose the algorithm had to classify two types of music, whereas all other music genres were excluded. This we did for each combination of the genres and for the seven predictors. Figure 7.3 shows a matrix with the discrimination performance. The numbers depicted in the fields

Table 7.6. Classification performance for classifying seven music genres.

percussiveness predictor	classification performance
1	22.96 ± 14.24%
2	24.65 ± 13.30%
3	30.65 ± 15.40%
4	33.51 ± 13.34%
5	22.80 ± 15.25%
6	31.94 ± 12.93%
7	34.44 ± 14.07%

Table 7.7. Accuracy of correct detections of a music genre out of the remaining six.

music genre	best perc. predictor	detection performance
jazz	7	70.41 ± 8.56%
folk	3	65.58 ± 11.8%
electronica	3	72.25 ± 10.86%
R&B	6	73.18 ± 7.58%
rock	7	72.46 ± 9.25%
reggae	7	80.5 ± 9.91%
vocal	4	84.25 ± 12.49%

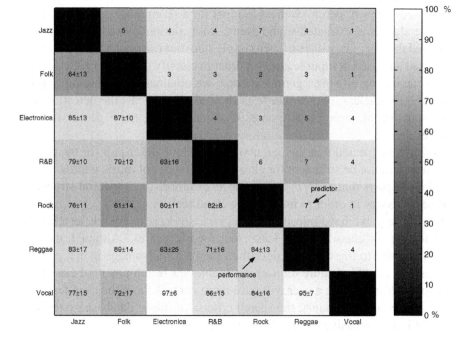

Figure 7.3. Accuracy of discriminating between two music genes. The values below main diagonal are the average accuracies, the numbers above the main diagonal denote the best predictors.

below the main diagonal are the average discrimination accuracy of each class pair, while the numbers above the main diagonal denote the best predictor for each class pair. With average values of about 60 % the genres Jazz, Folk and Rock are difficult to distinguish from each other on basis of the percussiveness predictions. In contrast several class pairs can be discriminated quite well (about 90%): Electronica vs. Folk, Electronica vs. Vocal, Reggae vs. Folk as well as Reggae vs. Vocal.

7.4 Summary

In an experimental study we tested percussiveness as a new feature for automatic classification of music.

In a first experiment we tested an algorithm that describes the envelope of a sound with a parametric three-phase description (Attack – Decay/Sustain – Release). Performing classification runs, we saw that some of these features were able to predict the percussiveness of single instrument sounds quite well (around 90% accuracy).

Encouraged by these findings we did a second experiment in which we tried to use predictions of percussiveness as input for a seven music genre classifier. First the algorithm extracted sound events from continuous audio streams and predicted their percussiveness. Then simple statistics on these predictions were computed and used as input features in order to classify the seven music genres. When classifying all seven genres at once, we obtained disappointingly poor results. Nevertheless, two further tests showed, that at least the detection and discrimination of some music genres is possible using these percussiveness features.

In fact we expected that the percussiveness as one descriptor for the timbre of sounds will not be sufficient for classifying all types of music genres. Additional descriptors that analyze other aspects of timbre (e.g. spectral structure) will be necessary in order to improve the classification.

In summary, we have introduced the *percussiveness* of sounds as a higher level feature for classifying (at least some) music genres. Although there is some necessity to improve the algorithm, the major advantage is that percussiveness is a tangible feature for users. It enables the development of classification applications, where a user will have more control about the used features. For instance he could adjust the search criteria of music database systems in terms of how percussive the desired music should be.

References

McKinney, M.F., D.J. Breebaart [2003]. Features for audio and music classification. *4th International Symposium on Music Information and Retrieval*, Baltimore, Maryland.

Jensen, K. [1999]. *Timbre Models of Musical Sounds*. PhD. Thesis, University of Copenhagen, DIKU Report 99/7.

Gouyon, F., and P. Herrera [2001]. Exploration of techniques for automatic labeling of audio drum tracks instruments. *Proceedings of MOSART Workshop on Current Research Directions in Computer Music*, Barcelona.

Moore, B.C.J., and B.R. Glasberg [1996]. A revision of Zwicker's loudness model. *Acta Acustica*, 82:335–345.

Schrader, J.E. [2003]. *Detecting and interpreting musical note onsets in polyphonic music*. MSc Thesis, Technische University Eindhoven, Eindhoven, The Netherlands.

Duda, R., and P. Hart [1973]. *Pattern Classification and Scene Analysis*. Wiley, New York.

Papoulis, A. [1991]. *Probability, Random Variables and Stochastic Processes*. McGraw-Hill, New York.

Chapter 8

EXTRACTING THE KEY FROM MUSIC

Steffen Pauws

Abstract   Extracting a sense of key from music audio is indispensable for various unequalled end-user applications dealing with music playback. This chapter
presents an audio key extraction algorithm that is based on models of human
auditory perception and music cognition. It is straightforward and has minimal computing requirements. First, it computes a chroma spectrum from short
non-overlapping time frames of audio; a chroma spectrum represents the spectral energies collected over all pitches that share the same chroma. This chroma
spectrum is compared with profiles for all existing 24 Western keys; a key profile represents the perceived stability of each chroma within a given key. The
key profile that compares best with the computed chroma spectrum is taken as
the most likely key. An evaluation with 237 CD recordings of Classical piano
sonatas indicated a classification accuracy of 86%. By considering keys that are
'friendly' to each other as equal keys, the accuracy is even 96%.

Keywords   Audio feature extraction, signal processing, music retrieval.

8.1 Introduction

Musical key is an important feature of Western (tonal) music. Knowing
the key of a piece of music is often required to do further music analysis,
segmentation, or classification. It provides a meaningful encoding of melodic
and harmonic events [Krumhansl, 1990]. It is also essential to some unequalled
end-user applications:

Advanced music playback.

> Mixing songs is a skillful art done by professional DJs to create seam-
> less transitions between songs. If two songs are mixed by aligning their
> beats (i.e., the perceived pulse in the music that makes us tap and dance),
> we talk about 'beat mixing'. If songs are mixed with respect to similar
> keys (or similar harmonic/chordal contexts), we talk about 'harmonic

119

Wim F.J. Verhaegh et al. (Eds.), Intelligent Algorithms in Ambient and Biomedical Computing, 119-132.
© 2006 *Springer. Printed in the Netherlands.*

mixing'. The latter is not pursued by all contemporary DJs as it requires a definite 'music-ear'. However, clashing keys in a mix can be startling, though do not sound as clumsy as misaligned beat structures do.

Ambiance creation.

The mode of a key (whether it is major or minor) is deemed to provide a specific emotional connotation, leaving all other music features intact [Kastner & Crowder, 1990]. This provides ample opportunities for ambiance creation using calm technology like ambient light colouring while music is played back.

For many recordings of (Historical) Classical music, the key is provided in the title of the piece. In all other cases, it is cumbersome to get at the key of a song recording. By mere listening, only musically well-trained people are able to identify a key in music. After hearing the first measure of a Bach composition, university majors were able to sing, though not name as that requires absolute pitch labelling, the scale of the key correctly in 75% of the times [Cohen, 1977]. To enable above-mentioned applications, online key extraction algorithms are required that work directly on the raw audio.

This chapter presents a key extraction algorithm, which has as input (PCM) audio data. It is based on human auditory models and music cognition and needs only a small amount of computing resources. In short, it computes a chroma spectrum from non-overlapping 100 msecs time frames of audio. A chroma spectrum represents the spectral energies collected over all pitches that share the same chroma. Then, the chroma spectrum is compared with the profiles for all 24 Western keys. A key profile represents the perceived stability of each chroma within the context of a particular musical key. The key profile that has maximum correspondence with the computed chroma spectrum is taken as the most likely key.

8.1.1   Related work

Many key extraction algorithms that are found in the literature work on symbolic data only (e.g., MIDI or notated music) by eliminating keys if the pitches are not contained in the key scales [Longuet-Higgins & Steedman, 1971], by looking for key-establishing harmonic aspects [Holtzman, 1977] or key-establishing aspects at accent locations [Chafe et al., 1982], by using key profile correlation [Krumhansl, 1990] extended with the role of subsidiary pitches and sensory memory [Huron & Parncutt, 1993], by searching for keys in the scalar and chordal domain in parallel [Vos & Van Geenen, 1996], by harmonic analysis [Temperley, 1997], by median filtering using an inter-key distance [Shmulevich & Yli-Harja, 2000], or by computing a inter-key distance using a geometric topology of tonality (i.e., the spiral array) [Chew, 2002].

Extracting key from music audio attracts more attention lately in the music cognition and music retrieval literature. They are based on modelling human tone center recognition [Leman, 1994], on a chroma spectrum representation as input to key profile correlation or a machine learning technique [Gómez & Herrera, 2004], on a constant Q transform and a fuzzy distance measure with a reference set [Purwins et al., 2000], or on a rule-based approach for rhythm structure and chord changes using a chroma spectrum representation [Shenoy et al., 2004]. Reported accuracies are in the range 59-64% for a classical music database of 217 titles [Gómez & Herrera, 2004] and 90% for a set of only 20 popular songs [Shenoy et al., 2004].

8.2 Musical pitch and key

Western music has specialized itself in the use of musical harmony, which refers to the accompaniment of a melody by chords or other melodies. Other music cultures have, for instance, strong origins in melody and rhythm. Therefore, the sense of a musical key is an important concept for composers and listeners, especially in (Historical) Classical music. But what is musical key and what is musical pitch?

Musical pitch is a categorical percept, which requires the perceptual pitch continuum expressed in Hertz-frequencies to be organized in discrete steps called semitones. To this end, the pitch continuum is first divided into octaves. An octave is an interval containing 12 semitones; two pitches that are an octave apart are a twofold of one another in frequency. The twelve positions within an octave are called *chromas* or *pitch classes*. They are notated by their standard note name (i.e., C, C♯, D, ⋯, B) or numbered from 1 to 12. Two musical pitches that are separated by an integral number of octaves share the same value of chroma. To convey both its chroma and octave, a musical pitch is often notated by its chroma and its octave number. For instance, C3, C4, and C5 share all the same chroma but occur in different octaves.

A tuning system provides frequencies to all musical pitches. In general, contemporary Western music uses equal temperament in which the size of a seminote is defined as a single frequency ratio within an octave, that is, $1 : 2^{1/12} \approx 1.05946$. Using A4 at 440 Hz as reference frequency[1] to anchor this tuning system, it is easy to calculate that a musical pitch that is i semitones apart from A4 has a frequency f of

$$f = 440 \cdot 2^{i/12}\text{Hz}. \tag{8.1}$$

---

[1]A4 means the musical pitch labelled as 'A' in the fourth octave. This comes down to key number 49 counted from the left on a grand piano. Since the end of the 17th century, the frequency of this pitch has ranged from 373.7 to 457.6 Hz. Only in 1939, A4 became 440 Hz known as the concert pitch.

In a piece of Western tonal music, there is a particular pitch that sounds most often and acts as the 'most stable' melodic base to return to. Together with this so-called tonic, six other pitches are chosen from the twelve pitches in an octave to construct a scale. These seven pitches are used as a basis for a composition. In a musically harmonic context, we do not hear these seven pitches as equal, let alone, the remaining five pitches in an octave. Put differently, the pitches are organized in a tonal hierarchy in which some pitches are 'more stable' than others [Krumhansl, 1990]. This organized set of seven pitches is called the musical key of a piece. In summary, a musical key defines the tonal material that is primarily used in a composition.

Most compositions have a main or 'home' key. Many composers (of Classical music) have their favourite key [Purwins et al., 2003]. For instance, Bach preferred A-minor, C-minor and C-major, Vivaldi preferred C-major and D-major and Chopin favored A♭-major. However, a composition rarely has a single key in its entirety. Typically, it starts and ends with the same key with some key changes (called modulations) in the middle; such a composition is called *monotonal*. If there were no key changes in the music, the music would be rather dull to listen to, not considering melodic, rhythmic and dynamic aspects. Key changes build up tension and expectations in the music.

Keys are characterized by their tonic and their mode. The mode defines what pitches are actually assigned to a key denoted as distances (i.e., semitone intervals) from the tonic. Contemporary Western music uses major mode and minor mode. For instance, the major key with C as tonic contains all white keys on the piano. Its minor variant exhanges two white piano keys for two black ones. It is this mode of a key that is believed to have specific emotional connotation [Kastner & Crowder, 1990], with minor mode referring to 'sad' and major mode referring to 'happy'.

As keys borrow seven notes from the same set of available twelve notes, keys are interrelated. Keys are said to be 'friendly' to each other if they share all or almost all pitches. 'Friendliness' means that some keys blend nicely, whereas others 'clash' when used simultaneously or successively (e.g., in a key change). Without going in-depth, a major key (e.g., C-major) is 'friendly' to its relative minor (A-minor), its dominant (G-major), and its subdominant (F-major), and, to a lesser extent, to its parallel (C-minor). Friendly keys are often a cause of confusion in music analysis by human experts and computer algorithms.

8.3   Method

In short, a chroma spectrum is computed over six octaves (i.e., 72 pitches) from A0 (27.5 Hz) to A6 (1760 Hz) using the raw audio data, which will be discussed in Section 3.1. Note that a grand piano starts at A0 (27.5 Hz) and ends at C8 (4186 Hz), so about 81% (72/88) of the keys on the piano are covered. Then, as will be explained in Section 3.2, this chroma spectrum is used as input to a key profile distance algorithm to get the musical key.

8.3.1 Chroma spectrum

As described in Section 2, a chroma represent one of the twelve note positions within an octave. By mapping the musical pitches along different octaves in a spectral representation to their respective chromas using the rules of A440-equal temperament (see Equation 8.1), we arrive at a chroma spectrum. In this way, the chroma spectrum summarizes the harmonic content of a music sample as a compact 12-element feature vector.

General model. A chroma c_i, where $i = 1, \ldots, 12$, represents a set of pitches $\{p_{ik} | k = 1, \ldots, K\}$, where K denotes the number of octaves, that have the same position i within a given octave, but that differ in octave number k. To arrive at a likelihood score for a single chroma, we have to collect the likelihood scores for all pitches sharing the same chroma in the music signal. The pitches can originate from any music instrument, from a melody, a chord, or musical background. We will extend on an existing spectral model that copes with the correlation in spectral energy between the pitches of a leading soloist and musical background [Shalev-Shwartz et al., 2002].

Let $O = o_1 \ldots o_T$ be a sequence of vectors of length T. Each vector represents a spectrum representation of the music signal over a short time frame. The problem can be formulated as finding the likelihood $P(o_t; c_i)$ of 'hearing' a chroma c_i in vector o_t.

We denote the spectral distribution of the music signal at frequency f as $S(f)$. Part of it consists of the spectral content of the chroma c_i to be found, denoted as $C(f)$. As an ideal simplification, we model a single tone of a music instrument as a harmonic series. Its spectrum contains high energy bursts at integral multiples np_{ik}, for integer n referring to the index of the harmonic. The spectral content of c_i can then be modelled as a combination of harmonic series for all pitches p_{ik} sharing c_i. In other words,

$$C(f) = \sum_{k=1}^{K} \sum_{n=1}^{N} G_{ik}(n) \delta(np_{ik} - f), \qquad (8.2)$$

where N denotes the number of harmonics, $G_{ik}(n)$ is the amplitude gain for the n-th harmonic of pitch p_{ik}, and $\delta()$ is Dirac's delta function, which is 1 at the origin and zero elsewhere. Note that we ignore the fact that the harmonics of different pitches may coincide.

Another part of $S(f)$ consists of the spectral content of the musical background. We assume that the spectral energy of this background affects the entire spectrum $S(f)$, though it matches the energy of the pitches to be found. Its spectral energy at frequency f is denoted as $\eta(f)$. The spectral content of the music signal at frequency f is then modelled as,

$$S(f) = \sum_{k=1}^{K} \sum_{n=1}^{N} G_{ik}(n)(\eta(f) + \delta(np_{ik} - f)). \tag{8.3}$$

With some rearrangement, the background energy level at frequency f, $\eta(f)$, is,

$$\eta(f) = \frac{S(f) - \sum_{k=1}^{K} \sum_{n=1}^{N} G_{ik}(n)\delta(np_{ik} - f)}{\sum_{k=1}^{K} \sum_{n=1}^{N} G_{ik}(n)}. \tag{8.4}$$

The characteristics of the musical background, $\eta(f)$, are unknown. A simple assumption is to model it as a random variable from a zero-mean multivariate Gaussian process with statistical independence at all frequencies f and equal variance v. Then, the joint conditional probability density function (pdf), given the unknown variables, is given by,

$$f(\eta|v) = \frac{1}{(2\pi v)^{L/2}} e^{-\frac{\|\eta\|^2}{2v}}, \tag{8.5}$$

where L denotes the spectrum resolution (or, the number of independent observations) and $\| \cdot \|^2$ is the l_2-norm.

The maximum likelihood (ML) estimator [Eliason, 1993] for the unknown variables can be obtained by maximizing the log-likelihood function with respect to the unknown variables:

$$(\hat{v}, \hat{G}_{ik}(n))_{\text{ML}} = \arg \max_{v, G_{ik}(n)} L_\eta(v, G_{ik}(n)) \text{ where} \tag{8.6}$$

$$L_\eta(v, G_{ik}(n)) \overset{\triangle}{=} -\frac{L}{2} \log(2\pi v) - \frac{\|\eta\|^2}{2v}. \tag{8.7}$$

By maximizing the log-likelihood function for the unknown background variance v, one obtains

$$\hat{v} = \frac{\|\eta\|^2}{L}. \tag{8.8}$$

A sensible choice for the unknown harmonic amplitudes $G_{ik}(n)$ is by letting them correspond with the spectral peaks $S(np_{ik})$ at the harmonics of the pitch frequency p_{ik}. This is sensible because we can expect that the musical background level $\eta(f)$ is relatively small compared to the energy at the harmonics of p_{ik}. By substituting the estimates for $G_{ik}(n)$ and v and using Equation 8.4, the log-likelihood score becomes

$$\begin{aligned}
L_\eta(\hat{v}, \hat{A}(n)) &= -\frac{L}{2}(\log(2\pi) + \log(\|\eta\|^2) - \log(L) + 1) \\
&= c + \frac{L}{2}\log\left(\frac{\|C\|^2}{\|N\|^2}\right),
\end{aligned} \tag{8.9}$$

where $\|C\|^2$ denotes the energy of the spectrum from Equation 8.2 in which the gains $G_{ik}(n)$ are substituted by their estimates, and $\|N\|^2$ denotes the energy in the musical background spectrum $N(f) = S(f) - C(f)$.

We consider the musical background only as a random part of the model. Consequently, the likelihood as provided by Equation 8.9 also constitutes the likelihood for finding the chroma in the spectrum. In other words,

$$P(o_t; c_i) \propto \frac{\|C\|^2}{\|N\|^2}. \tag{8.10}$$

The right-hand expression of Equation 8.10 represents the calculation of an element of the chroma spectrum vector corresponding to chroma c_i. The chroma spectrum of an observation sequence $O = o_1 \ldots o_T$ with multiple non-overlapping observation vectors is defined as the mean vector of all individual chroma spectra for each o_t.

Implementation. To identify the musical pitches more easily, the subharmonic sum spectrum [Hermes, 1988] is used as a spectral representation in Equation 8.10. All harmonics are resolved (or, folded) to their fundamental by harmonic compression (i.e., by multiplying the frequency scale by an integral factor). In other words,

$$H(f) = \sum_{n=1}^{N} h^{n-1} W(nf) A(nf), \tag{8.11}$$

where N is the number of harmonics, $h \leq 1$ is a factor controlling the contribution of each harmonic to its fundamental, $W(\cdot)$ is an auditory sensitivity filter, and $A(\cdot)$ is the amplitude spectrum.

In the calculation of Equation 8.11, the following properties are implemented for reducing computing time resources and increasing frequency resolution.

1. The input music signal is partitioned in non-overlapping time frames of 100 milliseconds.

2. The signal is low-pass filtered and downsampled to cut off spectral content above 5 kHz for performing harmonic compression over 6 octaves. It is assumed that harmonics above 5 kHz do not contribute significantly

to the pitches below 5 kHz, though we might miss highly-tuned instruments. A Hamming-window 1024-point FFT is used to compute the amplitude spectrum.

3. Spectral components (i.e., the peaks) are enhanced to cancel out spurious peaks that do not contribute to pitches.

4. Only a limited number of harmonically compressed spectra are added. We use $N = 15$. Spectral components at higher frequencies contribute less to pitch than spectral components at lower frequencies. We use $h = 0.75$.

5. Harmonic compression on a linear frequency scale is implemented as a harmonic shift in the logarithmic frequency scale by using $s = \log_2 f$. Instead of Equation 8.11, we use

$$H(s) = \sum_{n=1}^{N} h^{n-1} W(s + \log_2 n) A(s + \log_2 n). \qquad (8.12)$$

To achieve a higher frequency resolution, interpolation is used in the logarithmic frequency scale. In total, 171 ($\lceil 1024/6 \rceil$) points per octave are interpolated over 6 octaves by a cubic spline method. If we would use the original frequency resolution provided to us by FFT, we would get a frequency resolution of 9.77 Hz (10,000/1024). The lowest octave A0-A1 that we consider has a frequency range of 27.5 Hz, so this resolution would be far too low as it covers about 35% of this octave.

6. A weighting function is used to model the human auditory sensitivity for frequencies below 1250 Hz. We use a raised arc-tangent function.

The chroma spectrum for each frame is computed by locating and re-mapping the spectral regions in the harmonically compressed spectrum that correspond with each chroma in A-440 equal temperament. For the chroma C, this comes down to the spectral regions centered around the pitch frequencies for C1 (32.7 Hz), C2 (65.4 Hz), C3 (130.8 Hz), C4 (261.6 Hz), C5 (523.3 Hz) and C6 (1046.5 Hz). The width of each spectral region is taken as a half semi-tone around this center to reduce the effects of 'slightly mistuned' pitches. The amplitudes in all spectral regions are combined to form one chroma region. The chroma spectrum elements are computed using Equation 8.10. Adding and averaging the chroma spectra over all frames results in a chroma spectrum for the complete music sample (use Figure 8.1 and 8.2 for an example).

8.3.2 Maximum key-profile correlation

The maximum key-profile correlation (MKC) is an algorithm for finding the most prominent key in a music sample [Krumhansl & Kessler,

Figure 8.1. The first measure of the First Prelude in C major from Book I composed by J.S. Bach.

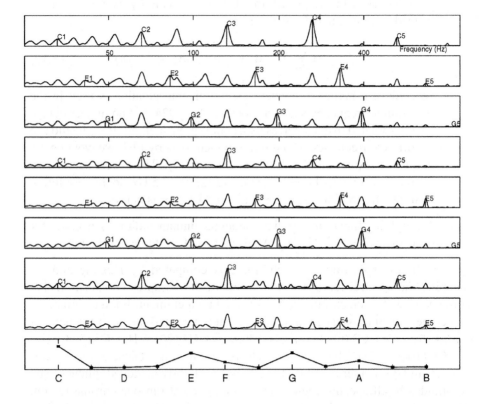

Figure 8.2. The first eight graphs present the harmonically compressed spectra for each note in the first measure of a piano performance of the First Prelude in C major from Book I composed by J.S. Bach. The frequency positions of the musical pitches over several octaves corresponding to chromas are marked in the spectra. The last graph at the bottom presents the chroma spectrum of all eight notes collecting the information in all eight spectra.

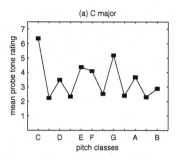 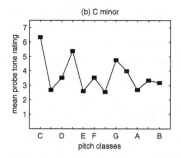

Figure 8.3. Mean probe tone rating (or key profiles) in the context of the key C major (a) and the key C minor (b).

1982; Krumhansl, 1990]. It has shown its value in research on psychological tonality measures for music [Takeuchi, 1994] and harmonic progression in improvised jazz music [Järvinen, 1995]. Originally, the algorithm was devised for symbolic encodings of music (i.e., MIDI, notated music). Here, it is used as a back-end to a signal processing step that works on raw audio data.

The MKC algorithm is based on key profiles that represent the perceived stability of each chroma within the context of a particular musical key. Krumhansl and Kessler [Krumhansl, 1990; Krumhansl & Kessler, 1982] derived the key profiles by a probe tone rating task. In this task, subjects were asked to rate, on a scale of 1 to 7, the suitability of various concluding pitches after they had listened to a preceding musical sample that established a particular key. The mean ratings represent the key profiles used in the current algorithm. These key profiles are shown for the keys C major and C minor in Figure 8.3. The graph indicates clearly that there are differences in the perceived stability of the chromas: highest ratings are given to the tonic (C), and the other two pitches of the triad (G, E), followed by the rest of pitches of the scale (F, A, D, B) to be concluded by the non-scale pitches (all sharps and flats).

Key profiles only depend on the relationship between a pitch and a tonal center and not on absolute pitches. Consequently, profiles for different major or minor keys are all transpositions of each other. For instance, the key profile for C major can be shifted six positions to arrive at a key profile for G major.

As discussed in Section 2, if music is composed in a particular key, the pitches are likely to be drawn from a single major or minor scale. Some pitches are more stable than others. The MKC algorithm is based on the assumption that the most stable chromas occur most often in a music sample. This is found to be true at least for Classical tonal compositions [Knopoff & Hutchinson, 1983]. The MKC algoritm computes the correlation (i.e., Pearson's product moment correlation) between the distribution of chroma occurrences in the musical sample and all 24 key profiles. Recall the chroma spectrum takes the role of this distribution of chroma occurences given as a vector with 12 elements. The key profile that provides the maximum correlation with the chroma spectrum is taken as the most probable key of the musical sample. The correlation value can be used as the salience of the perceived key or the degree of tonal structure of the music sample.

8.4 Evaluation

The evaluation of the algorithm consisted of an assessment of finding the correct key from a set of 237 performances of Classical piano sonatas on CD. The correct key was defined as the main key for which the musical composition was originally composed. Recall that music composers use various key modulating techniques to build up tension and relaxation in the music. However, many compositions start and end with the same key; these pieces are called *monotonal*. All recordings of the following CDs were used in the experiment.

J.S. Bach played by Rosalyn Tureck
   The Well-tempered Clavier Books I & II, 48 Preludes and Fugues
   Deutsche Grammophon, 1999.

J.S. Bach played by Jeno Jando
   The Well-tempered Clavier Book I, 24 Preludes and Fugues
   Naxos Classical, 1995.

D. Shostakovich played by Vladimir Askenazy
   24 Preludes & Fugues, op.87,
   Decca, 1999.

J. Brahms played by Glenn Gould
   The Glenn Gould Edition,
   Sony Classical, 1993.

F.F. Chopin played by Evgeny Kissin
   24 Preludes Op. 28, Sonate no. 2, Marche funebre / Polonaise op.53
   Sony Classical, 1999.

The original main key of the composition was compared with the extracted key from the CD-PCM data of the complete piano performances. In the left-hand graph of Figure 8.4, the results are shown in terms of percentage correct classification. In 86.1% of the recordings, the algorithm identified correctly the key in which the work was originally composed and performed. If we are only interested in the tonic of the key, we obtain the same accuracy, as there were no confusion between parallel keys. If we are only interested in the mode of the key, the accuracy is 92.4%. As discussed in Section 2, 'friendly' keys are often a source of confusion. If we consider the 'friendly' keys, that is, the exact key, its relative key, its dominant key (V), and its sub-dominant key (IV), as equal keys, the accuracy runs up to 96.2%. In other words, for only 3.8% (nine recordings out of a total of 237), the algorithm provided a key whose incorrectness could not be easily interpreted on music theoretic grounds.

In the right-hand graph of Figure 8.4, the same results are projected across composers. The year and time period when the pieces were composed are

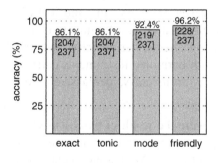
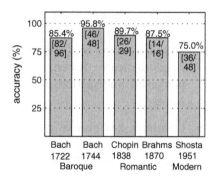

Figure 8.4. Key extraction accuracy: (Left-hand) Accuracy for finding the exact main key, the tonic of the key, the mode of the key, and all 'friendly' keys in 237 complete piano sonatas. 'Friendly' keys are the exact main key, its relative key, its dominant key (V), and its sub-dominant key (IV). (Right-hand) Accuracy for finding the exact main key for works of different composers.

shown as well. It is evident that the algoritm was less accurate for the works of Shostakovich (i.e., 75%). In addition, for five pieces, the algorithm identified a key which was not one of the 'friendly' keys. These works can still be considered highly tonal, but contain modern interpretations of harmony use because of its recency.

8.5 Conclusion

Musical key extraction from music audio is a prerequisite for matchless end-user applications like advanced music playback (e.g., automatic dj) and ambiance creation (e.g., music and lights). We presented an algorithm that correlates a chroma spectrum from audio data with profiles for all possible 24 Western keys. The key profile that has the highest correlation with the provided chroma spectrum is taken as the key of the musical fragment. The algorithm needs only minimum amount of computing necessities; it runs about 100 times real-time on a P4-2GHz platform.

The algorithm identifies correctly the exact main key in 86.1% of the cases by analyzing the complete CD recordings of Classical piano sonatas. If we assume exact, relative, dominant, sub-dominant and parallel keys as similar, it achieves a 96.2% accuracy. We have no data on recordings with other instrumentation or from other musical idioms.

The following points of the current algorithm need attention.

- Modelling the tone of an instrument as a harmonic series is highly idealized, accounting only for a single tone in a steady state. For instance, it does not account for instrument, playing, and tuning characteristics.

- A signal pre-processing stage might reveal fragments in a musical performance that contain key-relevant information and fragments that do not. This stage may check on masking effects, harmonicity, and transients to clearly discern fragments with harmonic instruments carrying perceived information on musical key from noisy, percussive instruments.

- Currently, key profiles are used that were the result of empirical work. Alternatively, the profiles can also be trained by a supervised machine learning method. The whole approach transforms then into a classification problem.

- Music perceptive and cognitive factors that establish a musical key at a human listener can be further integrated into the algorithm. Temporal, rhythmic and musical harmonic factors of pitches are not modelled, whereas it is known that the temporal order of pitches and the position of pitches in a metrical organization (e.g., the first beat, strong accents) influence the perception of a tonal center (i.e., the tonic of the key).

- Music theoretical and compositional constructs are not modelled in the algorithm. Composers use various key modulation techniques in which they signify how strong a new key will be established.

References

Chafe, C., B. Mont-Reynaud, and L. Rush [1982]. Toward an intelligent editor of digital audio: Recognition of musical constructs. *Computer Music Journal*, 6: 30–41.

Chew, E. [2002]. An algorithm for determining key boundaries. In: *Proceedings of the 2nd Intl Conference on Music and Artificial Intelligence*.

Cohen, A.J. [1977]. Tonality and perception: Musical scales prompted by excerpts from Das Wohl-temperierte Clavier of J.S.Bach. *Paper presented at the Second Workshop on Physical and Neuroppsychological Foundations of Music*, Ossiach, Austria.

Eliason, S.R. [1993]. *Maximum Likelihood Estimation: Logic and Practice*. SAGE Publications.

Hermes, D. [1988]. Measurement of pitch by subharmonic summation, *Journal of Acoustical Society of America*, 83(1): 257–264.

Holtzman, S.R. [1977]. A program for key determination, *Interface*, 6: 29–56.

Huron, D., and R. Parncutt [1993]. An improved model of tonality perception incorporating pitch salience and echoic memory, *Psychomusicology*, 12: 154–171.

Gómez, E., and P. Herrera [2004]. Estimating the tonality of polyphonic audio files: Cognitive versus machine learning modelling strategies, In: *Proceedings of 5th International Conference on Music Information Retrieval (ISMIR2004)*, Barcelona, Spain.

Järvinen, T. [1995] Tonal hierarchies in jazz improvisation, *Music Perception*, 12(4): 415–437.

Kastner, M.P., and R.G. Crowder [1990]. Perception of the major/minor distinction: IV. Emotional connations in young children, *Music Perception*, 8(2): 189–202.

Knopoff, L., and W. Hutchinson [1993]. Entropy as a measure of style: The influence of sample length. *Journal of Music Theory*, 27: 75–97.

Krumhansl, C.L. [1990]. *Cognitive Foundations of Musical Pitch*. Oxford Psychological Series, no. 17, Oxford University Press, New York.

Krumhansl, C.L., and E.J. Kessler [1982]. Tracing the dynamic changes in perceived tonal organization in a spatial representation of musical keys. *Psychological Review*, 89(4): 334–368.

Leman, M. [1994]. Schema-based tone center recognition of musical signals, *Journal of New Music Research*, 23: 169–204.

Longuet-Higgins, H.C., and M.J. Steedman [1971]. On interpreting Bach, *Machine Intelligence*, 6: 221–241.

Purwins, H., B. Blankertz, and K. Obermayer [2000]. A new method for tracking modulations in tonal music in audio data format, In: *International Joint Conference on Neural Network (IJCNN'00)*, 6: 270–275, IEEE Computer Society.

Purwins, H., T. Graepel, B. Blamkertz, and K. Obermayer [2003]. Correspondence analysis for visualizing interplay of pitch class, key and composer. In: *Perspectives in Mathematical Music Theory* E. Luis-Puebla, G. Mazzalo, and T. Noll (eds.).

Shalev-Shwartz, S., S. Dubnov, N. Friedman, and Y. Singer [2002]. Robust temporal and spectral modelling for query by melody, *Proceedings of SIGIR'02*, Tampere, Finland.

Shmulevich, I., and O. Yli-Harja [2000]. Localized key-finding: Algorithms and applications. *Music Perception*, 17(4): 531–544.

Takeuchi, A.H. [1994]. Maximum key-profile (MKC) as a measure of tonal structure in music. *Perception & Psychophysics*, 56(3): 335–346.

Temperly, D. [1997]. An algorithm for harmonic analysis. *Music Perception*, 15(1): 31–68.

Shenoy, A., R. Mohapatra, and Y. Wang [2004]. Key determination of acoustical musical signals, *Proceedings of IEEE Internal Conference on Multimedia and Expo (ICME 2004)*.

Vos, P.G., and E.W. van Geenen [1996]. A parallel processing key-finding model. *Music Perception*, 14(2): 185–224.

Chapter 9

APPROXIMATE SEMANTIC MATCHING
OF MUSIC CLASSES ON THE INTERNET

Zharko Aleksovski, Warner ten Kate, and Frank van Harmelen

Abstract   We address the problem of semantic matching, which concerns the search for semantic agreement between heterogeneous concept hierarchies. We propose a new approximation method to discover and assess the "strength" (preciseness) of a semantic match between concepts from two such concept hierarchies. We apply the method in the music domain, and present the results of preliminary tests on concept hierarchies from actual sites on the Internet.

Keywords   Music genre, music style, ontology, semantic web, semantic matching, approximation.

9.1   Introduction

The progress of information technology has made it possible to store and access large amounts of data. However, since people think in different ways and use different terminologies to store information, it becomes hard to search each other's data stores. With the advent of the Internet, which has enabled the integrated access of an ever-increasing number of such data stores, the problem becomes even more serious. The music domain is no exception. (We restrict ourselves to legal distributions.) The variety and size of offered content makes it difficult to find music of interest. It is often cumbersome to retrieve even a known piece of music.

Our ultimate goal is to improve this search for music on the Internet. We aim to use semantics in the retrieval process, which is conveyed in the Semantic Web. In this context we study the problem of semantic matching over different music provider's schemas. More specific, the problem is to find pairs of concepts (genres, styles, classes...) from different metadata schemas that have an equivalent meaning. It is not sufficient to use the concept labels only, since, for example, their position in the schemas influences their meaning as

133

Wim F.J. Verhaegh et al. (Eds.), Intelligent Algorithms in Ambient and Biomedical Computing, 133-147.

well. Figure 9.1 illustrates the problem with an example from existing music schemas.

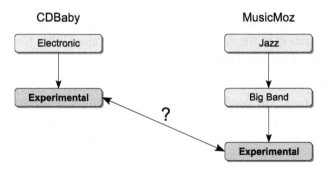

Figure 9.1. Two music genres. Although the labels are equivalent *Experimental*, they represent different classes.

The problem of finding music that fits a user's preferences is similar to the problem of matching the schemas of two different providers. In the latter case we usually need to find the pairs of concepts that have equivalent meaning. In the first case, we can regard the user's preferences as a concept description, and then the problem is to match this concept with those in the provider's terminology.

Being able to search for matches at the level of concepts, without using instances, is important. The search may use instances (artists, releases, tracks), of which there are various such approaches existing [Hayes & Cunningham, 2003; Ichise et al., 2003; Maynard et al., 2004]. However, when it comes to user preferences, people think of them semantically [Ten Kate, Ter Horst & Pauws, 2003], and these can be usually expressed in terms of concepts. Also, the provider sites publish their content in a structured way, organized in classes. Finally, the number of comparisons to be conducted reduces significantly when matching at the concept level.

We address the problem of matching between two different music concept hierarchies, which can be seen as metadata schemas as well (for survey on ontology matching techniques see the survey of Maynard et al. [2004]. We base ourselves on the approach proposed by Bouquet et al. [2003]. Main contributions of this chapter are the following.

- We propose a new method to find approximate mappings between concepts from two different concept hierarchies (see next section). Given two concepts from different concept hierarchies, our method checks whether the first concept is a subconcept of the second as described by Bouquet et al. [2003], but in addition, when that is not the case, it calculates "how strongly" the first concept can be considered a subconcept of the second. This is indicated by a value that we call sloppiness and ranges between 0.0 and 1.0 for each pair of concepts. The sloppiness indicates the error in the subsumption relation between the two concepts. Close to 0.0 means that most of the (semantic) content of the first concept is also present in the second concept, while values close to 1.0 indicate that there is no subsumption relation.

- We present first results from an analysis of the approximation method. In our study, we conducted experiments using actual data from music providers sites on the Internet. We extracted the music metadata schemas, which were underlying the navigation paths at the provider sites. We applied our approximation method on those schemas and compared them with the matches based on the instances (music artists) classified in the schemas. We discuss the problems we encountered in applying our method, and the level of correspondence observed between concepts and instances.

The rest of the chapter is structured as follows: In order to make the chapter self-contained, in Section 9.2 we discuss the approach of Bouquet et al. [2003], for the part relevant to this chapter. In Section 9.3 we discuss the present situation of music metadata schemas on the Internet. In Section 9.4 we introduce and explain our idea of approximate matching. In Section 9.5 we present some experimental results from applying the method. In Section 9.6 we discuss possible improvements on this work. Finally, Section 9.7 concludes the chapter with a brief summary.

9.2 Semantic coordination

We have taken the approach of Bouquet et al. [2003]; we summarize it briefly in this section, as relevant to our contribution. The goal is to find mappings between the concepts of two concept hierarchies.

Mapping means a relation between a pair of concepts from the two different hierarchies. There are five main types of mappings, which are subclass, superclass, equivalent, disjoint, and overlapping. For the current discussion a *concept hierarchy* can be thought of as a rooted tree where each node and each edge has a label, where a node represents a concept and an edge the subconcept relation between the concepts. It has the explicit purpose to provide an object classification.

The method compares nodes from two concept hierarchies. It proceeds in two phases. In the first phase, called explicitation, it creates a logic expression that represents each node. In the second phase it pairwise compares the nodes (logic expressions) for their relationship, in particular, whether one is subsumed by the other.

The first phase, next to the label of the nodes, the method accounts for the position of the nodes in the hierarchy. There are two main points in the phase:

- *Linguistic interpretation*: The senses that WordNet[1] returns on the words in node's label are combined as propositional terms to form the base of the logical formula. The formula represents all the possible linguistic interpretations of the label.

- *Contextualization*: The position of the nodes in the hierarchy is encoded in the logical formula. Each node's formula is considered in conjunction with its ancestor's formula, i.e. each node is assumed to be in the intersection with its ancestor. This makes sense because we expect a class to contain everything that its subclasses contain. In a similar way the disjointness of siblings in the hierarchy can be encoded into the formula.

In the second phase the method proceeds with:

- *Semantic comparison*: The so-obtained formulas from both hierarchies are evaluated for the five relationships. This is done by pair-wise combining the formulas in a grand formula that expresses the relationship to be tested. The test is performed by a SAT solver, which tests whether the grand formula is satisfiable.

For more details, see [Bouquet et al., 2003].

9.3   Internet music schemas

On the Internet, music metadata schemas mostly exist in the form of a navigation path when browsing through the music offered. A metadata schema is not always offered next to the music, but a visitor can interactively navigate through different pages that list the music offered. We consider this structure of navigation paths together with the labeling on the links and pages as the metadata schema of that provider.

9.3.1   Overview of the extracted data

After considering several music provider sites, we selected seven of them and extracted the schema, i.e. the navigation path, of each of them: CD-NOW(Amazon.com)[2], MusicMoz[3], Artist Direct Network[4], CD Baby[5], All

---

[1] http://wordnet.princeton.edu/
[2] http://www.cdnow.com
[3] http://musicmoz.org
[4] http://artistdirect.com
[5] http://www.cdbaby.com

Music Guide[6], Yahoo Launchcast[7] and Artist Gigs[8]. All of the schemas have a form of concept hierarchies that only used the subclass relationship. Sibling classes often have overlap (they are not disjoint). A general overview of the data is shown in Figure 9.2.

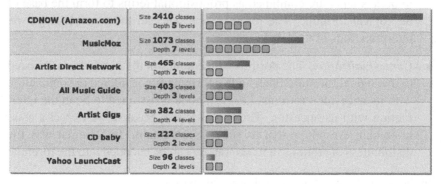

Figure 9.2. The extracted schemas from the Internet provider sites.

We simplified the extracted data. This included normalization of the labels, correcting typing mistakes, removing abbreviations and so on. Such changes were needed in order to make the data more suitable for our experiments.

Providers named the classes in their schemas following different criteria, and some even included classes whose meaning lies outside the music domain. Artist Direct Network tends to classify music by decades and has two levels depth in its schema. In general they follow the naming pattern that the first level classes are styles of music and the second level are decades. CDNOW does not strictly follow division by music-styles, but a big part of its schema has a structure adjusted for human-friendly browsing. They use classes named *Music Accessories, Vinyl releases,* which, while being a useful navigational path, do not represent a music style in our context.

Most of the labels in the schemas appeared to be of one of the following forms: style of music (the genre of the music like *Blues*), geographic region with music style (region where the music originates from, for example *American Blues*) and time or historical period when the music was created (decades like *90's*, named periods like *Baroque...*).

The labels usually consist of more than one word. In most cases, the intended meaning is either an intersection of the meanings of the separate words, or there are multiwords involved whose meaning must be considered as such. For example, the first case appears in the label *Chicago Blues*, and the second

[6]http://allmusic.com
[7]http://launch.yahoo.com
[8]http://www.artistgigs.com

case appears in *New Zealand Rock* where the meaning of the words *New* and *Zealand* can not be treated separately. The first case happens most often.

9.3.2   Fuzziness in music classification

Musical genres are not precisely defined, see e.g. [Aucouturier & Pachet, 2003; Pachet & Cazaly, 2000]. There are no objective criteria that sharply define music classes. As a result, different providers often classify the same music entities (artists, albums, songs...) differently. Widely used terms like *Pop* and *Rock* do not denote the same sets of artists at different portals. Such disagreement also appears for more specific styles of music like *Speed Metal*.

In our experiments, we compared the found matches between the classes in the schemas with the actual instance data in those classes. We restricted to the artists shared by MusicMoz and Artist Direct Network, i.e. artists that are present and classified in both portals[9]. In the sequel we refer to MusicMoz and Artist Direct Network as MM and ADN, respectively.

As Figure 9.3 shows, in the class named *Rock* (including its subclasses) in MM there are 471 shared classified artists, in ADN there are 245, and 196 shared artists are classified under *Rock* in both of them. Hence, from all the artists classified under *Rock* in at least one of the two portals, only about 38% is classified under *Rock* in both portals. This example shows that there is a high degree of fuzziness present in the music domain. This observation supports our expectation that the rigid exact-matching methods will not find matches that users would expect. Some form of flexibility is required in the matching, and approximating methods will appear to be more useful.

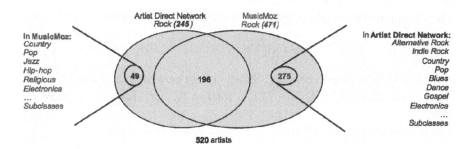

Figure 9.3.   The class *Rock* in ADN and MM.

---

[9]In MusicMoz there is a substantial number of annotated artists that are not classified in the music-style schema.

9.4 Approximate matching

In this section we explain how we extend the approach of Bouquet et al. [2003] with a form of approximation to deal with the impreciseness occurring in actual data. In our notation we use propositional logic, but we interpret the formulas as sets, and we use the operators union, intersection, and subset instead of the logical operators disjunction, conjunction, and implication, respectively. This set interpretation is justified with the fact that our formulas represent classes of music entities (artists, songs, releases...), so they actually are sets of these entities.

We focus on the problem of checking whether a subclass relation holds between two formulas. We first rewrite the formulas in normal forms. The left-hand formula is transformed into *disjunctive normal form* and the right-hand formula into *conjunctive normal form*. In this way, the subclass check can be split into a set of subproblems, each checking if one (left) disjunct is a subclass of a (right) conjunct. If all the subproblems are satisfied, the original problem is satisfied. In our approximation, we allow a few of the subproblems to be unsatisfiable, while still declaring the original problem satisfiable. The (relative) number of satisfiable subproblems is a measure of *how strongly* the subclass relation between the two given formulas hold. Below, we explain the approach more formally.

9.4.1 Normal forms

Given two propositional logic formulas A and B, the problem is to check whether the relation $A \subseteq B$ holds. We transform A into disjunctive normal form and B into conjunctive normal form.

The Disjunctive Normal Form (DNF) has the following form:

$$A = (A_1^1 \cap A_1^2 \cap \cdots \cap A_1^{n_1}) \cup (A_2^1 \cap A_2^2 \cap \cdots \cap A_2^{n_2}) \cup \cdots \cup (A_I^1 \cap A_I^2 \cap \cdots \cap A_I^{n_I}),$$

where each A_i^n is an atomic concept. Shortly it can be written as $A = A_1 \cup A_2 \cup \cdots \cup A_I$ where $A_i = (A_i^1 \cap A_i^2 \cap \cdots \cap A_i^{n_i})$ for $i = 1, \ldots, I$. Each A_i is called a *disjunct*.

The Conjunctive Normal Form (CNF) has the following form:

$$B = (B_1^1 \cup B_1^2 \cup \cdots \cup B_1^{m_1}) \cap (B_2^1 \cup B_2^2 \cup \cdots \cup B_2^{m_2}) \cap \cdots \cap (B_J^1 \cup B_J^2 \cup \cdots \cup B_J^{m_J}),$$

where each B_j^m is an atomic concept. Shortly it can be written as $B = B_1 \cap B_2 \cap \cdots \cap B_J$ where $B_j = (B_j^1 \cup B_j^2 \cup \cdots \cup B_j^{m_j})$ for $j = 1, \ldots, J$. Each B_j is called a *conjunct*.

Now, the problem to check whether $A \subseteq B$ can be written as

$$A_1 \cup A_2 \cup \cdots \cup A_I \subseteq B_1 \cap B_2 \cap \cdots \cap B_J.$$

This relation holds if and only if (iff) anything that belongs to some of the disjuncts A_i on the left-hand side also belongs to all of the conjuncts B_j on the right-hand side. Written formally:

$$A_1 \cup A_2 \cup \cdots \cup A_I \subseteq B_1 \cap B_2 \cap \cdots \cap B_J \Leftrightarrow \forall_{i=1,\ldots,I} \; \forall_{j=1,\ldots,J} \; A_i \subseteq B_j.$$

Hence, the problem whether $A \subseteq B$ is transformed into $I \cdot J$ number of subproblems of the following form:

$$\forall_{i,j} \; A_i \subseteq B_j. \tag{9.1}$$

One such a pair (A_i, B_j) will be referred to as a disjunct-conjunct pair.

Now we introduce the *idea of approximation*: the relation $A \subseteq B$ holds iff for all disjunct-conjunct pairs the subclass relation (9.1) holds. If (9.1) holds for most of the disjunct-conjunct pairs we say that the relation $A \subseteq B$ *almost holds*. Even more, we can express the strength at which the relation $A \subseteq B$ holds as the ratio between the number of disjunct-conjunct pairs that satisfy the subclass relations and the total number of pairs. We call this ratio the sloppiness and use the letter s to denote its value:

$$s(A \subseteq B) = \frac{|\{(i,j) \mid A_i \not\subseteq B_j\}|}{I \cdot J}.$$

Here $|\{(i,j) \mid A_i \not\subseteq B_j\}|$ denotes the number of disjunct-conjunct pairs that do not satisfy the subclass relation, I is the number of disjuncts in the DNF of A, and J is the number of conjuncts in the CNF of B.

Note that this method works on the concept level and can be applied when no information about the instances is available.

9.5 Experiment with approximate matching

In this section we summarize the results of experiments that we conducted using the approximate matching method. We used the metadata schemas extracted from ArtistDirectNetwork and MusicMoz.

The linguistic interpretation (i.e., the formulas build from the labels of the nodes) were obtained using simple techniques. For example, *Alternative Rock* was transformed into the following formula:

$$(Alternative \cap Rock) \cup Alternative\_Rock.$$

Special characters "&" and "/" were treated as logical union. For example, *Pop & Rock* was transformed into the formula $Pop \cup Rock$. No background knowledge was used. When using background knowledge, each atomic concept (e.g., *Alternative, Rock, Alternative_Rock*) should be replaced with the union of the different senses for that concept.

We made the assumption that concepts with the same label have the same meaning. When comparing the disjunct-conjunct relations we made a simplification: a disjunct A_i is considered to be a subclass of a conjunct B_j when some literal in the disjunct (which is an intersection of literals) is present in the conjunct (which is an union of literals). So, given a disjunct-conjunct pair:

$$A_i = (A_i^1 \cap A_i^2 \cap \cdots \cap A_i^{n_i}), B_j = (B_j^1 \cup B_j^2 \cup \cdots \cup B_j^{m_j}),$$

we say that $A_i \subseteq B_j$ if $A_i^n = B_j^m$ for some n and m. If no such pair is found, the disjunct A_i is not considered to be a subclass of the conjunct B_j. This simplification, however, may lead to some incorrect rejections of subclass relations. Also, more sophisticated techniques can be used to match the names [Bilenko et al., 2003].

9.5.1 Example of an approximate matching

Now we explain the process of approximate inferring an equivalence relation in detail. For the sake of the explanation we have chosen an example that produces simple formulas, however, in practice these formulas can grow bigger and be more complex.

In our example, consider the relation between two styles from ADN and MM that are named *Glam Rock* on both portals (Figure 9.4).

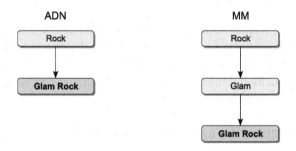

Figure 9.4. Glam Rock style from the schemas of ADN and MM.

The first step is to transform the concepts into formulas. We first transform the *Glam Rock* style from ADN. Note that *Glam Rock* is a substyle of *Rock* as shown in Figure 9.4. Also note that *Glam Rock* consists of two words. For the formula, we therefore have to take into account the separate meanings of those words (i.e., the intersection of their meanings), as well as those words constituting a single term (as is the case in "New Zealand"). Therefore the formula representing the meaning of *Glam Rock* from ADN is the following:

$$Glam\_Rock\_A = Rock \cap ((Glam \cap Rock) \cup Glam\_Rock).$$

This leads to the following normal forms:

$$Glam\_Rock\_DNF\_A = (Glam \cap Rock) \cup (Glam\_Rock \cap Rock), \quad (9.2)$$
$$Glam\_Rock\_CNF\_A = (Rock) \cap (Glam \cup Glam\_Rock). \quad (9.3)$$

Analogously, the "Glam Rock" style from MM is transformed into the formula:

$$Glam\_Rock\_B \quad = Rock \cap Glam \cap ((Glam \cap Rock) \cup Glam\_Rock)$$
$$= Rock \cap Glam.$$

The literal Glam_Rock in the formula is discarded because of the absorption rule [Mendelson, 1997]. This leads to the following normal forms:

$$Glam\_Rock\_DNF\_B \quad = \quad (Glam \cap Rock), \tag{9.4}$$
$$Glam\_Rock\_CNF\_B \quad = \quad (Rock) \cap (Glam). \tag{9.5}$$

The normal forms can be used to test the equivalence relation between the concepts *Glam_Rock_A* and *Glam_Rock_B*. We therefore have to check the subclass relation for those two concepts in both directions.

In order to check the subsumption *Glam_Rock_B* $\subseteq$ *Glam_Rock_A* the normal forms (9.3) and (9.4) are needed. *Glam_Rock_B* consists of only one disjunct, and *Glam_Rock_A* consists of two conjuncts. We therefore have to check two disjunct-conjunct pairs:

$(Glam \cap Rock) \subseteq (Rock)$ $-$ true (Rock is on both sides),
$(Glam \cap Rock) \subseteq (Glam \cup Glam\_Rock)$ $-$ true (Glam is on both sides).

Both disjunct-conjunct pairs satisfy the relation, so *Glam_Rock_B* $\subseteq$ *Glam_Rock_A* holds with a sloppiness of 0%.

In order to check the subsumption *Glam_Rock_A* $\subseteq$ *Glam_Rock_B* the normal forms (9.2) and (9.5) are needed. *Glam_Rock_A* consists of two disjuncts, and *Glam_Rock_B* consists of two conjuncts. We therefore have to check four disjunct-conjunct pairs:

$(Glam \cap Rock) \subseteq (Rock)$ $-$ true (Rock is on both sides),
$(Glam \cap Rock) \subseteq (Glam)$ $-$ true (Glam is on both sides),
$(Glam\_Rock \cap Rock) \subseteq (Rock)$ $-$ true (Rock is on both sides),
$(Glam\_Rock \cap Rock) \subseteq (Glam)$ $-$ false.

Three out of four disjunct-conjunct pairs satisfy the relation, however, one disjunct-conjunct pair does not. Hence, 25% of the disjunct-conjunct pairs do not satisfy the subsumption relation, and the relation *Glam_Rock_A* $\subseteq$ *Glam_Rock_B* therefore holds with a sloppiness of 25%.

When assessing the sloppiness in the equivalence relation between *Glam_Rock_A* and *Glam_Rock_B*, we take the maximum of the sloppiness values calculated in the two subsumptions. The equivalence relation between *Glam_Rock_A* and *Glam_Rock_B* therefore holds with a sloppiness of 25%.

9.5.2   Comparison with instance data

For our experiments we extracted real data from the Internet (Section 9.3). In the following, the results are presented that were obtained using the data sets MM and ADN (Table 9.1).

Table 9.1.   Size of the data in ArtistDirectNetwork and MusicMoz.

name	number of classes	number of artists	number of classified artists	number of shared classified artists
Artist Direct Network	465	16072	16072	1183
MusicMoz	1073	6451	2356	

Most of the shared classified artists are classified under *Rock*-related classes (e.g., *Alternative Rock, Glam Rock, Heavy Metal*). A significant limitation of our dataset is that the number of instances is of the same order as the number of classes.

The tests were performed to discover the equivalence matchings between the classes in both hierarchies, i.e., whether each is a subclass of the other. Different values for the sloppiness measure were used in the tests. In order to assess the success of the matching we introduce a value called significance, which we define as the cardinality ratio between the intersection and the union of the two classes. Formally:

$$significance(A \Leftrightarrow B) = \frac{|A \cap B|}{|A \cup B|}.$$

The significance is close to 0 when the two classes have no overlap, i.e., a relatively small amount of instances belong to their intersection. When the value is close to 1 (or 100%) then the two classes denote almost the same set of instances.

Figure 9.5 presents the average significance for different values of the sloppiness in case of equivalence testing between ADN and MM. Only classes that have at least 3 instances were observed, leaving to compare roughly 150 against 350 classes. The figure shows that the significance stays constant with increasing sloppiness before dropping down. On the other hand the number of matched equivalences was found to increase with sloppiness: from 18 matches at 0%, to 51 at 30%, to 140 at 45%, to 900 at 55%, where the onset is passed of exploding to all possible 43000 matches at 100%. This increase at constant significance suggests that the matches additionally found at first do represent correct matches. Above 40% incorrect matches prevail.

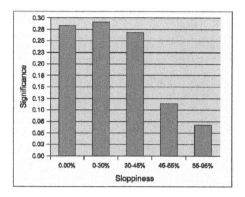

Figure 9.5.   Significance of matched equivalences between ADN and MM.

The relatively low value of the initial average significance reflects the presence of fuzziness, as discussed in Section 9.3.2. It is a notification that people have large deviation in the way they think about the music style names. It is stated by Aucouturier & Pachet [2003] that the music domain constantly evolves, and there is no centralized authority that can assign styles to the artists. They are classified in different ways, although the same name is given by the music providers.

Figure 9.6 shows the number of equivalence relations inferred given some value for the sloppiness parameter. The number of inferences increases when the sloppiness is increased. At the beginning, the number of inferences increases slowly. This is reasonable since a relatively small amount of pairs of classes from different sources should be considered to be equivalent or approximately equivalent. In general, most of the pairs of classes are not related at all, and adding sloppiness should not change this. Still, as said, more classes were found, and most of them were relevant, not altering the significance. From 50% toward the end, the number of inferences increases more rapidly. At 100% there is a "cliff", because all classes are considered to be equivalent with a sloppiness of 100%.

9.6   Future work

The presented general scheme of approximation can be improved in several directions. For example, not all disjunct-conjunct pairs are equally important in their contribution to the tested formulas. Disjuncts and conjuncts can have a different size, i.e., a different number of literals they contain. Literals may also have different size when it comes to the sets of instances they denote. Accounting for these differences, e.g., weighing may result in a more accurate sloppiness measure.

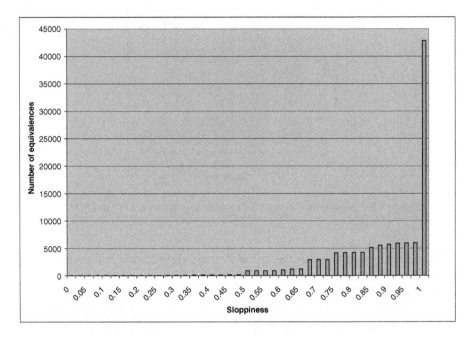

Figure 9.6. Number of equivalent relations inferred between ADN and MM using different sloppiness parameter.

Using background knowledge is another way to improve the mapping scheme. Given that two concepts are synonyms, they can be considered as equivalent in the matching process, and therefore provide a better match. Also, other relations, such as subclass between concepts, will boost the quality of the results. For example, using the fact that the Chicago region is part of America, the method can discover that *Blues, Chicago* is a subclass of *Blues, American*.

Prerequisite is the availability of the background knowledge. We are not aware of such an ontology existing in the music domain. One approach is to create one through knowledge discovery mechanisms. We conducted some preliminary experiments in which we considered two ways to extract relations between terms from the music domain. For the first we used The Free Dictionary[10] as a source, and in the second we used Google[11]. In The Free Dictionary we used as measure how strongly two terms are related, the co-occurrence of words between the pages that describe the terms. In the Google case, we assumed that related terms occur on the same pages; then, the number of Google

---

[10] http://www.thefreedictionary.com/
[11] http://www.google.com/

hits when querying for both terms relative to the number of hits when querying for each term separately, was used as strength measure for the term relation. The experiments produced useful results and we plan to continue in this direction in the future.

9.7 Conclusion

In this chapter, we have presented a new method to do approximate matching between classes from different concept hierarchies. We presented the results from applying this method to the music domain. The method is based on the approach of semantic models [Bouquet et al., 2003], and it discovers matches using logic inferencing.

We discussed the present problems in music artist classifications on the Internet, based on music content data extracted from Internet music providers. In the course of this analysis, we identified the need of integrating music content from different providers. Further, we discussed that fuzziness, as one of the main characteristics of the domain, makes the problem of matching music classes from different sources even more severe.

We applied our approximate matching method on music data extracted from the Internet. We presented and discussed the first results from these experiments. There is clear indication that the method helps to deal with this problem.

This is a preliminary work; additional research should focus not only upon implementing the suggested improvements and testing against other state-of-the-art methods, but also testing with richer data, and data from other domains. Due to the size limitations of the test data, in our study we couldn't assess the performance of the method accurately.

Acknowledgements

We would like to use this opportunity of thanking Heiner Stuckenschmidt for his useful feedback and fruitful discussions. Our thanks are also due to Aleksandar Pechkov for his feedback about the relation extraction from the Internet, and Perry Groot for his feedback and the translation into LaTeX.

References

Aucouturier, Jean-Julien, and Francois Pachet [2003]. Representing musical genre: A state of the art. *Journal of New Music Research 2003*, 32(1): 83–93.

Bilenko, Mikhail, Raymond Mooney, William Cohen, Pradeep Ravikumar, and Stephen Fienberg [2003]. Adaptive name matching in information integration. *IEEE Intelligent Systems*.

Bouquet, Paolo, Luciano Serafini, and Stefano Zanobini [2003]. Semantic coordination: A new approach and an application. In *Proc. of 2nd Int. Semantic Web Conf. (ISWC)*, Sanibel Island, Florida, USA, pages 130–145.

Hayes, Conor, and Padraig Cunningham [2003]. Context boosting collaborative recommendations. Technical Report TCD-CS-2003-26, Trinity College Dublin, Computer Science Department.

Ichise, Ryutaro, Hiedeaki Takeda, and Shinichi Honiden [2003]. Integrating multiple internet directories by instance-based learning. In *Proc. 18th Int. Joint Conf. on Artificial Intelligence (IJCAI)*, Acapulco, Mexico, pages 22–28.

Maynard, Diana, Giorgos Stamou, Heiner Stuckenschmidt, Ilya Zaihrayeu, Jesus Barrasa, Jerome Euzenat, Manfred Hauswirth, Marc Ehrig, Mustafa Jarrar, Paolo Bouquet, Pavel Shvaiko, Rose Dieng-Kuntz, Ruben Lara Hernandez, Sergio Tessaris, Sven Van Acker, and Thanh-Le Bach [2004]. State of the art on ontology alignment. Knowledge Web Deliverable D2.2.3, INRIA, Saint Ismier.

Mendelson, E. [1997]. *Introduction to Mathematical Logic*. Chapman & Hall.

Pachet, Francois, and Daniel Cazaly [2000]. A taxonomy of musical genres. In *Proc. Content-Based Multimedia Information Access (RIAO)*, Paris, France, pages 1238–1245.

ten Kate, Warner, Herman ter Horst, and Steffen Pauws [2003]. Semantics in media systems: Adapting machine operation to the human context. In *Proc.1 Int. WS on Socio-Cognitive Grids*, Santorini, Greece, pages 11–18.

Chapter 10

ONTOLOGY-BASED INFORMATION EXTRACTION FROM THE WORLD WIDE WEB

Jan Korst, Gijs Geleijnse, Nick de Jong, and Michael Verschoor

Abstract   We study possibilities to automatically extract information from the Internet, by structuring and combining data from web pages. The web pages are found with the use of a search engine and the information is structured by using ontologies. The ontologies are populated with the use of statistical and linguistic techniques.

   We present the results of a case study that is aimed at finding the names of famous persons. The results indicate that, even if we only use the summaries that Google provides of web pages, the approach results in a high precision and recall for the specific application.

Keywords   Information extraction, ontology, Google, World Wide Web, famous persons.

10.1   Introduction

In less than ten years time, the World Wide Web (WWW) has become the world's largest knowledge source. Although not all web pages contain correct information, we can benefit from the collective knowledge that it contains.

   This chapter describes methods to find, identify, structure, and combine the information available. More specifically, we investigate how information for a given knowledge domain can be extracted from the WWW. We assume that the structure of the knowledge domain is given by an ontology that specifies the type of objects (classes) and the relations (properties) between them. An example of a knowledge domain is topography, where the ontology specifies classes such as *country*, *region*, *city*, and *river*, and relations such as *capital-of (country, city)* and *lies-in (city, region)*. In addition, we assume that for some classes a number of example instances are given. The objective is to automatically extend the set of instances of each class and to determine the

Wim F.J. Verhaegh et al. (Eds.), Intelligent Algorithms in Ambient and Biomedical Computing, 149-167.
© 2006 *Springer. Printed in the Netherlands.*

properties of these instances. For example, given a few examples of countries, we want to find a complete list of all countries, identify large cities that lie in each country, and determine which countries are neighbors.

The extracted information can be used in various specific settings. For example, it can be used by a recommender system to acquire additional metadata to make meaningful recommendations for TV programs, by establishing links that would not have been found through the direct mapping of keywords representing a user's preferences to the metadata of TV programs. For example, if a user has expressed a preference for TV programs relating to Italy, then by using the extracted information the recommender system will be able to recognize terms as Tuscany, Rome, and Uffizi as relevant. Likewise, if the user has expressed a preference for TV programs relating to photography the system will be able to recognize the names of famous photographers as Cartier-Bresson and Moholy-Nagy. The extraction of information from the WWW as described in this chapter is complementary to the extraction of low-level features from the audio/video content itself that can also be used to intelligently reason about this content [Breebaart & McKinney, 2004].

In addition, the approach to automatically extract information from web pages can be instrumental in developing the semantic web [Berners-Lee, Hendler & Lassila, 2001]. This development aims at adding more structure to web pages, by using XML-based web languages as RDF(S) and OWL, such that machine interpretation of the content of web pages is facilitated. Unfortunately, the vast majority of web pages is not yet 'semantic-web'-enabled. The approach described in this chapter can be seen as a first step in automatically generating semantic web content from the ordinary WWW.

The main advantage of the use of the WWW as data source is that we need not restrict ourselves to using only a few expert sources. Instead, we can combine data from numerous pages to populate our ontology. In addition, errors that occur in some web pages can be filtered out if these are outweighed by enough correct web pages. Also, once we are able to automatically extract relevant information for a given knowledge domain, we can also automatically update it to add newly available information at regular intervals. In addition, it will be possible to adapt the database of available information to the changing preferences of the user(s).

The subject addressed in this chapter is referred to as information extraction or ontology population. It touches on multiple disciplines including information retrieval [Van Rijsbergen, 1979; Salton & McGill, 1983; Frakes & Baeza-Yates, 1992], natural language processing [Jurafsky & Martin, 2001], data mining [Fayyad, Piatetsky-Shapiro, Smith & Uthurusamy, 1996], web mining [Cooley, Mobasher & Srivastava, 1997], and other fields. It relates to areas as automatic thesaurus construction [Grefenstette, 1994] automatic term recognition [Frantzi, 1999] as well as to question-answering systems [Kwok,

Etzioni & Weld, 2001]. The problem has recently drawn quite some attention from these different fields and the approaches to tackle the problem are correspondingly quite diverse [Brin, 1998; Clerkin, Cunningham & Hayes, 2001; Faatz & Steinmetz, 2002]. Quite some papers propose methods that extensively apply linguistic techniques to parse natural language documents such as [Buchholtz, 2001]. One of the objectives of our work is to investigate to what extent one can extract information without extensive use of linguistic techniques.

The organization of this chapter is as follows. In Section 10.2 we formulate the problem of extracting information as an ontology completion problem, and we consider two important subproblems, called *similar instances problem* (SIP) and *similar statements problem* (SSP). In Section 10.3 we present a general solution approach that is applicable to both SIP and SSP. Next, in Section 10.4 we give a detailed discussion on a case study that aims at finding the names of famous people from the WWW. We present experimental results, and give an indication of the precision and recall. Finally, we end with concluding remarks in Section 10.5.

10.2 Problem definition

The subject of this chapter is to automatically instantiate the classes and properties in a given ontology that specifies an arbitrary knowledge domain. The ontology is assumed to consist of a set of classes C, a set of instances for each of these classes I, a set of properties P and a set of instances for each of these properties T, where a property is a relation, i.e., a subset of the Cartesian product of the instance sets of two given classes. The classes and properties together define the structure of the knowledge domain, the instances populate this structure. Before giving a formal problem definition, we first define the concepts of reference ontology and partial ontology.

Definition 10.1. *A reference ontology O is defined by a 4-tuple $O = (C, I, P, T)$, where*

$C = (c_1, c_2, \ldots, c_n)$ *is a set of n classes,*
  where c_i is the name of the i-th class,

$I = (I_1, I_2, \ldots, I_n)$, *where I_i is the set of instances of class c_i,*

$P = (p_1(c_{1,1}, c_{1,2}), p_2(c_{2,1}, c_{2,2}), \ldots, p_m(c_{m,1}, c_{m,2}))$ *is a set of m properties,*
  where p_j is the name of the j-th property
  and $c_{j,1}, c_{j,2} \in C$ are the names that identify the associated classes,

$T = (T_1, T_2, \ldots, T_m)$, *where T_j is the set of instances of property p_j,*
  i.e., the pairs $(a, b) \in I_{j,1} \times I_{j,2}$ for which property p_j holds.

Examples of classes in C are *country* and *city*, examples of properties in P are *capital-of (country, city)* and *lies-in (city, country)*. Of the reference ontology only the set C of classes and the set P of properties are explicitly given. The instance sets in I and T are only partially given. The problem is to extend the partially given ontology to the complete reference ontology.

Definition 10.2. *For a reference ontology $O = (C,I,P,T)$, a partial ontology O' is defined as a 4-tuple $O' = (C,I',P,T')$, where*

C and P are given by O,

$I' = (I'_1, I'_2, \ldots, I'_n)$, with $I'_i \subseteq I_i$ for $i = 1, \ldots, n$, and

$T' = (T'_1, T'_2, \ldots, T'_m)$, with $T'_j \subseteq T_j$ for $j = 1, \ldots, m$.

Now, the problem can be stated as follows.

Definition 10.3 (Ontology Completion Problem). *For a reference ontology $O = (C,I,P,T)$ describing a given knowledge domain, let $O' = (C,I',P,T')$ be a partial ontology. Extend O' to $O'' = (C,I'',P,T'')$ such that O'' approximates the reference ontology O as well as possible, that is, I''_i approximates I_i for each $c_i \in C$ and T''_j approximates T_j for each $p_j \in P$.*

The performance of an algorithm that aims at automatically extending a partial ontology by extracting information from web pages will obviously depend on the availability of web pages relating to the knowledge domain. To quantify 'approximating as well as possible' we use the common notions of precision and recall, where the precision and recall for a class $c_i \in C$ are given by

$$precision(c_i) = \frac{|I_i \cap I''_i|}{|I''_i|} \quad \text{and} \quad recall(c_i) = \frac{|I_i \cap I''_i|}{|I_i|}.$$

Precision gives the fraction of relevant instances in the total set of found instances and recall gives the fraction of found instances in the total set of relevant instances. Precision and recall for a property $p_j \in P$ are defined analogously.

The instance sets in I and T are typically not given explicitly. For some knowledge domains, human experts might even disagree on the exact content of the reference ontology. For example a list of the 1000 most important composers will be difficult to generate as musicologists will probably disagree on who should be included. However, disagreement will probably be only on a small fraction of the potential instances of a class. For our experiments, we assume that there is an undisputed knowledge source to determine the quality of our ontology completion algorithm.

Given the above definition of the ontology completion problem, we can now consider two subproblems that can be studied in isolation.

Definition 10.4 (Similar Instances Problem (SIP)). *Let $I_i' \subseteq I_i$ be given for a class $c_i \in C$. Extend I_i' to I_i'' such that $\alpha \cdot precision(c_i) + \beta \cdot recall(c_i)$ is maximized, for given $\alpha, \beta \in \mathbb{R}^+$.*

Definition 10.5 (Similar Statements Problem (SSP)). *Let $T_j' \subseteq T_j$ be given for a property $p_j \in P$. Extend T_j' to T_j'' such that $\alpha \cdot precision(p_j) + \beta \cdot recall(p_j)$ is maximized, for given $\alpha, \beta \in \mathbb{R}^+$.*

For SSP we can distinguish three variants. In the first variant, the complete instance sets $I_{1,j}$ and $I_{2,j}$ are given and the problem reduces to finding out which of the pairs (a, b) from $I_{1,j} \times I_{2,j}$ are instances of property p_j. In the second variant, only one of the instance sets is given completely. In the third variant, none of the instance sets are given completely. In the case study discussed in Section 10.4, we will concentrate on the second variant.

10.3 Solution approach

To handle subproblems SIP and SSP, we propose the following general solution approach. To illustrate the approach, let us assume that we want to extract the dates of birth of a given set of persons. The approach consists of the following four phases.

1. *Preselection.* First a preselection of potentially relevant web pages is made by issuing one or more queries to a given search engine. In our experiments we used Google as search engine. For the example, we may issue a query *"Charles Darwin was born on"*.

2. *Extraction.* The text of the web pages is scanned for occurrences of relevant phrases. If such an occurrence is found, then its context is extracted to identify potentially relevant terms. For the example, we may extract the text fragments that directly follow the phrase *"was born on"* and end with a 4-digit number expressing a year.

3. *Normalization.* Instances of a given class may occur in different textual formats. Multiple extracted text fragments that refer to the same instance are normalized to some canonical form. For the example, we may transform *Feb. 12, 1809, February 12th, 1809, 12 February, 1809,* and *the 12th of February 1809* to *February 12, 1809.*

4. *Filtering.* The list of potentially relevant terms found in the extraction phase will contain errors or incomplete instances. By using occurrence statistics and by performing additional checks, we want to filter these

out. For the example, we might discard the text fragment *this day in 1809* as incomplete, as it does not contain the name of a month. In general, the filtering phase will increase the algorithm's precision but it will potentially decrease recall.

For many applications, the above solution approach can be applied iteratively, where the terms extracted in one iteration can be used in the preselection phase of the next iteration [Geleijnse & Korst, 2005]. In the following section, we describe the four phases in more detail for a given case study.

10.4   Case study: Finding famous people on the Web

The goal of this case study is to extract the names of famous people from the Web and to determine the period in which they lived, restricting ourselves to persons that are born after the year 1000 and have already died. For this case study, the given partial ontology is given by $O' = (C, I', P, T')$, with

$C = (name, year-of-birth, year-of-death),$

$I' = (\emptyset, \{1000, \ldots, 1990\}, \{1015, \ldots, 2004\}),$

$P = (born-in(name, year-of-birth), died-in(name, year-of-death)),$ and

$T' = (\emptyset, \emptyset).$

We assume that a potentially famous person has at least reached the age of 15.

As we are interested in extracting an "eternal hall of fame", we restrict ourselves to persons that have already died. For a living person, the number of web pages that contain his or her name may vary considerably over time. And very often, fame turns out to be of a temporary nature.

Note that in many cases person names are not unique. A name may refer to multiple persons, e.g., *Theo van Gogh* refers to the brother of the painter Vincent van Gogh as well as to the movie director that has been murdered in Amsterdam in 2004. Since we restrict ourselves to persons that have already died, we can in addition use the years of birth and death to uniquely identify a person.

This section is organized as follows. We first present the preselection, extraction, normalization and filtering phases of our solution approach in more detail. Next, we present a possible measure of fame that allows us to rank the persons found. Finally, we describe the results of a number of tests that we have carried out to estimate recall and precision of the extraction and filtering algorithms.

10.4.1 Preselection

The first phase in our solution approach is to make a preselection of potentially relevant web pages by issuing appropriate queries to Google. The pages that Google returns can then be analyzed in the following phases.

Now, what would be an appropriate query to find the names of famous persons in web pages? One possibility is to look for text fragments as "*was born in*" or "*died in*" in web pages to analyze the text that directly precedes and succeeds these text fragments to identify person names and years, respectively. A drawback of these queries is that they are not specific enough. Google returns at most 1000 pages for each query, which considerably restricts the number of persons that we can find in this way. More specific queries are "*was born in x*" and "*died in x*", where x is a year chosen from the set of possible birth and death years. Although this considerably increases the potential number of persons that we can find in this way, these queries might still not be specific enough. For example, query "*was born in 1685*" will result in at most 1000 pages, that are likely to be dominated by Johann Sebastian Bach. Other less famous persons that are born in this year might not be found in this way.

To generate enough specific queries, we instead use the lifetime of persons to try to find the corresponding person names. To this end, we simply issue Google queries of the form

$$\text{``}(y_1 - y_2)\text{''}$$

with $y_1 \in [1000..1990]$, $y_2 - y_1 \in [15..110]$, and $y_2 \leq 2004$, resulting in a total of approximately 100,000 queries. In other words, we search for persons who were born during the last millennium and who died at an age between 15 and 110.

Since we are only interested in potentially famous people, we want to avoid web pages that offer genealogical data. To this end, we restrict ourselves to web pages that do not contain words as *genealogy* or *genealogie*.

10.4.2 Extraction

For each of these issued queries, we scan the (at most) 1000 excerpts that Google returns. In each of these excerpts, we determine the first occurrence of the queried pair of numbers. Since Google ignores non-alphanumeric characters, the queried pair of numbers may also occur as y_1, y_2 or as y_1/y_2. If the queried pair of numbers is in the intended context $(y_1 - y_2)$, i.e., if the numbers are separated by a hyphen and surrounded by brackets, then the words directly preceding this first occurrence are stored for later analysis, to a maximum of six words. In this way, we obtain for each queried pair of numbers up to 1000 short text fragments that potentially contain person names. In addition, for each of the stored text fragments, we trim potential prefixes that normally cannot be part of a name. For example, we delete all words that precede a full stop (except when preceded by a single capital letter), a colon, or a semicolon.

10.4.3 Normalization

We observe that a single person is often identified in different ways, e.g. *Johann Sebastian Bach, JS Bach, JOHANN SEBASTIAN BACH* and *Bach, Johann Sebastian* all refer to the same person. The latter variant is called an *inversion*. The latter two variants can be transformed into the first variant by substituting upper-case characters by lower-case ones and by adjusting the order of first and last names[1]. Of the words that only consist of multiple upper-case letters we transform the upper-case into lower-case letters, except for the first one (with some specific exceptions concerning initials, ordinal numbers of kings, queens, etc., composite names including hyphens or apostrophes, and Scottish and Irish names). Complicating factors in the identification of inversions are (*i*) that a comma between last name and first names is sometimes omitted and (*ii*) that many first names also occur as last names. An additional complication is that the first names of a name sometimes vary per language.

We assume the combination of last name and lifetime to be specific enough to uniquely identify famous persons. Clearly, examples can be found were multiple persons share the same last name and lifetime, especially when they have a common last name such as *Smith*. However, we assume the probability that two of these are both well known to be negligible.

In addition, we also want to filter out other multiple occurrences of the same person name. These occurrences are caused by variations in spelling of names and errors in the lifetimes. To this end, we carried out the following filtering steps.

1. *Keeping only the first-name variant that occurs most often.* For each last-name/lifetime combination, we often find different variants of first names preceding it. For example, *Bach (1685 – 1750)* is preceded by, e.g., *Johann Sebastian, JS*, and *Johann S*. Of all these variants we only keep the one that is found most often, i.e., the variant that occurs most often in the text fragments we found in the 1000 excerpts that Google returned on query *"(1685 – 1750)"*.

2. *Filtering out small variations in name.* If two names have exactly the same lifetime and the edit distance [Levenshtein, 1966; Gusfield, 1997] between these full names is less than a given threshold, then only the

---

[1] We use 'first names' to identify given names as *William* and *Johann Sebastian* and 'last names' to identify family names as *Shakespeare* and *Bach*, even though for some countries as Hungary and Japan it is more common to place the family name before the given name.

variant that is found most often is kept. As threshold we use an edit distance of two.

3. *Filtering out single errors in lifetimes.* If two names are completely identical but their lifetimes differ in only the year of birth *or* the year of death, then only the variant that is found most often is kept.

Experiments indicate that in this step we reduce the candidate set of names by approximately 25%.

10.4.4 Filtering

Not all text fragments we have found after the first three phases will be person names. Typically, historic periods, art styles, geographic names, etc. can also directly precede a time interval. Table 10.1 illustrates the difficulties in discriminating between person names and other text fragments. We note that *West Mae* is supposed to be an inversion of the name *Mae West* and that *Napoleon Hill* refers to a person as well as to a geographic location in the state Idaho (USA).

Table 10.1. Some examples to illustrate the difficulties in discriminating between persons names and other text fragments.

person name	non-person names
Art Blakey	Art Deco
West Mae	West Virginia
Amy Beach	Miami Beach
HP Lovecraft	HP Inkjet
Napoleon Hill	Napoleon Hill

To filter out non-person names, we distinguish the following three approaches.

1. *Looking for negative clues.* By using a list of words that typically do not occur in person names, we can filter out all text fragments that contain one of more of these words.

2. *Looking for positive clues.* By using a list of common first and last names, we can select only the text fragments that contain one or more of these names.

3. *Looking for further evidence.* By issuing additional Google queries ("X was born in"), one can build up further evidence on whether or not the candidate name refers to a person.

Experimental results indicate that the third approach seems to be less effective for this application, although it requires the least amount of specific knowledge, and as such is the most generally applicable one. Comparing the first two approaches, we observe that selecting on negative clues puts more emphasis on high recall, while selecting on positive clues puts more emphasis on high precision. Since we aim at generating a list of famous persons that is as complete as possible, we prefer to filter out on negative clues. Hence, we chose to use the first approach.

To obtain a list of words that typically do not occur in person names, we cannot simply use the words occurring in a dictionary, for the following two reasons. First of all, many common words that occur in a dictionary also occur in person names. Table 10.2 gives a number of examples. Furthermore, we do not want to restrict ourselves to web pages that are written in one or a few languages.

Table 10.2. Examples illustrating that common words as well as geographic names can be part of a person name.

Philip Glass	Jack London	Charles Herbert Best
Alonzo Church	Shirley Temple	Edwin Herbert Land
Nicci French	Dorothy Day	Bernard Law Montgomery
Christopher Love	Irving Berlin	Isambard Kingdom Brunel
Max Born	Sitting Bull	Francis de Sales
Vincent Voiture	Florence Nightingale	Hound Dog Taylor
Witte de With	Wilhelm Reich	Edward Mandell House

Instead, we semi-automatically generated a list of words that do not occur in person names, as follows. For the words that occur in the text fragments we have generated so far, we determined how often they occur in the text fragments. More specifically, we determined how often they occur with a capital and how often they occur without a capital. If words mostly occur without a capital, then this is a strong indication that they do not occur in person names. Of these candidates, we next checked by hand the ones that occurred most often in the text fragments. This resulted in a list of less than 2500 words, where different variants of the same word are counted separately. For example, *Archive*, *Archives*, *archive*, and *archives* are four of the 2500 words. Hence, these 2500 words represent only a very small fraction of the total number of words that occur in the various languages. Experiments indicate that in this step we reduce the candidate set of names by approximately 10%.

10.4.5 Measure of fame

To be able to rank the persons found, we use the Google page count (GPC) as our measure of fame, i.e., the number of web pages that contain the person's name as estimated by Google. Now, the question is which query we should issue to Google to determine the GPC of a person. The query should be neither too general nor too specific.

The problem is that very often person names are not unique. We have already observed that a full name is often not specific enough to uniquely identify a person but that we assume the combination of last name and years of birth and death to uniquely identify a person. Consequently, we issue the following query to determine the GPC:

$$\text{``last-name (year-of-birth – year-of-death)''},$$

where *last-name* is simply the last word in a candidate person name consisting of one or more words. In this way *Johann Sebastian Bach (1685 – 1750)*, *JS Bach (1685 – 1750)*, and *Bach (1685 – 1750)* are all covered by this same query. For kings, queens, popes, etc., we use the Latin ordinal number as last name. In this way *Charles V (1500 – 1558)*, *Carlos V (1500 – 1558)*, and *Karel V (1500 – 1558)* are all covered by query *"V (1500 – 1558)"*.

10.4.6 Experimental results

By carrying out the successive phases as described above, we obtained a list of approximately 450,000 candidate famous persons. For each of the last name/lifetime combinations found, we issued a query to Google to determine the corresponding GPC. The list was next sorted on descending GPC. In Table 10.3, we give the top 80 persons of the second millennium. Scanning the table, we make the following observations. First of all, we observe that the top is dominated by composers of western classical music. Their music is offered for sale by many Internet shops, which explains their relatively high GPC. On the whole, we observe that the top 80 indeed contains persons that are of international fame, with maybe two exceptions: Melvin Jones (founder of the International Lions Club Association) and John Bartlett (editor of *Familiar Quotations*). Their high GPC is the result of relatively few web sites. This suggests to base our measure of fame not (only) on number of web pages but (also) on the number of web sites. This is considered an issue for further research. In addition, we observe that Ronald Reagan has a relatively high score. In general, we observe that persons who have recently died, such as Ronald Reagan and Yasser Arafat receive a relatively high score. For example, for Yasser Arafat we observed a GPC of more than 80,000 some two weeks after his death in November 2004, while in January 2005 his GPC had dropped to 15,300. This suggests to not only exclude living persons from our list but also persons who have recently died.

In Tables 10.4–10.8, we present the top 30 persons found for the 11th, 13th, 15th, 17th, and 19th century, respectively. Assuming that on average a person establishes his/her fame at an age of around 30, we assign a person to the century in which he/she reaches an age of 30. However, for persons that died at an age of a, with $a < 50$, the year of his/her $\lfloor \frac{10+a}{2} \rfloor$-th anniversary is used. From these table we again conclude that indeed the presented approach is able to extract famous persons, even for the Late Middle Ages.

When considering the top of all ten centuries, one can make the following observations. The number of persons found per century increases considerably for the successive centuries. In addition, the average GPC of the top 30 increases considerably for the successive centuries. In the 17th, 18th, and 19th centuries, we observe that composers of western classical music dominate the top. In the 11th, 12th, and 13th centuries, we observe that kings and wise men from various religious backgrounds dominate the top.

Recall. We estimate the recall by choosing a diverse set of six books containing short biographies of persons whom we expect to find in our list. For each of these books, we determined for the persons that could potentially be found by our algorithm (i.e., the persons who are born in the intended time period and have died), the fraction of persons that are actually in our list. Table 10.9 gives an overview of the recall we obtained for each of the books. For further details on the books we refer to the list of references. We observe that the recall is close to one, for each of the six books, even for a more specialized topic as 17th century Dutch painters. Of the total 108 of these painters mentioned in the book, 106 were found. We note that of the 16 persons that did not appear in our list, there were 4 persons for which the books could not provide the lifetime.

In addition, we carried out some more tests. From these tests, we conclude that our list contains all 37 American presidents that are no longer alive. Also, our list contains all 6 rulers of the former USSR that passed away, and all 33 popes from Clement VIII who was elected pope in 1592 are in our list.

Precision. All kinds of imperfections can still be observed in our list, such as remaining inversions, missing parts of a name, and errors in lifetimes, although each of these occurs relatively infrequently. We concentrate on estimating the fraction of names that do not relate to persons. The corresponding precision that is obtained by the algorithm has been estimated as follows. We selected three decennia, namely 1220–1229, 1550–1559 and 1880–1889, and analyzed for each the candidate persons that were born in this decennium. For the first two decennia we analyzed the complete list, for decennium 1880–1889 we

Table 10.3. The 80 persons of the 2nd millenium that have the highest GPC.

2ND MILLENNIUM					
Johann Sebastian Bach	(1685–1750)	98700	Gabriel Faure	(1845–1924)	34200
Wolfgang Amadeus Mozart	(1756–1791)	88800	Felix Mendelssohn-Bartholdy	(1809–1847)	34000
Ludwig van Beethoven	(1770–1827)	80700	Benjamin Britten	(1913–1976)	33800
Albert Einstein	(1879–1955)	76800	Arnold Schoenberg	(1874–1951)	33300
Franz Schubert	(1797–1828)	69300	Camille Saint-Saens	(1835–1921)	32800
Johannes Brahms	(1833–1897)	65200	Mark Twain	(1835–1910)	32100
William Shakespeare	(1564–1616)	57300	Bela Bartok	(1881–1945)	32100
Franz Joseph Haydn	(1732–1809)	52900	Sigmund Freud	(1856–1939)	32000
Johann Wolfgang Goethe	(1749–1832)	52900	Domenico Scarlatti	(1685–1757)	31900
Charles Darwin	(1809–1882)	52000	Galileo Galilei	(1564–1642)	31900
Robert Schumann	(1810–1856)	51800	Arcangelo Corelli	(1653–1713)	31700
Leonardo da Vinci	(1452–1519)	50600	Georges Bizet	(1838–1875)	31200
Giuseppe Verdi	(1813–1901)	47900	Sergei Prokofiev	(1891–1953)	31200
Frederic Chopin	(1810–1849)	46800	Sergei Rachmaninov	(1873–1943)	31200
Antonio Vivaldi	(1678–1741)	46700	Francois Couperin	(1668–1733)	31100
Richard Wagner	(1813–1883)	44700	Charles Gounod	(1818–1893)	31000
Ronald Reagan	(1911–2004)	44300	Cesar Franck	(1822–1890)	30900
Franz Liszt	(1811–1886)	43700	Melvin Jones	(1879–1961)	30700
Claude Debussy	(1862–1918)	42300	Jean-Philippe Rameau	(1683–1764)	30600
Henry Purcell	(1659–1695)	41500	Carl Maria von Weber	(1786–1826)	29900
Voltaire	(1694–1778)	40300	Friedrich Nietzsche	(1844–1900)	29700
Immanuel Kant	(1724–1804)	40100	John Dowland	(1563–1626)	29300
James Joyce	(1882–1941)	39900	Paul Hindemith	(1895–1963)	29300
Friedrich Schiller	(1759–1805)	39900	George Bernard Shaw	(1856–1950)	29200
Georg Philipp Telemann	(1681–1767)	39800	Francis Bacon	(1561–1626)	29100
Antonin Dvorak	(1841–1904)	39200	Oscar Wilde	(1854–1900)	29000
Gustav Mahler	(1860–1911)	39000	Pablo Picasso	(1881–1973)	28900
Richard Strauss	(1864–1949)	38700	Jacques Offenbach	(1819–1880)	28900
Giacomo Puccini	(1858–1924)	38700	Samuel Barber	(1910–1981)	28200
Rene Descartes	(1596–1650)	38500	John Bartlett	(1820–1905)	28200
Maurice Ravel	(1875–1937)	37900	Jorge Luis Borges	(1899–1986)	28000
Winston Churchill	(1874–1965)	37900	Vincent van Gogh	(1853–1890)	28000
Gioacchino Rossini	(1792–1868)	36400	Mahatma Gandhi	(1869–1948)	28000
Hector Berlioz	(1803–1869)	35800	Bedrich Smetana	(1824–1884)	27700
Bertrand Russell	(1872–1970)	35000	Leos Janacek	(1854–1928)	27500
Anton Bruckner	(1824–1896)	34700	Ottorino Respighi	(1879–1936)	27400
Benjamin Franklin	(1706–1790)	34500	Henry I	(1100–1135)	27200
Napoleon Bonaparte	(1769–1821)	34500	Roland de Lassus	(1532–1594)	27100
Jean Sibelius	(1865–1957)	34300	Martin Luther	(1483–1546)	27000
Isaac Newton	(1642–1727)	34200	Anton Webern	(1883–1945)	26600

Table 10.4. The 30 persons of the 11th century that have the highest GPC.

11TH CENTURY					
Edward the Confessor	(1042–1066)	20300	Marpa	(1012–1097)	550
Anselm of Canterbury	(1033–1109)	3750	Alfonso VI	(1065–1109)	535
Shlomo Yitzchaki	(1040–1105)	2880	Vratislav II	(1061–1092)	518
king of Navarre	(1000–1035)	2550	Sima Guang	(1019–1086)	518
Su Shi	(1037–1101)	1510	Malcolm III	(1058–1093)	512
Omar Khayyam	(1048–1122)	1200	Tughril Beg	(1037–1063)	474
Heinrich IV	(1056–1106)	1050	Naropa	(1016–1100)	459
Heinrich II	(1002–1024)	952	William the Conqueror	(1027–1087)	446
Abu Hamid al-Ghazali	(1058–1111)	927	Henry I	(1068–1135)	417
Ramanuja	(1040–1137)	674	Henry III	(1017–1056)	375
Malcolm II	(1005–1034)	636	Robert I	(1032–1076)	352
Seton Kunrig	(1025–1113)	621	Isaac I Comnenus	(1007–1060)	336
Wladyslaw Herman	(1079–1102)	599	Constantine X Ducas	(1006–1067)	335
Milarepa	(1040–1123)	560	Alexius I Comnenus	(1057–1118)	331
Philip I	(1060–1108)	560	Romanus IV Diogenes	(1032–1072)	331

Table 10.5. The 30 persons of the 13th century that have the highest GPC.

13TH CENTURY					
Edward I	(1239–1307)	24100	Meister Eckhart	(1260–1328)	1040
Dante Alighieri	(1265–1321)	22300	Eihei Dogen	(1200–1253)	977
Alfonso X	(1221–1284)	15900	Ramon Llull	(1232–1316)	939
Thomas Aquinas	(1225–1274)	10200	Henry III	(1216–1272)	879
Giotto di Bondone	(1266–1337)	6050	Alexander III	(1249–1286)	790
Jalal al-Din Rumi	(1207–1273)	5000	Edward II	(1284–1327)	785
Saint Francis of Assisi	(1182–1226)	4580	Roger Bacon	(1214–1294)	781
Marco Polo	(1254–1324)	4530	Kublai Khan	(1215–1294)	778
Saint Bonaventure	(1217–1274)	3570	Friedrich II	(1194–1250)	747
Albertus Magnus	(1193–1280)	1640	Madhva	(1199–1278)	739
Philip II Augustus	(1180–1223)	1360	Alexander II	(1214–1249)	718
Cimabue	(1240–1302)	1320	Leonardo of Pisa	(1175–1250)	708
Nasreddin Hoca	(1208–1284)	1300	Louis IX	(1226–1270)	692
Dogen Zenji	(1200–1253)	1290	Robert Greathead	(1175–1253)	617
Snorri Sturluson	(1179–1241)	1180	Afonso III	(1248–1279)	605

analyzed only the first 1000 as well as the last 1000 names. This resulted in a precision of 0.94, 0.95, and 0.98, respectively. As the decennium of 1880–1889 resulted in considerably more names, we take a weighted average of these results. This yields an estimated precision for the complete list of 0.98.

Table 10.6. The 30 persons of the 15th century that have the highest GPC.

15TH CENTURY					
Leonardo da Vinci	(1452–1519)	50600	Benozzo Gozzoli	(1420–1497)	1560
Niccolo Machiavelli	(1469–1527)	13300	Guru Nanak	(1469–1539)	1530
duc de Bourbon	(1439–1503)	5330	Charles VI	(1380–1422)	1510
Guillaume Jouvenel des Ursins	(1401–1472)	5280	Andrea Mantegna	(1431–1506)	1480
Charles VII	(1403–1461)	5270	Donato Bramante	(1444–1514)	1470
Leon Battista Alberti	(1404–1472)	4010	Henry the Navigator	(1394–1460)	1450
Sandro Botticelli	(1445–1510)	3850	Amerigo Vespucci	(1454–1512)	1440
Hieronymus Bosch	(1450–1516)	3470	John Skelton	(1460–1529)	1420
Paolo Uccello	(1397–1475)	2890	Giovanni Pico della Mirandola	(1463–1494)	1330
Donatello	(1386–1466)	2880	Domenico Ghirlandaio	(1449–1494)	1290
Desiderius Erasmus	(1466–1536)	2490	Nikolaus von Kues	(1401–1464)	1280
Kaiser Friedrich III	(1440–1493)	2200	Enrique IV	(1454–1474)	1240
Tommaso Masaccio	(1401–1428)	2160	Friedrich III	(1415–1493)	1220
Alessandro di Mariano Filipepi	(1445–1510)	2030	Guillaume Dufay	(1400–1474)	1200
Henry VI	(1422–1461)	1960	Jeanne d'Arc	(1412–1431)	1180

Table 10.7. The 30 persons of the 17th century that have the highest GPC.

17TH CENTURY					
Henry Purcell	(1659–1695)	41500	Jonathan Swift	(1667–1745)	11800
Rene Descartes	(1596–1650)	38500	Baruch Spinoza	(1632–1677)	11500
Isaac Newton	(1642–1727)	34200	Christiaan Huygens	(1629–1695)	10900
Arcangelo Corelli	(1653–1713)	31700	Rembrandt van Rijn	(1606–1669)	10300
Francois Couperin	(1668–1733)	31100	Michel-Richard Delalande	(1657–1726)	8080
Alessandro Scarlatti	(1660–1725)	26400	Moliere	(1622–1673)	8080
Blaise Pascal	(1623–1662)	25800	Nicolas Poussin	(1594–1665)	7770
Jean-Baptiste Lully	(1632–1687)	25800	John Dryden	(1631–1700)	7760
John Locke	(1632–1704)	25400	Louis XIV	(1643–1715)	7300
Johannes Kepler	(1571–1630)	24900	Johannes Vermeer	(1632–1675)	7210
Gottfried Wilhelm Leibniz	(1646–1716)	22600	Heinrich Schuetz	(1585–1672)	7180
Thomas Hobbes	(1588–1679)	16500	Pierre de Fermat	(1601–1665)	7160
John Donne	(1572–1631)	15200	Jean de la Fontaine	(1621–1695)	7040
Peter Paul Rubens	(1577–1640)	14200	Johann Pachelbel	(1653–1706)	7000
John Milton	(1608–1674)	13700	Orlando Gibbons	(1583–1625)	6560

Regarding the precision of the properties *born-in* and *died-in*, we make the following observations. Considering the list of 450,000 potential instances that our algorithm found for this property, we observe that 235 were found with a GPC of at least 10,000 and 2450 were found with a GPC of at least 1000.

Table 10.8. The 30 persons of the 19th century that have the highest GPC.

19TH CENTURY					
Ludwig van Beethoven	(1770–1827)	80700	Hector Berlioz	(1803–1869)	35800
Franz Schubert	(1797–1828)	69300	Anton Bruckner	(1824–1896)	34700
Johannes Brahms	(1833–1897)	65200	Jean Sibelius	(1865–1957)	34300
Charles Darwin	(1809–1882)	52000	Gabriel Faure	(1845–1924)	34200
Robert Schumann	(1810–1856)	51800	Felix Mendelssohn-Bartholdy	(1809–1847)	34000
Giuseppe Verdi	(1813–1901)	47900	Camille Saint-Saens	(1835–1921)	32800
Frederic Chopin	(1810–1849)	46800	Mark Twain	(1835–1910)	32100
Richard Wagner	(1813–1883)	44700	Sigmund Freud	(1856–1939)	32000
Franz Liszt	(1811–1886)	43700	Georges Bizet	(1838–1875)	31200
Claude Debussy	(1862–1918)	42300	Charles Gounod	(1818–1893)	31000
Antonin Dvorak	(1841–1904)	39200	Cesar Franck	(1822–1890)	30900
Gustav Mahler	(1860–1911)	39000	Carl Maria von Weber	(1786–1826)	29900
Richard Strauss	(1864–1949)	38700	Friedrich Nietzsche	(1844–1900)	29700
Giacomo Puccini	(1858–1924)	38700	George Bernard Shaw	(1856–1950)	29200
Gioacchino Rossini	(1792–1868)	36400	Oscar Wilde	(1854–1900)	29000

Table 10.9. Recall for six popular scientific editions.

book	nr. of candidates	nr. found	recall
The Science Book	156	147	0.94
The Art Book	358	353	0.99
The Dutch Painters: 100 17th Century Masters	108	106	0.98
Philosophy: 100 Essential Thinkers	78	78	1.00
Herinneringen in Steen	195	195	1.00
Scientists and Inventions	154	154	1.00

Clearly, the probability that instances with a high GPC contain spelling errors in last name or lifetime is quite low, since accidental spelling errors in the last name or in the lifetime will result in a low GPC. Indeed, we found that the accuracy of our results was better than that of the information in some of the books. Especially *The Art Book* contained several errors in the lifetimes and even spelling errors in the last name of two of the artists.

There are two sources of errors influencing the precision of the properties *born-in* and *died-in* that we want to mention explicitly. For kings, queens, emperors, etc., the period in which they reigned is also found quite often in combination with their name. If lifetime and period of reign end with the same year, and if the person name is found more often in combination with the period of reign, then the name/lifetime combination might even be filtered out. But since the period of reign is always a subinterval of the corresponding lifetime, very often ending in the same year, this could be handled more carefully. We consider this an issue for further research. Another source of errors that we observe, albeit very infrequently, is the occurrence of living persons in our list. Very often this is caused by a publication that is devoted to a specific period from the life of that person, where the name of the person and the period is mentioned explicitly in the title.

A more detailed analysis of the precision is required to get a more accurate estimate of the fraction of errors that are caused by the various error sources.

10.5   Concluding remarks

The presented case study on finding famous persons indicates that non-trivial results can be obtained by extracting information from the WWW, even if we restrict ourselves to using the excerpts that Google provides. The case study shows that high recall and precision can be obtained. Currently, we are extending the case study to also automatically extract for example the corresponding nationality and professions of persons. In this way we can automatically extract subsets of persons such as 17th century Dutch painters as well as further details that are profession dependent.

The case study supports current social studies, where it is argued that collective knowledge can sometimes be more powerful than individual knowledge [Surowiecki, 2004]. The obtained list of persons seems to extend far beyond the list of well-known persons that are known to a single person.

Experiments as the one presented above show that the approach described in this chapter provides an effective method to extract information. However, adapting the approach to a knowledge domain of interest still requires manual effort to e.g. identify text fragments that can best be applied in the preselection step. Moreover, manual effort is required to adapt the normalization and filtering phases. Automatic adaption to a knowledge domain by e.g. automatic detection of appropriate text fragments that express relevant properties is considered an important next step in our research.

Acknowledgments

We thank our colleagues Verus Pronk and Herman ter Horst for their valuable comments on a draft version of this chapter.

References

The Art Book, Phaidon Press, 1994, Dutch Edition by Waanders Uitgevers, 1997.

Berners-Lee, T., J. Hendler, and O. Lassila [2001]. The semantic web. *Scientific American*, May 2001.

Breebaart, J., and M.F. McKinney [2004]. Features for audio classification. in: W. Verhaegh, E. Aarts, and J. Korst (Eds.), *Algorithms in Ambient Intelligence*, Kluwer Academic Publishers.

Brin, S. [1998]. Extracting patterns and relations from the World Wide Web. *Proceedings WebDB Workshop at EDBT'98*.

Buchholtz, S. [2001]. Using grammatical relations, answer frequencies and the World Wide Web for TREC question answering. *Proceedings 10th Text Retrieval Conference, TREC'01*, Gaithersburg, MD, November 13–16.

Clerkin, P., P. Cunningham, and C. Hayes [2001]. Ontology discovery for the semantic web using hierarchical clustering. *Proceedings 1st Workshop on Semantic Web Mining, WS'01*, Freiburg, Germany.

Cooley, R., B. Mobasher, and J. Srivastava [1997]. Web mining and pattern discovery on the Word Wide Web. *Proceedings 9th IEEE International Conference on Tools with Artificial Intelligence, ICTAI'97*, Newport Beach, CA, pages 558–567.

Faatz, A., and R. Steinmetz [2002]. Ontology enrichment with texts from the WWW. *Proceedings 2nd Workshop on Semantic Web Mining, WS'02*, Helsinki, Finland.

Feldman, A., and P. Ford [1979]. *Scientists and Inventions*. Aldus Books, London.

Frakes, W.B., and R. Baeza-Yates [1992]. *Information Retrieval: Data Structures and Algorithms*, Prentice-Hall.

Frantzi, K.T., S. Ananiadou, and J. Tsujii [1999]. Classifying technical terms. *Proceedings ICCC/IFIP 3rd Conference on Electronic Publishing*, Ronneby, Sweden, pages 144–155.

Geleijnse, G., and J. Korst [2005]. Automatic ontology population by Googling. *Proceedings of the 17th Belgian-Dutch Conference on Artificial Intelligence*, Brussels, Belgium.

Grefenstette, G. [1994]. *Explorations in Automatic Thesaurus Discovery*. Kluwer Academic Publishers, 1994.

Gusfield, D. [1997]. *Algorithms on Strings, Trees, and Sequences*. Cambridge University Press.

Hearst, M. [1992]. Automatic acquisition of hyponyms from large text corpora. *Proceedings of the 14th International Conference on Computational Linguistics*, Nantes, France, pages 539–545.

Jurafsky, D., and J.H. Martin [2001]. *Speech and Language Processing: An introduction to natural language processing, computational linguistics and speech recognition*. Prentice-Hall.

Kwok, C., O. Etzioni, D. Weld [2001]. Scaling question answering to the Web. *Proceedings of WWW10*, Hong Kong, 2001.

Lassila, O., and R.R. Swick [1999]. Resource description framework (RDF) model and syntax specification, recommendation. W3C, February 1999, http://www.w3.org/TR/1999/RECrdf-syntax-19990222.

Levenshtein, V.I. [1966]. Binary codes capable of correcting insertions and reversals. *Sov. Phys. Dokl.*, 10:707–710.

Reeth, A. van, G. Peeters, B. Büch, and T. Weerheijm [1988]. *Herinneringen in Steen*, (Dutch for: Memories in Stone). De Haan/Unieboek, Houten.

Rijsbergen, C.J. van [1979]. *Information Retrieval*. Buttersworths.

Salton, G., and M.J. McGill [1983]. *Introduction to Modern Information Retrieval*. McGraw-Hill.

The Science Book. Weidenfeld & Nicolson, London, 2001.

Stokes, P. [2003]. *Philosophy: 100 Essential Thinkers*. Enchanted Lion Books.

Surowiecki, J. [2004]. *The Wisdom of Crowds*, Doubleday.

Verschoor, M.P.F. [2004]. *Automatic ontology-driven extraction and structuring of information from the Internet*. M.Sc. Thesis, Eindhoven University of Technology, Eindhoven, The Netherlands.

Wikimedia Foundation. *Wikipedia, the free encyclopedia*. http://en.wikipedia.org/

Wright, C. [1978]. *The Dutch Painters: 100 Seventeenth Century Masters*. Barron's Publishers, London.

Chapter 11

PRIVACY PROTECTION IN COLLABORATIVE FILTERING BY ENCRYPTED COMPUTATION

Wim F.J. Verhaegh, Aukje E.M. van Duijnhoven, Pim Tuyls, and Jan Korst

Abstract   We present a method to protect users' privacy in collaborative filtering by performing the computations on encrypted data. We focus on the commonly-used memory-based approach, and show that the two main steps in collaborative filtering, being the determination of similarities and the prediction of ratings, can be performed on encrypted profiles. We discuss both user-based and item-based collaborative filtering, and for a number of variants of the similarity measures and prediction formulas described in literature, we show how they can be computed using encrypted data only. Although we consider collaborative filtering in this chapter, the techniques of comparing profiles using encrypted data only is useful in a much wider range of applications.

Keywords   Collaborative filtering, privacy, encryption.

11.1 Introduction

One of the key characteristics of ambient intelligence [Aarts & Marzano, 2003] is personalization, which ensures tailored applications and services to users. In order to realize personalization, electronic systems need user profiles, indicating the specific characteristics and preferences of users. If such personalization is realized by a stand-alone device, there is no issue, but if the personalization is offered by a service on the internet, the privacy of users may be at stake. Although most internet services, such as Amazon, have a privacy statement on their web site, users may be reluctant to give their personal data away, for several reasons. First, they may not trust every server. Secondly, they may trust the server, but do not want to run the risk that it gets hacked. Thirdly, the server may be reliable, but if it goes bankrupt, the user profiles represent valuable information that may be sold to third parties. Of course the privacy concern depends on the kind of data, e.g. preferences for books may be less sensitive information than users' medical records.

169

Wim F.J. Verhaegh et al. (Eds.), Intelligent Algorithms in Ambient and Biomedical Computing, 169-184.
© 2006 Springer. Printed in the Netherlands.

In order to protect the users' privacy, we investigate the possibilities that encryption techniques offer. The idea is that a user only releases personal information in an encrypted form, and that all the computations necessary for the personalization are done on the encrypted data. In the end, the user will receive an encrypted personalization result, which he can decrypt and use.

In this chapter we show how the above can be realized for recommendation services based on collaborative filtering [Herlocker et al., 1999; Shardanand & Maes, 1995], as a first personalization application that we select. Collaborative filtering is a well-known technique to recommend e.g. new music or books to users, and helps users in coping with the overload of content that is available through the internet. Based on a user's likes and dislikes for previously encountered content, it estimates to what extent he would like other content that is available. To this end, collaborative filtering uses the preferences of a community of users.

We can distinguish two global types of collaborative filtering approaches: memory-based [Herlocker et al., 1999] and model-based [Canny, 2002]. Memory-based collaborative filtering is the most commonly used approach. In this approach, which is a *lazy learning* approach in machine learning terms, the preferences (in the form of ratings for content) of a community of users are collected at a web server. Then, a similarity measure is computed between each pair of users based on the content they jointly rated. Next, recommendations for a particular user can be made by considering users that are similar to him, and checking for content that they liked but that has not yet been rated by the user or that is not yet in the user's collection.

Model-based approaches pursue a more active learning strategy. First, the collected preference data is processed to build a model of the users' profile space. For instance, Canny [2002] describes a factor-analysis approach, which first distills a basis of user preference profiles and expresses the individual users' profiles in terms of this basis. Next, this model is used to make predictions.

As mentioned, we want to develop a system that prevents any information about a user's preferences to become known to others. This not only means that we want to keep the user's ratings for items secret, but even the information of what items he has rated. Furthermore, we do not even want to reveal this kind of information anonymously, as we do not want to run the risk that the identity of the user is traced back somehow, after which his data is in the clear. Finally, as similarities between users also give information about a user's preferences, we also want to keep this data secret.

In addition to the above requirements from a user's perspective, we add the requirement that the server should maintain some control over the service, i.e., it should not be possible for a user to trivially retrieve valuable gathered data to set up a recommendation service too.

Whereas Canny [2002] focuses on model-based collaborative filtering, we discuss in this chapter how the more commonly used memory-based collaborative filtering technique can be performed on encrypted data. This holds for all variants of similarity measures and prediction formulas that we describe.

The remainder of this chapter is organized as follows. First, in Section 11.2 we discuss the procedures and formulas behind memory-based collaborative filtering, where we distinguish user-based and item-based approaches. Next, in Section 11.3 we briefly describe the proposed encryption system and its beneficial properties. Then, we discuss how the above requirements can be met, by describing how to perform the collaborative-filtering computations on encrypted data for the user-based and item-based approaches in Sections 11.4 and 11.5, respectively.

Although we focus in this chapter on encryption of preference information in collaborative filtering, the techniques we present are applicable in a much broader context, as many more ambient intelligence applications will use some form of matching profiles. Also these applications may be much better accepted by users if private information can be protected. We will however not elaborate on this.

11.2   Memory-based collaborative filtering

Most memory-based collaborative filtering approaches work by first determining similarities between users, by comparing their jointly rated items. Next, these similarities are used to predict the rating of a user for a particular item, by interpolating between the ratings of the other users for this item. Typically, all computations are performed by the server, upon a user request for a recommendation.

Next to the above approach, which is called a *user-based* approach, one can also follow an *item-based* approach. Then, first similarities are determined between items, by comparing the ratings they have gotten from the various users, and next the rating of a user for an item is predicted by interpolating between the ratings that this user has given for the other items.

Before discussing the formulas underlying both approaches, we first introduce some notation. We assume a set U of users and a set I of items. Whether a user $u \in U$ has rated item $i \in I$ is indicated by a boolean variable b_{ui} which equals one if the user has done so and zero otherwise. In the former case, also a rating r_{ui} is given, e.g. on a scale from 1 to 5. The set of users that have rated an item i is denoted by U_i, and the set of items that have been rated by a user u is denoted by I_u.

11.2.1   The user-based approach

User-based algorithms are probably the oldest and most widely used collaborative filtering algorithms [Breese, Heckerman & Kadie, 1998; Herlocker et al., 1999; Resnick et al., 1994; Sarwar et al., 2000]. As described above, there are two main steps: determining similarities and calculating predictions. For both we discuss commonly used formulas, of which we show later that they all can be computed on encrypted data.

Similarity measures.   Quite a number of similarity measures have been presented in the literature before. We distinguish three kinds: correlation measures, distance measures, and counting measures.

Correlation measures.   A common similarity measure used in literature is the so-called *Pearson correlation coefficient* (see e.g. [Sarwar et al., 2000]), given by[1]

$$s(u,v) = \frac{\sum_{i \in I_u \cap I_v}(r_{ui} - \bar{r}_u)(r_{vi} - \bar{r}_v)}{\sqrt{\sum_{i \in I_u \cap I_v}(r_{ui} - \bar{r}_u)^2 \sum_{i \in I_u \cap I_v}(r_{vi} - \bar{r}_v)^2}}, \tag{11.1}$$

where $\bar{r}_u$ denotes the average rating of user u for the items he has rated. The numerator in this equation gets a positive contribution for each item that is either rated above average by both users u and v, or rated below average by both. If one user has rated an item above average and the other user below average, we get a negative contribution. The denominator in the equation normalizes the similarity, to fall in the interval $[-1, 1]$, where a value 1 indicates complete correspondence and -1 indicates completely opposite tastes.

Related similarity measures are obtained by replacing $\bar{r}_u$ in (11.1) by the middle rating (e.g. 3 if using a scale from 1 to 5) or by zero. In the latter case, the measure is called vector similarity or cosine, and if all ratings are non-negative, the resulting similarity value will then lie between 0 and 1.

Distance measures.   Another type of measures is given by distances between two users' ratings, such as the mean-square difference [Shardanand & Maes, 1995] given by

$$\frac{\sum_{i \in I_u \cap I_v}(r_{ui} - r_{vi})^2}{|I_u \cap I_v|}, \tag{11.2}$$

or the normalized Manhattan distance [Aggarwal et al., 1999] given by

$$\frac{\sum_{i \in I_u \cap I_v}|r_{ui} - r_{vi}|}{|I_u \cap I_v|}. \tag{11.3}$$

---

[1] Note that if $I_u \cap I_v = \emptyset$, then the similarity $s(u,v)$ is undefined, and it should be discarded in the prediction formulas (11.9)–(11.11).

Such a distance is zero if the users rated their overlapping items identically, and larger otherwise. A simple transformation converts a distance into a measure that is high if users' ratings are similar and low otherwise.

Counting measures. Counting measures are based on counting the number of items that two users rated (nearly) identically. A simple counting measure is the majority voting measure [Nakamura & Abe, 1998] of the form

$$s(u,v) = (2-\gamma)^{c_{uv}} \gamma^{d_{uv}}, \tag{11.4}$$

where γ is chosen between 0 and 1, $c_{uv} = |\{i \in I_u \cap I_v \mid r_{ui} \approx r_{vi}\}|$ gives the number of items rated 'the same' by u and v, and $d_{uv} = |I_u \cap I_v| - c_{uv}$ gives the number of items rated 'differently'. The relation $\approx$ may here be defined as exact equality, but also nearly-matching ratings may be considered sufficiently equal.

Another counting measure is given by the weighted kappa statistic [Cohen, 1968], which is defined as the ratio between the observed agreement between two users and the maximum possible agreement, where both are corrected for agreement by chance. More formally, the measure is given by

$$s(u,v) = \frac{o_{uv} - e_{uv}}{1 - e_{uv}}. \tag{11.5}$$

Here, o_{uv} is the observed fraction of agreement, given by

$$o_{uv} = \frac{\sum_{i \in I_u \cap I_v} w(r_{ui}, r_{vi})}{|I_u \cap I_v|}, \tag{11.6}$$

where weights $w(x,y)$, with $0 \leq w(x,y) = w(y,x) \leq 1$ and $w(x,x) = 1$, indicate the degree of correspondence between ratings x and y. The offset e_{uv} is the expected fraction of agreement, and is given by

$$e_{uv} = \sum_{x \in X} \sum_{y \in X} p_u(x) p_v(y) w(x,y), \tag{11.7}$$

where X is the set of possible ratings, and $p_u(x)$ is the fraction of items that u has given a rating x, i.e.,

$$p_u(x) = \frac{|\{i \in I_u \mid r_{ui} = x\}|}{|I_u|}. \tag{11.8}$$

Prediction formulas. The second step in collaborative filtering is to use the similarities to compute a prediction for a certain user-item pair. Also for this step several variants exist. For all formulas, we assume that there are users that have rated the given item; otherwise no prediction can be made.

Weighted sums.   The first prediction formula, as used by Herlocker et al. [1999], is given by

$$\hat{r}_{ui} = \bar{r}_u + \frac{\sum_{v \in U_i} s(u,v)(r_{vi} - \bar{r}_v)}{\sum_{v \in U_i} |s(u,v)|}. \tag{11.9}$$

So, the prediction is the average rating of user u plus a weighted sum of deviations from the averages. In this sum, all users are considered that have rated item i. Alternatively, one may restrict them to users that also have a sufficiently high similarity to user u, i.e., we sum over all users in $U_i(t) = \{v \in U_i \mid s(u,v) \geq t\}$ for some threshold t.

An alternative, somewhat simpler prediction formula is given by

$$\hat{r}_{ui} = \frac{\sum_{v \in U_i} s(u,v) r_{vi}}{\sum_{v \in U_i} |s(u,v)|}. \tag{11.10}$$

Note that if all ratings are positive, then this formula only makes sense if all similarity values are non-negative, which may be realized by choosing a non-negative threshold.

Maximum total similarity.   A second type of prediction formula is given by choosing the rating that maximizes a kind of total similarity, as is done in the majority voting approach, given by

$$\hat{r}_{ui} = \arg \max_{x \in X} \sum_{v \in U_i^x} s(u,v), \tag{11.11}$$

where $U_i^x = \{v \in U_i \mid r_{vi} \approx x\}$ is the set of users that gave item i a rating similar to value x. Again, the relation $\approx$ may be defined as exact equality, but also nearly-matching ratings may be allowed. Also in this formula one may use $U_i(t)$ instead of U_i to restrict oneself to sufficiently similar users.

Time complexity.   The time complexity of user-based collaborative filtering is as follows.

For the first step, there are $O(|U|^2)$ pairs of users between which a similarity has to be computed. Similarity measures (11.1)–(11.4) require sums of $O(|I|)$ items, hence giving a total time complexity of $O(|U|^2|I|)$ to determine all similarities. The computation of the weighted kappa statistic (11.5) requires per pair of users $O(|I|)$ steps to compute (11.6) and $O(|X|^2)$ steps to compute (11.7), where for the latter one needs to compute (11.8) in $O(|I|)$ steps once per user and per value in X. So, this gives a total time complexity of $O(|U|^2|I| + |U|^2|X|^2 + |U||I||X|)$ for the kappa statistic. As $|X|$ is typically bounded by a small constant, say between 5 and 10, this reduces to the same time complexity $O(|U|^2|I|)$ as for the other measures.

If for all users all items with a missing rating are to be given a prediction, then this requires $O(|U||I|)$ predictions to be computed. Prediction formulas (11.9) and (11.10) can be computed in $O(|U|)$ steps, where (11.9) requires $O(|I|)$ steps once per user u to compute his average ratings $\bar{r}_u$. So, this gives a total complexity of $O(|U|^2|I|)$ to compute all predictions. Prediction formula (11.11) however requires $O(|U||X|)$ steps per prediction, thereby giving a total complexity of $O(|U|^2|I||X|)$ to compute all predictions. Again, if $|X|$ is bounded by a constant, this time complexity reduces to $O(|U|^2|I|)$.

11.2.2 The item-based approach

As mentioned, item-based algorithms [Karypis, 2001; Sarwar et al., 2001] first compute similarities between items, e.g. by using a similarity measure

$$s(i,j) = \frac{\sum_{u \in U_i \cap U_j}(r_{ui} - \bar{r}_u)(r_{uj} - \bar{r}_u)}{\sqrt{\sum_{u \in U_i \cap U_j}(r_{ui} - \bar{r}_u)^2 \sum_{u \in U_i \cap U_j}(r_{uj} - \bar{r}_v)^2}}. \tag{11.12}$$

Note that the exchange of users and items as compared to (11.1) is not complete, as still the average rating $\bar{r}_u$ is subtracted from the ratings. The reason to do so is that this subtraction compensates for the fact that some users give higher ratings than others, and there is no need for such a correction for items.

The standard item-based prediction formula to be used for the second step is given by

$$\hat{r}_{ui} = \bar{r}_i + \frac{\sum_{j \in I_u} s(i,j)(r_{uj} - \bar{r}_j)}{\sum_{j \in I_u} |s(i,j)|}. \tag{11.13}$$

The other similarity measures and prediction formulas we presented for the user-based approach can in principle also be turned into item-based variants, but we will not show them here.

Also in the time complexity for item-based collaborative filtering the roles of users and items interchange as compared to the user-based approach, as expected. For the first step, $O(|I|^2)$ similarity measures (11.12) have to be computed, each of which takes $O(|U|)$ steps. The prediction formula (11.13) requires $O(|I|)$ steps for each user and each item, where the average rating $\bar{r}_i$ takes $O(|U|)$ steps once per item i. As a result, the total time complexity is given by $O(|U||I|^2)$.

If the number $|U|$ of users is much larger than the number $|I|$ of items, the time complexity of the item-based approach is favorable over that of user-based collaborative filtering. Another advantage in this case is that the similarities are generally based on more elements, which gives more reliable measures. A further advantage of item-based collaborative filtering, as argued by Sarwar et al. [2001], is that correlations between items may be more stable than correlations between users, but we will not elaborate on this.

11.3   Encryption

In the next section we show how the presented formulas for collaborative filtering can be computed on encrypted ratings. Before doing so, we present the encryption system we use, and the specific properties it possesses that allow for the computation on encrypted data.

11.3.1   A public-key cryptosystem

The cryptosystem we use is the public-key cryptosystem presented by Paillier [1999]. We will not describe it in full detail, for which we refer to the chapter, but we briefly describe how data is encrypted.

First, encryption keys are generated. To this end, two large primes p and q are chosen randomly, and we compute $n = pq$ and $\lambda = \text{lcm}(p-1, q-1)$. Furthermore, a *generator* g is computed from p and q (for details, see [Paillier, 1999]). Now, the pair (n, g) forms the public key of the cryptosystem, which is sent to everyone, and λ forms the private key, to be used for decryption, which is kept secret.

Next, a sender who wants to send a message $m \in \mathbb{Z}_n = \{0, 1, \ldots, n-1\}$ to a receiver with public key (n, g) computes a ciphertext $\varepsilon(m)$ by

$$\varepsilon(m) = g^m r^n \bmod n^2, \tag{11.14}$$

where r is a number randomly drawn from $\mathbb{Z}_n^* = \{x \in \mathbb{Z} \mid 0 < x < n \wedge \gcd(x, n) = 1\}$. This r prevents decryption by simply encrypting all possible values of m (in case it can only assume a few values) and comparing the end result. The Paillier system is hence called a *randomized* encryption system.

Decryption of a ciphertext $c = \varepsilon(m)$ is done by computing

$$m = \frac{L(c^\lambda \bmod n^2)}{L(g^\lambda \bmod n^2)} \bmod n,$$

where $L(x) = (x-1)/n$ for any $0 < x < n^2$ with $x \equiv 1 \pmod{n}$. During decryption, the random number r cancels out.

Note that in the above cryptosystem the messages m are integers. Nevertheless, rational values are possible by multiplying them by a sufficiently large number and rounding off [Fouque, Stern & Wackers, 2002]. For instance, if we want to use messages with two decimals, we simply multiply them by 100 and round off. Usually, the range $\mathbb{Z}_n$ is large enough to allow for this multiplication.

11.3.2 Properties

The above presented encryption scheme has the following nice properties. The first one is that

$$\varepsilon(m_1)\varepsilon(m_2) \equiv g^{m_1} r_1^n g^{m_2} r_2^n \equiv g^{(m_1+m_2)}(r_1 r_2)^n \equiv \varepsilon(m_1 + m_2) \pmod{n^2},$$

which allows us to compute sums on encrypted data. Secondly,

$$\varepsilon(m_1)^{m_2} \equiv (g^{m_1} r_1^n)^{m_2} \equiv g^{m_1 m_2}(r_1^{m_2})^n \equiv \varepsilon(m_1 m_2) \pmod{n^2},$$

which allows us to compute products on encrypted data. An encryption scheme with these two properties is called a *homomorphic* encryption scheme. The Paillier system is one homomorphic encryption scheme, but more ones exist.

We can use the above properties to calculate sums of products, as required for the similarty measures and predictions, using

$$\prod_j \varepsilon(a_j)^{b_j} \equiv \prod_j \varepsilon(a_j b_j) \equiv \varepsilon(\sum_j a_j b_j) \pmod{n^2}. \tag{11.15}$$

So, using this, two users a and b can compute an inner product between a vector of each of them in the following way. User a first encrypts his entries a_j and sends them to b. User b then computes (11.15), as given by the left-hand term, and sends the result back to a. User a next decrypts the result to get the desired inner product. Note that neither user a nor user b can observe the data of the other user; the only thing user a gets to know is the inner product.

A final property we want to mention is that

$$\varepsilon(m_1)\varepsilon(0) \equiv g^{m_1} r_1^n g^0 r_2^n \equiv g^{m_1}(r_1 r_2)^n \equiv \varepsilon(m_1) \pmod{n^2}.$$

This action, which is called *(re)blinding*, can be used also to avoid a trial-and-error attack as discussed above, by means of the random number $r_2 \in \mathbb{Z}_n^*$. We will use this in Section 11.4.2.

11.3.3 A threshold version of the cryptosystem

The Paillier cryptosystem can also be implemented in a threshold version [Fouque, Poupard & Stern, 2000], in which the decryption key is shared among a number l of users, and a ciphertext can only be decrypted if more than a threshold t of users cooperate. In this version, the generation of the keys is somewhat more complicated, as well as the decryption mechanism. Without further going into details, for which we refer to Fouque, Poupard & Stern [2000], we briefly discuss the decryption procedure in the threshold cryptosystem. For this, first a subset of at least $t + 1$ users is chosen that will be involved in the decryption. Next, each of these users receives the ciphertext and computes a decryption share, using his own share of the key. Finally, these decryption shares are combined to compute the original message. As long as at least

$t+1$ users have combined their decryption share, the original message can be reconstructed.

11.4 Encrypted user-based algorithm

Having all ingredients in place, we now explain how memory-based collaborative filtering can be performed on encrypted data, in order to compute a prediction $\hat{r}_{ui}$ for a certain user u and item i. Note that although the computations are done on encrypted data, the outcome is of course identical to that of the original collaborative filtering algorithm.

We consider a setup as depicted in Figure 11.1, where user u communicates with other users v through a server. Furthermore, each user has generated his own key, and has published the public part of it. As we want to compute a prediction for user u, the steps below will use the keys of u.

Figure 11.1.   The setup for the user-based algorithm.

11.4.1 Computing similarities on encrypted data

First we take the similarity computation step, for which we start with the Pearson correlation given in (11.1). Although we already explained in Section 11.3 how to compute an inner product on encrypted data, we have to resolve the problem that the iterator i in the sums in (11.1) only runs over $I_u \cap I_v$, and this intersection is not known to either user. Therefore, we first introduce

$$
q_{ui} = \begin{cases} r_{ui} - \bar{r}_u & \text{if } b_{ui} = 1, \text{ i.e., user } u \text{ rated item } i, \\ 0 & \text{otherwise,} \end{cases}
$$

and rewrite (11.1) into

$$
s(u,v) = \frac{\sum_{i \in I} q_{ui} q_{vi}}{\sqrt{\sum_{i \in I} q_{ui}^2 b_{vi} \sum_{i \in I} q_{vi}^2 b_{ui}}}. \tag{11.16}
$$

The idea that we used is that any $i \notin I_u \cap I_v$ does not contribute to any of the three sums because at least one of the factors in the corresponding term will be zero. Hence, we have rewritten the similarity into a form consisting of three inner products, each between a vector of u and one of v.

The protocol now runs as follows. First, user u calculates encrypted entries $\varepsilon(q_{ui})$, $\varepsilon(q_{ui}^2)$, and $\varepsilon(b_{ui})$ for all $i \in I$, using (11.14), and sends them to the server. The server forwards these encrypted entries to each other user $v_1, \ldots, v_m$. Next, each user v_j, $j = 1, \ldots, m$, computes $\varepsilon(\sum_{i \in I} q_{ui} q_{vji})$, $\varepsilon(\sum_{i \in I} q_{ui}^2 b_{vji})$, and $\varepsilon(\sum_{i \in I} q_{vji}^2 b_{ui})$, using (11.15), and sends these three results back to the server, which forwards them to user u. User u can decrypt the total of $3m$ results and compute the similarities $s(u, v_j)$, for all $j = 1, \ldots, m$. Note that user u now knows similarity values with the other m users, but he need not know who each user $j = 1, \ldots, m$ is. The server, on the other hand, knows who each user $j = 1, \ldots, m$ is, but it does not know the similarity values.

For the other similarity measures, we can also derive computation schemes using encrypted data only. For the mean-square distance, we can rewrite (11.2) into

$$\frac{\sum_{i \in I_u \cap I_v} (r_{ui}^2 - 2 r_{ui} r_{vi} + r_{vi}^2)}{|I_u \cap I_v|} = \frac{\sum_{i \in I} r_{ui}^2 b_{vi} + 2 \sum_{i \in I} r_{ui}(-r_{vi}) + \sum_{i \in I} r_{vi}^2 b_{ui}}{\sum_{i \in I} b_{ui} b_{vi}},$$

(11.17)

where we additionally define $r_{ui} = 0$ if $b_{ui} = 0$ in order to have well-defined values. So, this distance measure can also be computed by means of four inner products.

The computation of normalized Manhattan distances is somewhat more complicated. Given the set X of possible ratings, we first define for each $x \in X$,

$$b_{ui}^x = \begin{cases} 1 & \text{if } b_{ui} = 1 \wedge r_{ui} = x, \\ 0 & \text{otherwise}, \end{cases}$$

and

$$a_{ui}^x = \begin{cases} |r_{ui} - x| & \text{if } b_{ui} = 1, \\ 0 & \text{otherwise}. \end{cases}$$

Now, (11.3) can be rewritten into

$$\frac{\sum_{i \in I} \sum_{x \in X} b_{ui}^x a_{vi}^x}{\sum_{i \in I} b_{ui} b_{vi}} = \frac{\sum_{x \in X} \sum_{i \in I} b_{ui}^x a_{vi}^x}{\sum_{i \in I} b_{ui} b_{vi}}. \tag{11.18}$$

So, the normalized Manhattan distance can be computed from $|X| + 1$ inner products. Furthermore, for the numerator a user v can compute $\prod_{x \in X} \varepsilon(\sum_{i \in I} b_{ui}^x a_{vi}^x) \equiv \varepsilon(\sum_{x \in X} \sum_{i \in I} b_{ui}^x a_{vi}^x)$, and send this result, together with the encrypted denominator, back to user u.

The majority-voting measure can also be computed in the above way, by defining

$$a_{ui}^x = \begin{cases} 1 & \text{if } b_{ui} = 1 \wedge r_{ui} \approx x, \\ 0 & \text{otherwise}. \end{cases} \tag{11.19}$$

Then, c_{uv} used in (11.4) is given by

$$c_{uv} = \sum_{x \in X} \sum_{i \in I} b_{ui}^x a_{vi}^x, \tag{11.20}$$

which can again be computed in a way as described above. Furthermore,

$$d_{uv} = \sum_{i \in I} b_{ui} b_{vi} - c_{uv}.$$

Finally, we consider the weighted kappa measure (11.5). Again, o_{uv} can be computed by defining

$$a_{ui}^x = \begin{cases} w(x, r_{ui}) & \text{if } b_{ui} = 1, \\ 0 & \text{otherwise,} \end{cases}$$

and then calculating

$$o_{uv} = \frac{\sum_{x \in X} \sum_{i \in I} b_{ui}^x a_{vi}^x}{\sum_{i \in I} b_{ui} b_{vi}}. \tag{11.21}$$

Furthermore, e_{uv} can be computed in an encrypted way if user u encrypts $p_u(x)$ for all $x \in X$ and sends them to each other user v, who can then compute

$$\prod_{x \in X} \prod_{y \in Y} \varepsilon(p_u(x))^{p_v(y)w(x,y)} \equiv \varepsilon(e_{uv}), \tag{11.22}$$

and send this back to u for decryption.

11.4.2　Computing predictions on encrypted data

For the second step of collaborative filtering, user u can calculate a prediction for item i in the following way. First, we rewrite the quotient in (11.9) into

$$\frac{\sum_{v \in U} s(u,v) q_{vi}}{\sum_{v \in U} |s(u,v)| b_{vi}}. \tag{11.23}$$

So, first user u encrypts $s(u, v_j)$ and $|s(u, v_j)|$ for each other user v_j, $j = 1, \ldots, m$, and sends them to the server. The server then forwards each pair $\varepsilon(s(u, v_j)), \varepsilon(|s(u, v_j)|)$ to the respective user v_j, who computes $\varepsilon(s(u, v_j))^{q_{v_j i}} \varepsilon(0) \equiv \varepsilon(s(u, v_j) q_{v_j i})$ and $\varepsilon(|s(u, v_j)|)^{b_{v_j i}} \varepsilon(0) \equiv \varepsilon(|s(u, v_j)| b_{v_j i})$, where he uses reblinding to prevent the server from getting knowledge from the data going back and forth to user v_j by trying a few possible values. Each user v_j next sends the results back to the server, which then computes

$$\prod_{j=1}^{m} \varepsilon(s(u, v_j) q_{v_j i}) \equiv \varepsilon(\sum_{j=1}^{m} s(u, v_j) q_{v_j i})$$

and

$$\prod_{j=1}^{m} \varepsilon(|s(u, v_j)| b_{v_j i}) \equiv \varepsilon(\sum_{j=1}^{m} |s(u, v_j)| b_{v_j i}),$$

and sends these two results back to user u. User u can then decrypt these messages and use them to compute the prediction. The simple prediction formula of (11.10) can be handled in a similar way.

The maximum total similarity prediction as given by (11.11) can be handled as follows. First, we rewrite

$$\sum_{v \in U_i^x} s(u, v) = \sum_{j=1}^{m} s(u, v_j) a_{v_j i}^x,$$ (11.24)

where $a_{v_j i}^x$ is as defined by (11.19). Next, user u encrypts $s(u, v_j)$ for each other user v_j, $j = 1, \ldots, m$, and sends them to the server. The server then forwards each $\varepsilon(s(u, v_j))$ to the respective user v_j, who computes $\varepsilon(s(u, v_j))^{a_{v_j i}^x} \varepsilon(0) \equiv \varepsilon(s(u, v_j) a_{v_j i}^x)$, for each rating $x \in X$, using reblinding. Next, each user v_j sends these $|X|$ results back to the server, which then computes

$$\prod_{j=1}^{m} \varepsilon(s(u, v_j) a_{v_j i}^x) \equiv \varepsilon(\sum_{j=1}^{m} s(u, v_j) a_{v_j i}^x),$$ (11.25)

for each $x \in X$, and sends the $|X|$ results to user u. Finally, user u decrypts these results and determines the rating x that has the highest result.

11.4.3   Time complexity revisited

The effect of encryption on the time complexity of computing the similarities and predictions, is as follows.

The time complexity to compute (11.16) and (11.17) is determined by the server, which has to forward for each user u, $O(|I|)$ encrypted items to all users v_j. This can be done in a total of $O(|U|^2|I|)$ steps, which equals the time complexity of the unencrypted case. For (11.18) and (11.20), $O(|I||X|)$ encrypted items have to be forwarded for each user u to all users v_j, giving a total of $O(|U|^2|I||X|)$ steps, which is a factor $O(|X|)$ more than in the unencrypted case. Finally, for the weighted kappa statistic we first need $O(|I||X|)$ steps per pair u, v to compute o_{uv} in (11.21), and secondly we need to compute e_{uv}. The latter takes for each u, $O(|U||X|)$ steps for the server to forward the encrypted items $\varepsilon(p_u(x))$ to all other users v_j, and $O(|X|^2)$ steps for all users v_j to compute (11.22). Note that the latter users can do so in parallel, and hence the total time complexity of computing all kappa statistics is given by $O(|U|^2|I||X| + |U|^2|X| + |U||X|^2)$.

For the prediction formulas, encryption does not have an effect on the time complexity. Formulas (11.23) and the encrypted version of (11.10) still require $O(|U|)$ steps per user u and item i for the work done by the user u himself and the server, giving a total time complexity of $O(|U|^2|I|)$. The time complexity to compute (11.24) in an encrypted way is determined by the server, which has to calculate $O(|X|)$ products (11.25) over $O(|U|)$ entries for each user u and item i, giving a total time complexity of $O(|U|^2|I||X|)$.

Although the time complexity is not much affected, we note that the run time of course will increase, because of the more demanding computations on encrypted data. This overhead is determined by the length of the encryption key.

11.5 Encrypted item-based algorithm

Also item-based collaborative filtering can be done on encrypted data, using the threshold system of Section 11.3.3. The general working of the item-based approach is slightly different than the user-based approach, as first the server determines similarities between items, and next uses them to make predictions.

So, first the server considers each pair $i, j \in I$ of items to determine their similarity as given in (11.12). Again, we first have to resolve the problem that the iterator, u in this case, does not run over the entire set U. To this end, we rewrite (11.12) into

$$\frac{\sum_{u \in U} q_{ui} q_{uj}}{\sqrt{\sum_{u \in U} q_{ui}^2 b_{uj} \sum_{u \in U} q_{uj}^2 b_{ui}}}.$$

So, we have to compute three sums, where each user $u \in U$ can compute his own contributions to them. Each of the three sums can be computed in the following way. First, the users compute their contributions, encrypt them using the threshold encryption scheme, and send the encrypted contributions to the server. The server multiplies all encrypted contributions, and gets in this way an encrypted version of the respective sum. Next, the server sends this encrypted sum back to the users, who each compute their decryption share of it. These decryption shares are sent back to the server, and if the server has received more shares than the threshold, it can decrypt the respective sum.

Next, we consider the prediction formula (11.13). For this, first the server computes the average rating $\bar{r}_i$ of each item i, which can be written as

$$\bar{r}_i = \frac{\sum_{u \in U} r_{ui}}{\sum_{u \in U} b_{ui}},$$

where again we define $r_{ui} = 0$ if $b_{ui} = 0$. The two sums in this quotient can again be computed by the server in the same way as described above.

Secondly, user u can compute a prediction for item i if we rewrite (11.13) into

$$\frac{\sum_{j \in I} b_{uj} |s(i,j)| \bar{r}_i + \sum_{j \in I} r_{uj} s(i,j) + \sum_{j \in I} b_{uj} s(i,j)(-\bar{r}_j)}{\sum_{j \in I} b_{uj} |s(i,j)|}. \tag{11.26}$$

Inspecting the four sums in this equation reveals that they are inner products between vectors of u and of the server. Note that the similarity values and item averages are valuable data to the server, so it does not want to share this with the users. So, user u encrypts his entries b_{uj} and r_{uj} and sends them to the server. The server then computes the sums in the encrypted domain, and sends the results back to u. User u then can decrypt the sums, and compute the prediction. Note that the server need not send all four encrypted sums back, as it can compute the denominator in (11.26) in the encrypted domain. It then only sends the encrypted denomimator and numerator back to u.

The time complexity of encrypted item-based collaborative filtering is not affected by encryption, apart from the computational overhead of computing with encrypted numbers. For the similarity measure, the server has to collect $O(|U|)$ contributions for three sums, for each pair i, j, giving a total time complexity of $O(|U||I|^2)$. Securely computing the average scores $\bar{r}_i$ requires $O(|U|)$ steps per item i, giving $O(|U||I|)$ in total. Finally, computing the predictions in an encrypted way takes $O(|I|)$ steps per user u and item i, giving also a total of $O(|U||I|^2)$.

11.6 Conclusion

We have shown how collaborative filtering can be done on encrypted data. In this way, sensitive information about a user's preferences, as discussed in the introduction, is kept secret, as it only leaves the user's system in an encrypted form. We have listed a number of variants of the similarity measures and prediction formulas described in literature, and showed for each of them how they can be computed using encrypted data only, without affecting their results.

Compared to the original set-up of collaborative filtering, the new set-up requires a more active role of the users' devices. This means that instead of a (single) server that runs an algorithm, we now have a system running a distributed algorithm, where all the nodes are actively involved in parts of the algorithm. The time complexity of the algorithm basically stays the same, except for an additional factor $|X|$ (typically between 5 and 10) for some similarity measures, and the overhead from computing with large encrypted numbers.

Although we showed that collaborative filtering can in principle be done on encrypted data, there are a few more issues to be resolved for a practical implementation. For instance, one should take into account the computational and communication overhead required due to the encryption and decryption of data. Furthermore, the system should be made robust against more complex forms of attacks, e.g. an attack where a user repeatedly computes similarities to other users, each time using a profile with only one item. These issues are topic of further research.

Although we only discussed collaborative filtering, the technique of computing similarities between profiles on encrypted data only is interesting for other applications as well, such as user matching and service discovery. In the vision of ambient intelligence, where much more (sensitive) profiling will be used in the future, this may play a crucial role in getting these applications accepted by a wide audience. This is also topic of further research.

References

Aarts, E., and S. Marzano [2003]. *The New Everyday: Visions of Ambient Intelligence.* 010 Publishing, Rotterdam, The Netherlands.

Aggarwal, C., J. Wolf, K.-L. Wu, and P. Yu [1999]. Horting hatches an egg: A new graph-theoretic approach to collaborative filtering. *Proceedings ACM KDD'99 Conference*, pages 201-212.

Breese, J., D. Heckerman, and C. Kadie [1998]. Empirical analysis of predictive algorithms for collaborative filtering. *Proceedings 14th Conference on Uncertainty in Artificial Intelligence*, pages 43–52.

Canny, J. [2002]. Collaborative filtering with privacy via factor analysis. *Proceedings ACM SIGIR'02*, pages 238–245, 2002.

Cohen, J. [1968]. Weighted kappa: Nominal scale agreement with provision for scaled disagreement or partial credit. *Psychological Bulletin*, 70:213–220.

Fouque, P.-A., G. Poupard, and J. Stern [2000]. Sharing decryption in the context of voting or loteries. *Proceedings of the 4th International Conference on Financial Cryptography, Lecture Notes in Computer Science*, 1962:90–104.

Fouque, P.-A., J. Stern, and J.-G. Wackers [2002]. Cryptocomputing with rationals. *Proceedings of the 6th International Conference on Financial Cryptography, Lecture Notes in Computer Science*, 2357:136–146.

Herlocker, J., J. Konstan, A. Borchers, and J. Riedl [1999]. An algorithmic framework for performing collaborative filtering. *Proceedings ACM SIGIR'99*, pages 230–237.

Karypis, G. [2001]. Evaluation of item-based top-n recommendation algorithms. *Proceedings 10th Conference on Information and Knowledge Management*, pages 247–254.

Nakamura, A., and N. Abe [1998]. Collaborative filtering using weighted majority prediction algorithms. *Proceedings 15th International Conference on Machine Learning*, pages 395–403.

Paillier, P. [1999]. Public-key cryptosystems based on composite degree residuosity classes. *Proceedings Advances in Cryptology – EUROCRYPT'99, Lecture Notes in Computer Science*, 1592:223–238.

Resnick, P., N. Iacovou, M. Suchak, P. Bergstrom, and J. Riedl [1994]. GroupLens: An open architecture for collaborative filtering of netnews. *Proceedings ACM CSCCW'94 Conference on Computer-Supported Cooperative Work*, pages 175–186.

Sarwar, B., G. Karypis, J. Konstan, and J. Riedl [2000]. Analysis of recommendation algorithms for e-commerce. *Proceedings 2nd ACM Conference on Electronic Commerce*, pages 158–167.

Sarwar, B., G. Karypis, J. Konstan, and J. Riedl [2001]. Item-based collaborative filtering recommendation algorithms. *Proceedings 10th World Wide Web Conference (WWW10)*, pages 285–295.

Shardanand, U., and P. Maes [1995]. Social information filtering: Algorithms for automating word of mouth. *Proceedings CHI'95*, pages 210–217.

van Duijnhoven, A.E.M. [2003]. Collaborative filtering with privacy. Master's thesis, Technische Universiteit Eindhoven.

Part III

TECHNOLOGY

Chapter 12

A FIRST LOOK AT THE MINIMUM DESCRIPTION LENGTH PRINCIPLE

Peter D. Grünwald

Abstract   This is an informal overview of Rissanen's Minimum Description Length (MDL) Principle. We provide an entirely non-technical introduction to the subject, focussing on conceptual issues.

Keywords   Machine learning, statistics, statistical learning, model selection, universal coding, prediction, information theory, data compression, minimum description length.

12.1   Introduction and overview

How does one decide among competing explanations of data given limited observations? This is the problem of *model selection*. It stands out as one of the most important problems of inductive and statistical inference. The Minimum Description Length (MDL) Principle is a relatively recent method for inductive inference that provides a generic solution to the model selection problem. MDL is based on the following insight: any regularity in the data can be used to *compress* the data, i.e., to describe it using fewer symbols than the number of symbols needed to describe the data literally. The more regularities there are, the more the data can be compressed. Equating 'learning' with 'finding regularity', we can therefore say that the more we are able to compress the data, the more we have *learned* about the data. Formalizing this idea leads to a general theory of inductive inference with several attractive properties:

1. *Occam's razor.* MDL chooses a model that trades-off goodness-of-fit on the observed data with 'complexity' or 'richness' of the model. As such, MDL embodies a form of Occam's razor, a principle that is both intuitively appealing and informally applied throughout all the sciences.

Wim F.J. Verhaegh et al. (Eds.), Intelligent Algorithms in Ambient and Biomedical Computing, 187-213.
© 2006 *Springer. Printed in the Netherlands.*

2. *No overfitting, automatically.* MDL procedures *automatically* and *in-herently* protect against overfitting and can be used to estimate both the parameters and the structure (e.g., number of parameters) of a model. In contrast, to avoid overfitting when estimating the structure of a model, traditional methods such as maximum likelihood must be *modified* and extended with additional, typically *ad hoc* principles.

3. *Bayesian interpretation.* MDL is closely related to Bayesian inference, but avoids some of the interpretation difficulties of the Bayesian approach[1], especially in the realistic case when it is known a priori to the modeler that none of the models under consideration is true. In fact:

4. *No need for 'underlying truth'.* In contrast to other statistical methods, MDL procedures have a clear interpretation independent of whether or not there exists some underlying 'true' model.

5. *Predictive interpretation.* Because data compression is formally equivalent to a form of probabilistic prediction, MDL methods can be interpreted as searching for a model with good predictive performance on *unseen* data.

Here we introduce the MDL Principle in an entirely non-technical way, concentrating on its most important applications, model selection and avoiding overfitting. We will make quite a few claims that we do not substantiate, and we will touch upon quite a few notions and ideas that we cannot discuss in detail. In all such cases, we refer the reader to our extensive tutorial [Grünwald, 2005], where the MDL Principle is discussed and motivated in much greater, and technical, detail.

Contents.   In Section 12.2 we discuss the relation between learning and data compression. Section 12.3 introduces model selection and outlines a first, 'crude' version of MDL that can be applied to model selection. Section 12.4 indicates how these 'crude' ideas need to be refined to tackle small sample sizes and differences in model complexity between models with the same number of parameters. Section 12.5 discusses the philosophy underlying MDL, and considers its relation to Occam's razor. Sections 12.7 and 12.8 briefly discuss the past, present and future of MDL. All this is summarized in Section 12.9.

---

[1] See [Grünwald, 2005, Section 2.9.2].

12.2 The fundamental idea: Learning as data compression

We are interested in developing a method for *learning* the laws and regularities in data. The following example will illustrate what we mean by this and give a first idea of how it can be related to descriptions of data.

Regularity .... Consider the following three sequences. We assume that each sequence is 10000 bits long, and we just list the beginning and the end of each sequence.

$$00010001000100010001 \ \dots \ 00010001000100010001 \quad (12.1)$$
$$01110100110100100110 \ \dots \ 10101110101110110010110 \quad (12.2)$$
$$00011000001010100000 \ \dots \ 00100010001000000100011 \quad (12.3)$$

The first of these three sequences is a 2500-fold repetition of 0001. Intuitively, the sequence looks regular; there seems to be a simple 'law' underlying it; it might make sense to conjecture that future data will also be subject to this law, and to predict that future data will behave according to this law. The second sequence has been generated by tosses of a fair coin. It is intuitively speaking as 'random as possible', and in this sense there is no regularity underlying it. Indeed, we cannot seem to find such a regularity either when we look at the data. The third sequence contains approximately four times as many 0s as 1s. It looks less regular, more random than the first; but it looks less random than the second. There is still some discernible regularity in these data, but of a statistical rather than of a deterministic kind. Again, noticing that such a regularity is there and predicting that future data will behave according to the same regularity seems sensible.

...and compression. We claimed that any regularity detected in the data can be used to *compress* the data, i.e. to describe it in a short manner. Descriptions are always relative to some *description method* which maps descriptions D' in a unique manner to data sets D. A particularly versatile description method is a general-purpose computer language like C or PASCAL. A description of D is then any computer program that prints D and then halts. Let us see whether our claim works for the three sequences above. Using a language similar to Pascal, we can write a program

```
for i = 1 to 2500; print '0001'; next; halt
```

which prints sequence (1) but is clearly a lot shorter. Thus, sequence (1) is indeed highly compressible. On the other hand, perhaps not surprisingly, it turns out [Grünwald, 2005, Section 2.2] that if one generates a sequence like

(2) by tosses of a fair coin, then with extremely high probability, the shortest program that prints (2) and then halts will look something like this:

```
print '0111010011010000101010........10111011000101100010'; halt
```

This program's size is about equal to the length of the sequence. Clearly, it does nothing more than repeat the sequence.

The third sequence lies in between the first two: generalizing $n = 10000$ to arbitrary length n, [Grünwald, 2005] shows that the first sequence can be compressed to $O(\log n)$ bits; with overwhelming probability, the second sequence cannot be compressed at all; and the third sequence can be compressed to some length αn, with $0 < \alpha < 1$.

Example 12.1 (Compressing various regular sequences). The regularities underlying sequences (1) and (3) were of a very particular kind. To illustrate that *any* type of regularity in a sequence may be exploited to compress that sequence, we give a few more examples:

- *The number* π. Evidently, there exists a computer program for generating the first n digits of π – such a program could be based, for example, on an infinite series expansion of π. This computer program has constant size, except for the specification of n which takes no more than $O(\log n)$ bits. Thus, when n is very large, the size of the program generating the first n digits of π will be very small compared to n: the π-digit sequence is deterministic, and therefore extremely regular.

- *Physics data.* Consider a two-column table where the first column contains numbers representing various heights from which an object was dropped. The second column contains the corresponding times it took for the object to reach the ground. Assume both heights and times are recorded to some finite precision. In Section 12.3 we illustrate that such a table can be substantially compressed by first describing the coefficients of the second-degree polynomial H that expresses Newton's law; then describing the heights; and then describing the deviation of the time points from the numbers predicted by H.

- *Natural language.* Most sequences of words are not valid sentences according to the English language. This fact can be exploited to substantially compress English text, as long as it is syntactically mostly correct: by first describing a grammar for English, and then describing an English text D with the help of that grammar [Grünwald, 1996], D can be described using much less bits than are needed without the assumption that word order is constrained.

12.2.1 Kolmogorov complexity and ideal MDL

To formalize our ideas, we need to decide on a description method, that is, a formal language in which to express properties of the data. The most general choice is a general-purpose[2] computer language such as C or PASCAL. This choice leads to the definition of the *Kolmogorov Complexity* [Li & Vitányi, 1997] of a sequence as the length of the shortest program that prints the sequence and then halts. The lower the Kolmogorov complexity of a sequence, the *more regular* it is. This notion seems to be highly dependent on the particular computer language used. However, it turns out that for every two general-purpose programming languages A and B and every data sequence D, the length of the shortest program for D written in language A and the length of the shortest program for D written in language B differ by no more than a constant c, which does not depend on the length of D. This so-called *invariance theorem* says that, *as long as the sequence D is long enough*, it is not essential which computer language one chooses, as long as it is general-purpose. Kolmogorov complexity was introduced, and the invariance theorem was proved, independently by Kolmogorov [1965], Chaitin [1969] and Solomonoff [1964]. Solomonoff's paper, called *A Theory of Inductive Inference*, contained the idea that the ultimate model for a sequence of data may be identified with the shortest program that prints the data. Solomonoff's ideas were later extended by several authors, leading to an 'idealized' version of MDL [Solomonoff, 1978; Li & Vitányi, 1997; Gács, Tromp, & Vitányi, 2001]. This idealized MDL is very general in scope, but not practically applicable, for the following two reasons:

- *Uncomputability.* It can be shown that there exists no computer program that, for every set of data D, when given D as input, returns the shortest program that prints D [Li & Vitányi, 1997].

- *Arbitrariness/dependence on syntax.* In practice we are confronted with small data samples for which the invariance theorem does not say much. Then the hypothesis chosen by idealized MDL may depend on arbitrary details of the syntax of the programming language under consideration.

12.2.2 Practical MDL

Like most authors in the field, we concentrate here on non-idealized, practical versions of MDL that explicitly deal with the two problems mentioned above. The basic idea is to scale down Solomonoff's approach so that it does become applicable. This is achieved by using description methods that are less

---

[2]By this we mean that a universal Turing Machine can be implemented in it [Li & Vitányi, 1997].

expressive than general-purpose computer languages. Such description methods *C* should be restrictive enough so that for any data sequence *D*, we can always compute the length of the shortest description of *D* that is attainable using method *C*; but they should be general enough to allow us to compress many of the intuitively 'regular' sequences. The price we pay is that, using the 'practical' MDL Principle, there will always be some regular sequences which we will not be able to compress. But we already know that there can be *no* method for inductive inference at all which will always give us all the regularity there is — simply because there can be no automated method which for any sequence *D* finds the shortest computer program that prints *D* and then halts. Moreover, it will often be possible to guide a suitable choice of *C* by a priori knowledge we have about our problem domain. For example, below we consider a description method *C* that is based on the class of all polynomials, such that with the help of *C* we can compress all data sets which can meaningfully be seen as points on some polynomial.

12.3 MDL and model selection

Let us recapitulate our main insights so far:

MDL: The basic idea
The goal of statistical inference may be cast as trying to find regularity in the data. 'Regularity' may be identified with 'ability to compress'. MDL combines these two insights by *viewing learning as data compression*: it tells us that, for a given set of hypotheses $\mathcal{H}$ and data set *D*, we should try to find the hypothesis or combination of hypotheses in $\mathcal{H}$ that compresses *D* most.

This idea can be applied to all sorts of inductive inference problems, but it turns out to be most fruitful in (and its development has mostly concentrated on) problems of *model selection* and, more generally, dealing with *overfitting*. Here is a standard example (we explain the difference between 'model' and 'hypothesis' after the example).

Example 12.2 (Model selection and overfitting). Consider the points in Figure 12.1. We would like to learn how the *y*-values depend on the *x*-values. To this end, we may want to fit a polynomial to the points. Straightforward linear regression will give us the leftmost polynomial — a straight line that seems overly simple: it does not capture the regularities in the data well. Since for any set of *n* points there exists a polynomial of the $(n-1)$-st degree that goes exactly through all these points, simply looking for the polynomial with the least error will give us a polynomial like the one in the second picture. This

Figure 12.1.   A simple, a complex and a trade-off (3rd degree) polynomial.

polynomial seems overly complex: it reflects the random fluctuations in the data rather than the general pattern underlying it. Instead of picking the overly simple or the overly complex polynomial, it seems more reasonable to prefer a relatively simple polynomial with small but nonzero error, as in the rightmost picture. This intuition is confirmed by numerous experiments on real-world data from a broad variety of sources [Rissanen, 1989; Vapnik, 1998; Ripley, 1996]: if one naively fits a high-degree polynomial to a small sample (set of data points), then one obtains a very good fit to the data. Yet if one *tests* the inferred polynomial on a second set of data coming from the same source, it typically fits this test data very badly in the sense that there is a large distance between the polynomial and the new data points. We say that the polynomial *overfits* the data. Indeed, all model selection methods that are used in practice either implicitly or explicitly choose a trade-off between goodness-of-fit and complexity of the models involved. In practice, such trade-offs lead to much better predictions of test data than one would get by adopting the 'simplest' (one degree) or most 'complex[3]' ($n-1$-degree) polynomial. MDL provides one particular means of achieving such a trade-off.

It will be useful to make a precise distinction between 'model' and 'hypothesis'. This is done in the box on the following page. In our terminology, the problem described in Example 12.2 is a 'hypothesis selection problem' if we are interested in selecting both the degree of a polynomial and the corresponding parameters; it is a 'model selection problem' if we are mainly interested in selecting the degree.

To apply MDL to polynomial or other types of hypothesis and model selection, we have to make precise the somewhat vague insight 'learning may be viewed as data compression'. This can be done in various ways. In this section, we concentrate on the earliest and simplest implementation of the idea. This is

---

[3]Strictly speaking, in our context it is not very accurate to speak of 'simple' or 'complex' polynomials; instead we should call the *set* of first degree polynomials 'simple', and the *set* of 100-th degree polynomials 'complex'.

Models vs. hypotheses

We use the phrase *point hypothesis* to refer to a *single* probability distribution or function. An example is the polynomial $5x^2 + 4x + 3$. A point hypothesis is also known as a 'simple hypothesis' in the statistical literature.

We use the word *model* to refer to a family (set) of probability distributions or functions with the same functional form. An example is the set of all second-degree polynomials. A model is also known as a 'composite hypothesis' in the statistical literature.

We use *hypothesis* as a generic term, referring to both point hypotheses and models.

---

the so-called *crude*[4] *two-part code* version of MDL:

Crude, two-part version of MDL principle (informally stated)

Let $\mathcal{H}^{(1)}, \mathcal{H}^{(2)}, \ldots$ be a list of candidate models (e.g., $\mathcal{H}^{(k)}$ is the set of k-th degree polynomials), each containing a set of point hypotheses (e.g., individual polynomials). The best point hypothesis $H \in \mathcal{H}^{(1)} \cup \mathcal{H}^{(2)} \cup \ldots$ to explain the data D is the one which minimizes the sum $L(H) + L(D|H)$, where

- $L(H)$ is the length, in bits, of the description of the hypothesis; and

- $L(D|H)$ is the length, in bits, of the description of the data when encoded with the help of the hypothesis.

The best *model* to explain D is the smallest model containing the selected H.

Example 12.3 (Polynomials, continued). In our previous example, the candidate hypotheses were polynomials. We can describe a polynomial by describing its coefficients in a certain precision (number of bits per parameter).

---

[4]The terminology 'crude MDL' is not standard. It is introduced here for pedagogical reasons, to make clear the importance of having a single, unified principle for designing codes. It should be noted that Rissanen's and Barron's early theoretical papers on MDL already contain such principles, albeit in a slightly different form than in their recent papers. Early practical applications [Grünwald, 1996] often do use *ad hoc* two-part codes which really are 'crude' in the sense defined here.

Thus, the higher the degree of a polynomial or the precision, the more[5] bits we need to describe it and the more 'complex' it becomes. A description of the data 'with the help of' a hypothesis means that the better the hypothesis fits the data, the shorter the description will be. A hypothesis that fits the data well gives us a lot of *information* about the data. Such information can always be used to compress the data [Grünwald, 2005]. Intuitively, this is because we only have to code the *errors* the hypothesis makes on the data rather than the full data. In our polynomial example, the better a polynomial H fits D, the fewer bits we need to encode the discrepancies between the actual y-values y_i and the predicted y-values $H(x_i)$. We can typically find a very complex point hypothesis (large $L(H)$) with a very good fit (small $L(D|H)$). We can also typically find a very simple point hypothesis (small $L(H)$) with a rather bad fit (large $L(D|H)$). The sum of the two description lengths will be minimized at a hypothesis that is quite (but not too) 'simple', with a good (but not perfect) fit.

12.4 Crude and refined MDL

Crude MDL picks the H minimizing the sum $L(H) + L(D|H)$. To make this procedure well-defined, we need to agree on precise definitions for the codes (description methods) giving rise to lengths $L(D|H)$ and $L(H)$. We now discuss these codes in more detail. We will see that the definition of $L(H)$ is problematic, indicating that we somehow need to 'refine' our crude MDL Principle.

12.4.1 Crude MDL: Definition of $L(D|H)$

Consider a two-part code as described above, and assume for the time being that all H under consideration define probability distributions; for example, $D = (x_1, \ldots, x_n)$ may be a binary sequence of given length n and $\mathcal{H}^{(k)}$ may stand for the class of k-th order Markov chain distributions, where the probability that the i-th bit is a 1 can depend on the previous k outcomes[6]. Each $H \in \mathcal{H}^{(k)}$ is a specific k-th order Markov chain distribution, which can be identified with a 2^k-dimensional parameter vector $\vec{\theta} = (\theta_{[0\ldots0]}, \theta_{[0\ldots1]}, \ldots, \theta_{[1\ldots1]})$, each parameter $\theta_{a_1 \ldots a_k}$ corresponding to the probability that x_i is a 1, given that the previous k bits $x_{i-1}, \ldots, x_{i-k}$ were equal to $a_1, \ldots, a_k$ respectively.

It turns out that for probabilistic hypotheses (Markov chains or otherwise), there is only one reasonable choice for the code $L(D \mid H)$. It is the so-called

---

[5] See the previous note.

[6] Note that in our previous example, $\mathcal{H}^{(k)}$ was a set of polynomials and thus did not directly define a distribution; we deal with this issue in Example 12.5.

Shannon-Fano code, satisfying, for all data sequences D,

$$L(D|H) = \lceil -\log P(D|H) \rceil, \tag{12.4}$$

where $P(D|H)$ is the probability mass of D according to H. It follows from the *Kraft inequality* [Cover & Thomas, 1991] – a cornerstone of information theory – that for every H that defines a probability distribution $P(\cdot \mid H)$, a code with lengths (12.4) must exist. Here, as in the sequel, log is logarithm to the base two, and codelengths are measured in bits. $\lceil \cdot \rceil$ means 'rounded up to the nearest integer'. In practice, we conveniently ignore rounding; the error this introduces in practice is always negligible. As a result, we can simply write $L(D|H) = -\log P(D|H)$. With such a code, the higher the probability H assigns to D, the shorter the codelength of D. This is what we mean by 'encoding D with the help of H': we use a code such that the better H *fits* D (i.e., the better H expresses the properties of D), the shorter the codelength.

It follows that, for given data D, the crude two-part MDL Principle tells us to pick the hypothesis $H \in \mathcal{H}^{(1)} \cup \mathcal{H}^{(2)} \cup \ldots$ that minimizes the sum

$$-\log P(D \mid H) + L(H). \tag{12.5}$$

Example 12.4 (Real-valued data). The use of the Shannon-Fano code with lengths $-\log P(D \mid H)$ is appropriate if H directly induces a probability *mass* function on the data. For example, this is the case if D is a sequence of 0s and 1s. If the x_i in $D = (x_1, \ldots, x_n)$ are real-valued, then, typically, every single D will have probability mass 0. To deal with that case, we first note that in practice, data that are analyzed on a computer are always discretized to some precision. Then each observed point x_i should really be interpreted as a small interval containing x_i. Such an interval will have probability mass > 0 according to H, so that $P(D|H)$, defined as the probability mass of the n *intervals* around the observed values $x_1, \ldots, x_n$, will be larger than 0 after all. If one discretizes at a fine enough precision, then the codelength $-\log P(D|H)$ of a discretized sequence $D = (x_1, \ldots, x_n)$ must be approximately equal to $-\log f(D|H) + c \cdot n$. Here f is the *probability density function* of data D according to distribution $P(\cdot \mid H)$, c is some number depending on the precision, but not depending on H. Since the additional term $c \cdot n$ does not depend on H, it plays no role in the minimization (12.5) and can be dropped. For this reason, in the case of real-valued data, (12.5) can be restated as: pick the H minimizing

$$-\log f(D \mid H) + L(H) \tag{12.6}$$

where f is the density of D given H. The form of $P(\cdot \mid H)$ and/or $f(\cdot \mid H)$ depends on the problem at hand – for, say, Gaussian mixture estimation, it will look very different than for Markov chain order selection. We now provide a concrete example.

Example 12.5 (Non-probabilistic hypotheses). Example 12.2, where $\mathcal{H}^{(k)}$ represented the class of k-th degree polynomials, introduces a further complication: the elements $H \in \mathcal{H}^{(k)}$ are functions rather than distributions. We deal with such cases by 'turning each H into a distribution', typically by making two additional assumptions: (1) $(x_1, y_1), \ldots, (x_n, y_n)$ are i.i.d. (independently and identically distributed); and (2), rather than expressing $Y = H(X)$, the function H really expresses that

$$Y = H(X) + Z, \tag{12.7}$$

where Z is a normally distributed noise term with mean 0. With these additional assumptions, the density $f(D|H)$ appearing in (12.6) is well-defined, and crude MDL can be applied. Let us evaluate this $f(D|H)$ in more detail. The assumption (12.7) entails that, for *given* x_i, the corresponding y_i has a normal density with mean 0 and some given variance σ^2. Therefore, the conditional density $f(y_i \mid x_i, H)$ is given by

$$f(y_i|x_i, H) = \frac{1}{\sqrt{2\pi\sigma^2}} e^{\frac{-(y_i - H(x_i))^2}{2\sigma^2}}. \tag{12.8}$$

The joint density $f(D|H)$ must satisfy

$$
\begin{aligned}
f(D|H) &= f(x_1, \ldots, x_n; y_1, \ldots, y_n | H) \\
&= f(y_1, \ldots, y_n | x_1, \ldots, x_n, H) f(x_1, \ldots, x_n) \\
&= \prod_{i=1}^{n} f(y_i|x_i, H) f(x_i),
\end{aligned} \tag{12.9}
$$

where the last step follows by our assumption that the data are i.i.d. Together (12.8) and (12.9) give

$$f(D|H) = \prod_{i=1}^{n} \frac{1}{\sqrt{2\pi\sigma^2}} e^{\frac{-(y_i - H(x_i))^2}{2\sigma^2}} f(x_i), \tag{12.10}$$

such that

$$-\ln f(D|H) = \frac{n}{2\sigma^2} \sum (y_i - H(x_i))^2 + \frac{n}{2} \ln 2\pi\sigma^2 - \sum \ln f(x_i). \tag{12.11}$$

The sum of the $\ln f(x_i)$ terms does not depend on H. Therefore it plays no role in the minimization (12.6) and can be dropped from the equation: only the conditional densities $f(y_i \mid x_i, H)$ are relevant to determine the MDL hypothesis. By (12.11), we see that with the assumption of normally distributed noise Z in (12.7), crude MDL (12.6) becomes a *penalized least-squares criterion*: we pick the H that minimizes a compromise between its fit on the data (measured by squared error) and its description length $L(H)$. In practice, the variance σ^2 will often be unknown. In that case, we can make it part of the hypothesis and try to learn it from the data. (12.6) then becomes: pick the H and σ^2 minimizing

$$-\log f(D \mid H, \sigma^2) + L(H, \sigma^2).$$

where $-\log f(D \mid H, \sigma^2)$ is given by (12.11). Because σ^2 has now become a part of our hypothesis, we must encode it explicitly in the first part of our code.

Consistency (asymptotic convergence).   To make crude MDL precise, we still need to define a code for hypotheses H, leading to codelengths $L(H)$. Broadly speaking, it turns out that *no matter what code we use* for encoding hypothesis, as long as it remains fixed (does not change when more data becomes available), the resulting procedure (expressed in (12.5) or (12.6)) is *statistically consistent* [Barron & Cover, 1991]. This means that *if* nature happens to generate data sequences by sampling from some distribution $P(\cdot \mid H)$ for a H which we can represent ($H \in \mathcal{H}^{(1)} \cup \mathcal{H}^{(2)} \cup \ldots$), then, eventually, we fill find this H with probability tending to 1. In our polynomial example, suppose data are sampled from a, say, 10-th degree polynomial H such that $L(H) < \infty$ ("we can represent H"), and then Gaussian noise with unit variance is added. Then, if the available sample D is small, MDL will, with high probability, select a polynomial of degree lower than 10. As more data becomes available, MDL will select a polynomial of degree slightly larger or smaller, but close to 10. And from some sample size on, there will be no more fluctuations and MDL will keep selecting a degree-10 polynomial, no matter how large the sample becomes [Grünwald, 2005].

12.4.2   Definition of $L(H)$: A problem for crude MDL

The statistical consistency result that we just presented is encouraging: it tells us that crude MDL is not a completely arbitrary procedure: eventually it will converge to the 'right' hypothesis, if such an hypothesis exists. This stands in contrast to some other model selection methods (such as leave-one-out cross-validation), which may select an overly complex hypothesis forever, even in the large sample limit.[7] But although consistency may be a necessary requirement of a good hypothesis selection method, it is certainly not *sufficient*: since the description length $L(H)$ of any fixed point hypothesis H can be very large under one code, but quite short under another, our procedure is still quite arbitrary, and we can give no guarantees at all on how *fast* MDL will converge. What we seek is an additional principle for determining $L(H)$, which allows

---

[7]This is explained in detail in the FAQ of the neural-nets newsgroup, see www.faqs.org/faqs/ai-faq/neural-nets/part3/section-12.html

us to find *reasonable* approximations of the data generating machinery based on *small* samples. In the first publications on MDL [Rissanen, 1978; Rissanen, 1983], it was advocated to choose some sort of *minimax code* for H, minimizing, in some precisely defined sense, the shortest worst-case total description length $L(H) + L(D|H)$, where the worst-case is over all possible data sequences. Thus, the MDL Principle is employed at a 'meta-level' to choose a code for H. However, this code requires a cumbersome discretization of the model space H, which is not always feasible in practice. Alternatively, Barron [1985] encoded H by the shortest computer program that, when input D, computes $P(D|H)$. While it can be shown that this leads to similar codelengths, it is computationally problematic. Later, Rissanen [1984] realized that these problems could be side-stepped by using a *one-part* rather than a *two-part code*. This development culminated in 1996 in a completely precise prescription of MDL for many, but certainly not all practical situations [Rissanen, 1996]. We call this modern version of MDL *refined MDL*.

12.4.3 Refined MDL

In refined MDL, we associate a code for encoding D *not with a single $H \in H$*, but with the full model H. Thus, given model H, we encode data not in two parts but we design a single *one-part code* with lengths $\bar{L}(D|H)$. This code is designed such that *whenever there is a member $H \in H$ that fits the data well, in the sense that $L(D \mid H)$ is small, then the codelength $\bar{L}(D|H)$ will also be small*. Codes with this property are called *universal codes* in the information-theoretic literature [Barron, Rissanen & Yu, 1998]. Among all such universal codes, we pick the one that is *minimax optimal*. The resulting universal code, defined and explained in detail by Grunwald [2005] , is known in the information-theoretic literature as the *Shtarkov* or *NML (Normalized Maximum Likelihood)* code.

To give an example, the set $H^{(3)}$ of third-degree polynomials is associated with a code with lengths $\bar{L}(\cdot \mid H^{(3)})$ such that, the better the data D are fit by the best-fitting third-degree polynomial, the shorter the codelength $\bar{L}(D \mid H)$. $\bar{L}(D \mid H)$ is called the *stochastic complexity* of the data given the model.

Parametric complexity. The second fundamental concept of refined MDL is the *parametric complexity* of a parametric model H, denoted by $\text{COMP}_n(H)$. This quantity, which depends on the sample size n, is a measure of the 'richness' of model H, indicating its ability to fit random data. To see how it relates to stochastic complexity, let, for given data D, $\hat{H}$ denote the distribution in H which maximizes the probability, and hence minimizes the codelength $L(D \mid \hat{H})$ of D. It turns out that

$$\text{stochastic complexity of } D \text{ given } H = L(D \mid \hat{H}) + \text{COMP}_n(H). \quad (12.12)$$

Consider a parametric statistical model $\mathcal{H}$ with k degrees of freedom and parameter sets restricted to a compact (closed and bounded) set $\Theta \subset \mathbb{R}^k$. In that case, under weak conditions on the parameterization (the mapping from parameters Θ to corresponding distributions $P(\cdot \mid \theta)$), $\text{COMP}_n(\mathcal{H})$ is well-defined and given by

$$\text{COMP}_n(\mathcal{H}) = \log \sum_{D' \in X^n} P(D' \mid \hat{H}(D')), \qquad (12.13)$$

where $\hat{H}(D)$ is the distribution in $\mathcal{H}$ that best fits (maximizes the probability of) data D, and the data D' in the denominator ranges over the set X^n of all possible samples of size n. Thus, the complexity is the logarithm of the sum over *all* data sequences of length n, of the probability of that data sequence according to the distribution in the model that best fits that particular data sequence. Clearly, if $\mathcal{H}$ contains just one distribution, $\hat{H}$ is the same for all D, and the complexity is 0. If $\hat{H}$ contains many distributions that are *substantially different* in the sense that they assign large probability to disjunct events, then the complexity is large.

To get more insight into (12.13), we note that, for sufficiently "regular" parametric statistical models $\text{COMP}_n(\mathcal{H})$ can be asymptotically expressed as follows:

$$\text{COMP}_n(\mathcal{H}) = \frac{k}{2} \log \frac{n}{2\pi} + \log \int_{\Theta} \sqrt{\det I(\theta)} d\theta + o(1). \qquad (12.14)$$

Here k is the number of parameters, 'det' stands for determinant and $I(\theta)$ is the $k \times k$ *Fisher information matrix* (for a definition see [Grünwald, 2005] or any basic textbook on statistics). $o(1)$ is a term which goes to 0 with increasing n. We see from (12.14) that the parametric complexity depends on the number of degrees of freedom in the model $\mathcal{H}$, where the influence of degree k grows logarithmically in n. Yet, it also depends on the *geometrical structure* of $\mathcal{H}$ [Myung, Balasubramanian & Pitt, 2000]. This is expressed by the third term. Since this term does not grow with n, its contribution becomes negligible for very large n; yet for practically relevant sample sizes its influence on the complexity can be substantial – see Example 12.6.

Refined MDL model selection. Refined MDL model selection between two parametric models (such as the models of first and second degree polynomials) now proceeds by selecting the model such that the stochastic complexity (12.12) of the given data D is smallest. Although we used a one-part code to encode data, refined MDL model selection still involves a trade-off between two terms: a goodness-of-fit term $L(D \mid \hat{H})$ and a complexity term $\text{COMP}_n(\mathcal{H})$. Typically, the first term $L(D \mid \hat{H})$ grows linearly in n, so that for very large samples, if only two models are compared, then the goodness-of-fit solely determines what model is chosen. But in practice, we often compare a countably infinite number of models. In that case, we consider models with arbitrarily large values of k. Then – although the procedure has to be slightly modified in this situation, see below – the complexity plays a role in determining what model is chosen, no matter how large the sample size n is.

Since in refined MDL we do not explicitly encode hypotheses H any more, there is no arbitrariness any more: it can be shown that $COMP_n(\mathcal{H})$ is independent of the arbitrary parameterization in which we represent our models. The resulting procedure can be interpreted in several different ways, some of which provide us with rationales for MDL beyond the pure coding interpretation [Grünwald, 2005]:

- *Counting/differential geometric interpretation.* The parametric complexity of a model is the logarithm of the number of *essentially different, distinguishable* point hypotheses within the model.

- *Two-part code interpretation.* For large samples, the stochastic complexity can be interpreted as a two-part codelength of the data after all, where hypotheses H are encoded with a special code that works by first discretizing the model space $\mathcal{H}$ into a set of 'maximally distinguishable hypotheses', and then assigning equal codelength to each of these.

- *Bayesian interpretation.* In many cases, refined MDL model selection coincides with Bayes factor model selection based on a *non-informative prior* such as *Jeffreys' prior* [Bernardo & Smith, 1994]. This prior was introduced by Jeffreys [1946] as a prior representing the elusive notion of 'prior ignorance about the parameter in a model', and it has the special property that it is invariant under continuous 1-to-1 reparameterizations of the model.

- *Prequential interpretation.* Refined MDL model selection can be interpreted as selecting the model with the best predictive performance when sequentially predicting *unseen* test data [Grünwald, 2005]. This makes it an instance of Dawid's [1984] *prequential* model validation and also relates it to *cross-validation* methods.

Refined MDL allows us to compare models of different functional form. It even accounts for the phenomenon that different models with the same number of parameters may not be equally 'complex'.

Example 12.6. Consider two models from psychophysics describing the relationship between physical dimensions (e.g., light intensity) and their psychological counterparts (e.g. brightness) [Myung, Balasubramanian & Pitt, 2000]: $y = ax^b + Z$ (Stevens' model) and $y = a\ln(x+b) + Z$ (Fechner's model) where Z is a normally distributed noise term. Both models have two free parameters; nevertheless, it turns out that in a sense, Stevens' model is more *flexible* or *complex* than Fechner's. Roughly speaking, this means there are a lot more data patterns that can be *explained* by Stevens' model than can be explained by Fechner's model. Myung et at. [2000] generated many samples of size 4 from Fechner's model, using some fixed parameter values. They then fitted both models to each sample. In 67% of the trials, Stevens' model fitted the data better than Fechner's, even though the latter generated the data. Indeed, in refined MDL, the 'complexity' associated with Stevens' model is much larger than the complexity associated with Fechner's, and if both models fit the data equally well, MDL will prefer Fechner's model.

Warning. Summarizing, refined MDL removes the arbitrary aspect of crude, two-part code MDL and associates parametric models with an inherent 'complexity' that does not depend on any particular description method for hypotheses. We should, however, warn the reader that we only discussed a special, simple situation in which we compared a finite number of parametric models that satisfy certain regularity conditions. These conditions are quite strong – in fact, for most models defined on infinite sample spaces, the parametric complexity is only defined if we restrict the parameter space to a bounded set. This is the case, for example, for the regression models of Examples 12.5 and 12.6. Such a boundedness assumption is often quite unnatural, and in that case, our 'refined' ideas have to be extended. This can be done in a number of ways, and to date, it is not clear which, or whether any, of these methods is in any sense optimal. Reassuringly though, in practice these different extensions tend to give similar results [De Rooij & Grünwald, 2005]. Also, if we compare an infinite number of models, then the refined ideas have to be slightly extended [Grünwald, 2005]. We then obtain a 'general' refined MDL Principle, which employs a combination of one-part and two-part codes.

12.5   The MDL philosophy

The first central MDL idea is that every regularity in data may be used to compress that data; the second central idea is that learning can be equated with finding regularities in data. Whereas the first part is relatively straightforward, the second part of the idea implies that *methods for learning from data must have a clear interpretation independent of whether any of the models under consideration is 'true' or not.* Quoting J. Rissanen [1989], the main originator of MDL:

> "We never want to make the false assumption that the observed data actually were generated by a distribution of some kind, say Gaussian, and then go on to analyze the consequences and make further deductions. Our deductions may be entertaining but quite irrelevant to the task at hand, namely, to learn useful properties from the data."
>
> *Jorma Rissanen [1989]*

Based on such ideas, Rissanen has developed a radical philosophy of learning and statistical inference that is considerably different from the ideas underlying mainstream statistics, both frequentist and Bayesian. We now describe this philosophy in more detail

Regularity as compression.   According to Rissanen, the goal of inductive inference should be to 'squeeze out as much regularity as possible' from the given data. The main task for statistical inference is to distill the meaningful information present in the data, i.e. to separate structure (interpreted as the regularity, the 'meaningful information') from noise (interpreted as the 'accidental information'). For the three sequences of Example 12.2.0.0, this would amount to the following: the first sequence would be considered as entirely regular and 'noiseless'. The second sequence would be considered as entirely random - all information in the sequence is accidental, there is no structure present. In the third sequence, the structural part would (roughly) be the pattern that 4 times as many 0s than 1s occur; given this regularity, the description of exactly which of all sequences with four times as many 0s than 1s occurs, is the accidental information.

Models as languages.   Rissanen interprets models (sets of hypotheses) as nothing more than languages for describing useful properties of the data – a model $\mathcal{H}$ is *identified* with its corresponding universal code $\bar{L}(\cdot \mid \mathcal{H})$. Different individual hypotheses within the models express different regularities in the data, and may simply be regarded as *statistics*, that is, summaries of certain regularities in the data. *These regularities are present and meaningful independently of whether some $H^* \in \mathcal{H}$ is the 'true state of nature' or not.* Suppose that the model $\mathcal{H}$ under consideration is probabilistic. In traditional theories, one typically assumes that some $P^* \in \mathcal{H}$ generates the data, and then 'noise' is defined as a random quantity relative to this P^*. In the MDL view 'noise' is defined relative to the model $\mathcal{H}$ as the residual number of bits needed to encode the data once the model $\mathcal{H}$ is given. Thus, noise is *not* a random variable: it is a function only of the chosen model and the *actually observed data*. Indeed, there is no place for a 'true distribution' or a 'true state of nature' in this view – there are only models and data. To bring out the difference to the ordinary statistical viewpoint, consider the phrase 'these experimental data are quite noisy'. According to a traditional interpretation, such a statement means that the data were generated by a distribution with high variance. According to the MDL philosophy, such a phrase means only that the data are not compressible with the currently hypothesized model – as a matter of principle, it can *never* be ruled out that there exists a different model under which the data are very compressible (not noisy) after all!

We have only the data. Many (but not all[8]) other methods of inductive infer-ence are based on the idea that there exists some 'true state of nature', typically a distribution assumed to lie in some model $\mathcal{H}$. The methods are then designed as a means to identify or approximate this state of nature based on as little data as possible. According to Rissanen[9], such methods are fundamentally flawed. The main reason is that the methods are designed under the assumption that the true state of nature is in the assumed model $\mathcal{H}$, which is often not the case. Therefore, *such methods only admit a clear interpretation under assumptions that are typically violated in practice.* Many cherished statistical methods are designed in this way – we mention hypothesis testing, minimum-variance un-biased estimation, several non-parametric methods, and even some forms of Bayesian inference [Grünwald, 2005, Section 2.9]. In contrast, MDL has a clear interpretation which *depends only on the data*, and not on the assumption of any underlying 'state of nature'.

Example 12.7 (Models that are wrong, yet useful). Even though the mod-els under consideration are often wrong, they can nevertheless be very *useful*. Examples are the successful 'Naive Bayes' model for spam filtering, Hidden Markov Models for speech recognition (is speech a stationary ergodic process? probably not), and the use of linear models in econometrics and psychology. Since these models are evidently wrong, it seems strange to base inferences on them using methods that are designed under the assumption that they contain the true distribution. To be fair, we should add that domains such as spam filtering and speech recognition are not what the fathers of modern statistics had in mind when they designed their procedures – they were usually think-ing about much simpler domains, where the assumption that some distribution $P^* \in \mathcal{H}$ is 'true' may not be so unreasonable.

MDL and consistency. Let $\mathcal{H}$ be a probabilistic model, such that each $P \in \mathcal{H}$ is a probability distribution. Roughly, a statistical procedure is called *consistent* relative to $\mathcal{H}$ if, for all $P^* \in \mathcal{H}$, the following holds: suppose data are distributed according to P^*. Then given enough data, the learning method will learn a good approximation of P^* with high probability. Many traditional statistical methods have been designed with consistency in mind [Grünwald, 2005].

The fact that in MDL, we do not assume a true distribution may suggest that we do not care about statistical consistency. But this is not the case: we would still like our statistical method to be such that in the *idealized* case where one

---

[8]For example, cross-validation cannot easily be interpreted in such terms of 'a method hunting for the true distribution'.

[9]The present author's own views are somewhat milder in this respect, but this is not the place to discuss them.

of the distributions in one of the models under consideration actually generates the data, our method is able to identify this distribution, given enough data. If even in the idealized special case where a 'truth' exists within our models, the method fails to learn it, then we certainly cannot trust it to do something reasonable in the more general case, where there may not be a 'true distribution' underlying the data at all. So: consistency *is* important in the MDL philosophy, but it is used *as a sanity check (for a method that has been developed without making distributional assumptions) rather than as a design principle*.

In fact, mere consistency is not sufficient. We would like our method to converge to the imagined true P^* *fast*, based on as small a sample as possible. Two-part code MDL with 'clever' codes achieves good rates of convergence in this sense (Barron and Cover [1991], complemented by Zhang [2004], show that in many situations, the rates are *minimax optimal*). The same seems to be true for refined one-part code MDL [Barron, Rissanen & Yu, 1998], although there is at least one surprising exception where inference based on the NML and Bayesian universal model behaves abnormally – see [Csiszár & Shields, 2000] for the details.

Summarizing this section, the MDL philosophy is quite agnostic about whether any of the models under consideration is 'true', or whether something like a 'true distribution' even exists. Nevertheless, it has been suggested [Webb, 1996; Domingos, 1999] that MDL embodies a naive belief that 'simple models' are 'a priori more likely to be true' than complex models. Below we explain why such claims are mistaken.

12.6   MDL and Occam's razor

When two models fit the data equally well, MDL will choose the one that is the 'simplest' in the sense that it allows for a shorter description of the data. As such, it implements a precise form of Occam's razor – *even though as more and more data becomes available, the model selected by MDL may become more and more 'complex'!* Occam's razor is sometimes criticized for being either (1) arbitrary or (2) false [Webb, 1996; Domingos, 1999]. Do these criticisms apply to MDL as well?

'Occam's razor (and MDL) is arbitrary'.   Because 'description length' is a syntactic notion it may seem that MDL selects an arbitrary model: different codes would have led to different description lengths, and therefore, to different models. By changing the encoding method, we can make 'complex' things 'simple' and vice versa. This overlooks the fact we are not allowed to use just any code we like! 'Refined' MDL tells us to use a specific code, independent of any specific parameterization of the model, leading to a notion of complexity that can also be interpreted without any reference to 'description lengths' but in terms of accumulated prediction errors [Grunwàld, 2005].

'Occam's razor is false'. It is often claimed that Occam's razor is false – we often try to model real-world situations that are arbitrarily complex, so why should we favor simple models? In the words of G. Webb[10]: 'What good are simple models of a complex world?'

The short answer is: even if the true data generating machinery is very complex, it may be a good strategy to prefer simple models for small sample sizes. Thus, MDL (and the corresponding form of Occam's razor) is a *strategy* for inferring models from data ("choose simple models at small sample sizes"), not a statement about how the world works ("simple models are more likely to be true") – indeed, a strategy cannot be true or false, it is 'clever' or 'stupid'. And the strategy of preferring simpler models is clever even if the data generating process is highly complex, as illustrated by the following example:

Example 12.8 ('Infinitely' complex sources). Suppose that data are subject to the law $Y = g(X) + Z$ where g is some continuous function and Z is some noise term with mean 0. If g is not a polynomial, but X only takes values in a finite interval, say $[-1, 1]$, we may still approximate g arbitrarily well by taking higher and higher degree polynomials. For example, let $g(x) = \exp(x)$. Then, if we use MDL to learn a polynomial for data $D = ((x_1, y_1), \ldots, (x_n, y_n))$, the degree of the polynomial $\ddot{f}^{(n)}$ selected by MDL at sample size n will increase with n, and with high probability, $\ddot{f}^{(n)}$ converges to $g(x) = \exp(x)$ in the sense that $\max_{x \in [-1,1]} |\ddot{f}^{(n)}(x) - g(x)| \to 0$. Of course, if we had better prior knowledge about the problem we could have tried to learn g using a model class $\mathcal{M}$ containing the function $y = \exp(x)$. But in general, both our imagination and our computational resources are limited, and we may be forced to use imperfect models.

If, based on a small sample, we choose the best-fitting polynomial $\hat{f}$ within the set of *all* polynomials, then, even though $\hat{f}$ will fit the data very well, it is likely to be quite unrelated to the 'true' g, and $\hat{f}$ may lead to disastrous predictions of future data. The reason is that, for small samples, the set of all polynomials is very large compared to the set of possible data patterns that we might have observed. Therefore, any particular data pattern can only give us very limited information about which high-degree polynomial best approximates g. On the other hand, if we choose the best-fitting $\hat{f}^{\circ}$ in some much smaller set such as the set of second-degree polynomials, then it is highly probable that the prediction quality (mean squared error) of $\hat{f}^{\circ}$ on future data is about the same as its mean squared error on the data we observed: the size (complexity) of

---

[10]Quoted with permission from KDD Nuggets 96:28, 1996.

the contemplated model is relatively small compared to the set of possible data patterns that we might have observed. Therefore, the particular pattern that we do observe gives us a lot of information on what second-degree polynomial best approximates g.

Thus, (a) $\hat{f}^{\circ}$ typically leads to better predictions of future data than $\hat{f}$; and (b) unlike $\hat{f}$, $\hat{f}^{\circ}$ is *reliable* in that it gives a correct impression of how good it will predict future data *even if the 'true' g is 'infinitely' complex*. This idea does not just appear in MDL, but is also the basis of Vapnik's [1998] Structural Risk Minimization approach and many standard statistical methods for non-parametric inference. In such approaches one acknowledges that the data generating machinery can be infinitely complex (e.g., not describable by a finite degree polynomial). Nevertheless, it is still a good strategy to approximate it by simple hypotheses (low-degree polynomials) as long as the sample size is small. Summarizing:

The inherent difference between under- and overfitting

If we choose an overly simple model for our data, then the best-fitting point hypothesis within the model is likely to be almost the best predictor, within the simple model, of future data coming from the same source. If we overfit (choose a very complex model) and there is noise in our data, then, *even if the complex model contains the 'true' point hypothesis*, the best-fitting point hypothesis within the model is likely to lead to very bad predictions of future data coming from the same source.

This statement is very imprecise and is meant more to convey the general idea than to be completely true. It becomes provably true if we use MDL's measure of model complexity; we measure prediction quality by logarithmic loss; and we assume that one of the distributions in $\mathcal{H}$ actually generates the data [Grünwald, 2005].

12.7 History

The MDL Principle has mainly been developed by J. Rissanen in a series of papers starting with [Rissanen, 1978]. It has its roots in the theory of *Kolmogorov* or *algorithmic* complexity [Li & Vitányi, 1997], developed in the 1960s by Solomonoff [1964], Kolmogorov [1965] and Chaitin [1966, 1969]. Among these authors, Solomonoff (a former student of the famous philosopher of science, Rudolf Carnap) was explicitly interested in inductive inference. The 1964 paper contains explicit suggestions on how the underlying ideas could be made practical, thereby foreshadowing some of the later work on two-part MDL. While Rissanen was not aware of Solomonoff's work at the time, Kolmogorov's [1965] paper did serve as an inspiration for Rissanen's [1978] development of MDL.

Another important inspiration for Rissanen was Akaike's [1973] AIC method for model selection, essentially the first model selection method based on information-theoretic ideas. Even though Rissanen was inspired by AIC, both the actual method and the underlying philosophy are substantially different from MDL.

MDL is much closer related to the *Minimum Message Length Principle*, developed by Wallace and his co-workers in a series of papers starting with the ground-breaking [Wallace & Boulton, 1968]; other milestones are [Wallace & Boulton, 1975] and [Wallace & Freeman, 1987]. Remarkably, Wallace developed his ideas without being aware of the notion of Kolmogorov complexity. Although Rissanen became aware of Wallace's work before the publication of [Rissanen, 1978], he developed his ideas mostly independently, being influenced rather by Akaike and Kolmogorov. Indeed, despite the close resemblance of both methods in practice, the underlying philosophy is quite different [Grünwald, 2005].

The first publications on MDL only mention two-part codes. Important progress was made by Rissanen [1984] , in which prequential codes are employed for the first time and [Rissanen, 1987], introducing the Bayesian mixture codes into MDL. This led to the development of the notion of stochastic complexity as the shortest codelength of the data given a model [Rissanen, 1986; Rissanen, 1987]. However, the connection to Shtarkov's *normalized maximum likelihood code* was not made until 1996, and this prevented the full development of the notion of 'parametric complexity'. In the mean time, in his impressive Ph.D. thesis, Barron [1985] showed how a specific version of the two-part code criterion has excellent frequentist statistical consistency properties. This was extended by Barron and Cover [1991] who achieved a breakthrough for two-part codes: they gave clear prescriptions on how to design codes for hypotheses, relating codes with good minimax codelength properties to rates of convergence in statistical consistency theorems. Some of the ideas of Rissanen [1987] and Barron and Cover [1991] were, as it were, unified when Rissanen [1996] introduced a new definition of stochastic complexity based on the *normalized maximum likelihood code*. The resulting theory was summarized for the first time by Barron, Rissanen and Yu [1998], and is called 'refined MDL' in the present overview.

12.8   Challenges for MDL: The road ahead

Having presented the history of MDL, we now take a brief look at the present and the future. While in many cases, the methods described here perform very well[11], there are also cases where they perform suboptimally compared to other state-of-the-art methods. Often this is due to one of two reasons:

- An asymptotic formula like (12.14) was used and the sample size was not large enough to justify this [Navarro, 2004].

- $COMP_n(\mathcal{H})$ was undefined for the models under consideration, and this was solved by cutting off the parameter ranges at *ad hoc* values [De Rooij & Grünwald, 2005].

In these cases the problem probably lies with the use of invalid approximations rather than with the MDL idea itself. More research is needed to find out when the asymptotics and other approximations can be trusted, and what is the 'best' way to deal with undefined $COMP_n(\mathcal{H})$. For the time being, we suggest to avoid using the asymptotic approximation (12.14) whenever possible, and to never cut off the parameter ranges at arbitrary values – instead, if $COMP_n(\mathcal{H})$ becomes infinite, then some of the alternative MDL methods [Grünwald, 2005, Section 2.7] should be used. Such methods will typically lead to well-behaved inference methods, comparable in quality to other state-of-the-art model selection techniques [De Rooij & Grünwald, 2005]; however, more research is needed (and is currently performed [Liang & Barron, 2005]) to determine exactly what type of method should be used in what context.

MDL and misspecification. Unfortunately, there is a class of problems where MDL is problematic in a more fundamental sense. Namely, if none of the distributions under consideration represents the data generating machinery very well, then both MDL and Bayesian inference may sometimes do a bad job in finding the 'best' approximation within this class of not-so-good hypotheses. This has been observed in practice [Clarke, 2002; Pednault, 2003]. Grunwald and Langford [2004] show that MDL can behave quite unreasonably for some classification problems in which the true distribution is not in $\mathcal{H}$. This is a bit ironic, since MDL was explicitly designed *not* to depend on the untenable assumption that some $P^* \in \mathcal{H}$ generates the data. But empirically we find that while it generally works quite well if some $P^* \in \mathcal{H}$ generates the data, it may sometimes fail if this is not the case. The author's own current research is focussed on extending MDL and its underlying principles to better deal with

---

[11]We mention [Hansen & Yu, 2000; Hansen & Yu, 2001] reporting excellent behavior of MDL in regression contexts; and [Allen, Madani & Greiner, 2003; Kontkanen, Myllymäki, Silander & Tirri, 1999; Modha & Masry, 1998] reporting excellent behavior of predictive (prequential) coding in Bayesian network model selection and regression. Also, 'objective Bayesian' model selection methods are frequently and successfully used in practice [Kass & Wasserman, 1996]. Since these are based on non-informative priors such as Jeffreys', they often coincide with a version of refined MDL and thus indicate successful performance of MDL.

'misspecification', and to establish precise conditions on the environment under which MDL can be expected to lead to reasonable inferences.

12.9 Summary, conclusion and further reading

We discussed how regularity is related to data compression, and how MDL employs this connection by viewing learning in terms of data compression. One can make this precise in several ways; in *idealized* MDL one looks for the shortest program that generates the given data. This approach is not feasible in practice, and here we concern ourselves with *practical* MDL. Practical MDL comes in a crude version based on two-part codes and in a modern, more refined version based on the concept of *universal coding*.

MDL methods are mostly applied to model selection but can also be used for other problems of inductive inference. In contrast to most existing statistical methodology, they can be given a clear interpretation irrespective of whether or not there exists some 'true' distribution generating data – inductive inference is seen as a search for regular properties in (interesting statistics of) the data, and there is no need to assume anything outside the model and the data. In contrast to what is sometimes thought, there is *no* implicit belief that 'simpler models are more likely to be true' – MDL does embody a preference for 'simple' models, but this is best seen as a strategy for inference that can be useful even if the environment is not simple at all.

MDL is a versatile method for inductive inference: it can be interpreted in at least four different ways, all of which indicate that it does something reasonable. It is typically asymptotically consistent, achieving good rates of convergence. All this strongly suggests that it is a good method to use in practice. Practical evidence shows that in many contexts it is, in other contexts its behavior can be problematic. In the author's view, the main challenge for the future is to improve MDL for such cases, by somehow extending and further refining MDL procedures in a non ad-hoc manner. I am confident that this can be done, and that MDL will continue to play an important role in the development of statistical, and more generally, inductive inference.

Further reading. Grunwald [2005] provides an 80-pages tutorial on MDL methods. It discusses all the concepts that we mentioned in this chapter in great conceptual and technical detail. Good other places to start further exploration of MDL are [Barron, Rissanen & Yu, 1998] and [Hansen & Yu, 2001]. Both papers provide excellent introductions, but they are geared towards a more specialized audience of information theorists and statisticians, respectively. Also worth reading is Rissanen's [1989] monograph. While outdated as an introduction to MDL *methods*, this famous 'little green book' still serves as a great introduction to Rissanen's radical but appealing *philosophy*, which is described very eloquently.

References

Akaike, H. [1973]. Information theory as an extension of the maximum likelihood principle. In B.N. Petrov and F. Csaki (Eds.), *Second International Symposium on Information Theory*, Budapest, pages 267–281. Akademiai Kiado.

Allen, T.V., O. Madani, and R. Greiner [2003]. Comparing model selection criteria for belief networks. Under submission.

Barron, A., and T. Cover [1991]. Minimum complexity density estimation. *IEEE Transactions on Information Theory*, 37(4):1034–1054.

Barron, A.R. [1985]. *Logically Smooth Density Estimation*. PhD thesis, Department of EE, Stanford University, Stanford, CA.

Barron, A.R., J. Rissanen, and B. Yu [1998]. The Minimum Description Length Principle in coding and modeling. *IEEE Transactions on Information Theory*, 44(6):2743–2760. Special Commemorative Issue: Information Theory: 1948-1998.

Bernardo, J., and A. Smith [1994]. *Bayesian Theory*. John Wiley.

Chaitin, G. [1966]. On the length of programs for computing finite binary sequences. *Journal of the ACM*, 13:547–569.

Chaitin, G. [1969]. On the length of programs for computing finite binary sequences: statistical considerations. *Journal of the ACM*, 16:145–159.

Clarke, B. [2002]. Comparing Bayes and non-Bayes model averaging when model approximation error cannot be ignored. Under submission.

Cover, T., and J. Thomas [1991]. *Elements of Information Theory*. Wiley Interscience, New York.

Csiszár, I., and P. Shields [2000]. The consistency of the BIC Markov order estimator. *The Annals of Statistics*, 28:1601–1619.

Dawid, A. [1984]. Present position and potential developments: Some personal views, statistical theory, the prequential approach. *Journal of the Royal Statistical Society, Series A*, 147(2):278–292.

de Rooij, S., and P. Grünwald [2005]. An empirical study of mdl model selection with infinite parametric complexity. Technical report, CORR ArXiv. see http://arxiv.org/abs/cs/0501028.

Domingos, P. [1999]. The role of Occam's razor in knowledge discovery. *Data Mining and Knowledge Discovery*, 3(4):409–425.

Gács, P., J. Tromp, and P. Vitányi [2001]. Algorithmic statistics. *IEEE Transactions on Information Theory*, 47(6):2464–2479.

Grünwald, P.D. [1996]. A minimum description length approach to grammar inference. In G.S.S. Wermter and E. Riloff (Ed.), *Connectionist, Statistical and Symbolic Approaches to Learning for Natural Language Processing, Lecture Notes in Artificial Intelligence*, 1040:203–216.

Grünwald, P.D. [2005]. MDL tutorial. In P.D. Grünwald, I.J. Myung, and M.A. Pitt (Eds.), *Advances in Minimum Description Length: Theory and Applications*. MIT Press.

Grünwald, P.D., and J. Langford [2004]. Suboptimal behaviour of Bayes and MDL in classification under misspecification. In *Proceedings of the Seventeenth Annual Conference on Computational Learning Theory (COLT' 04)*.

Hansen, M., and B. Yu [2000]. Wavelet thresholding via MDL for natural images. *IEEE Transactions on Information Theory*, 46:1778–1788.

Hansen, M., and B. Yu [2001]. Model selection and the principle of minimum description length. *Journal of the American Statistical Association*, 96(454):746–774.

Jeffreys, H. [1946]. An invariant form for the prior probability in estimation problems. *Proceedings of the Royal Statistical Society (London) Series A*, 186:453–461.

Kass, R., and L. Wasserman [1996]. The selection of prior distributions by formal rules. *Journal of the American Statistical Association*, 91:1343–1370.

Kolmogorov, A. [1965]. Three approaches to the quantitative definition of information. *Problems Inform. Transmission*, 1(1):1–7.

Kontkanen, P., P. Myllymäki, T. Silander, and H. Tirri [1999]. On supervised selection of Bayesian networks. In K. Laskey and H. Prade (Eds.), *Proceedings of the 15th International Conference on Uncertainty in Artificial Intelligence (UAI'99)*. Morgan Kaufmann Publishers

Li, M., and P. Vitányi [1997]. *An Introduction to Kolmogorov Complexity and Its Applications* (revised and expanded second ed.). Springer-Verlag, New York.

Liang, F., and A. Barron [2005]. Exact minimax predictive density estimation and MDL. In P.D. Grünwald, I.J. Myung, and M.A. Pitt (Eds.), *Advances in Minimum Description Length: Theory and Applications*. MIT Press.

Modha, D.S., and E. Masry [1998]. Prequential and cross-validated regression estimation. *Machine Learning*, 33(1):5–39.

Myung, I.J., V. Balasubramanian, and M.A. Pitt [2000]. Counting probability distributions: Differential geometry and model selection. *Proceedings of the National Academy of Sciences USA*, 97:11170–11175.

Navarro, D. [2004]. Misbehaviour of the Fisher information approximation to Minimum Description Length. Under submission.

Pednault, E. [2003]. Personal communication.

Ripley, B. [1996]. *Pattern Recognition and Neural Networks*. Cambridge University Press.

Rissanen, J. [1978]. Modeling by the shortest data description. *Automatica*, 14:465–471.

Rissanen, J. [1983]. A universal prior for integers and estimation by minimum description length. *The Annals of Statistics*, 11:416–431.

Rissanen, J. [1984]. Universal coding, information, prediction and estimation. *IEEE Transactions on Information Theory*, 30:629–636.

Rissanen, J. [1986]. Stochastic complexity and modeling. *The Annals of Statistics*, 14:1080–1100.

Rissanen, J. [1987]. Stochastic complexity. *Journal of the Royal Statistical Society, series B*, 49:223–239. Discussion: pages 252–265.

Rissanen, J. [1989]. *Stochastic Complexity in Statistical Inquiry*. World Scientific Publishing Company.

Rissanen, J. [1996]. Fisher information and stochastic complexity. *IEEE Transactions on Information Theory*, 42(1):40–47.

Solomonoff, R. [1964]. A formal theory of inductive inference, part 1 and part 2. *Information and Control*, 7:1–22, 224–254.

Solomonoff, R. [1978]. Complexity-based induction systems: comparisons and convergence theorems. *IEEE Transactions on Information Theory*, 24:422–432.

Vapnik, V. [1998]. *Statistical Learning Theory*. John Wiley.

Wallace, C., and D. Boulton [1968]. An information measure for classification. *Computing Journal*, 11:185–195.

Wallace, C., and D. Boulton [1975]. An invariant Bayes method for point estimation. *Classification Society Bulletin*, 3(3):11–34.

Wallace, C., and P. Freeman [1987]. Estimation and inference by compact coding. *Journal of the Royal Statistical Society, Series B*, 49:240–251. Discussion: pages 252–265.

Webb, G. [1996]. Further experimental evidence against the utility of Occam's razor. *Journal of Artificial Intelligence Research*, 4:397–417.

Zhang, T. [2004]. On the convergence of MDL density estimation. *Proceedings of the Seventeenth Annual Conference on Computational Learning Theory (COLT' 04), Lecture Notes in Artificial Intelligence*, 3120.

Chapter 13

SEMANTIC WEB ONTOLOGIES AND ENTAILMENT: COMPLEXITY ASPECTS

Herman J. ter Horst

Abstract   This chapter presents an overview of results relating to computational complexity of reasoning with Semantic Web ontologies. An overview of the completeness results that form the basis for these complexity results is also given. We prove NP-completeness of two standard entailment relations, simple entailment and RDFS (RDF Schema) entailment. These two entailment relations are in P if the target graph is assumed to contain no variables (blank nodes). We show that these results also apply to two stronger entailment relations, D^* entailment and pD^* entailment, which extend RDFS entailment to reasoning with datatypes and to reasoning with a subset of OWL (the Web Ontology Language), respectively. These results make use of deductive closure graphs that can be computed in polynomial time. We present new bounds on the size of these closure graphs.

Keywords   Ontology, reasoning, computational complexity, RDF, OWL.

13.1   Introduction

The W3C's Semantic Web vision [Berners-Lee et al., 2001] opens up new possibilities for intelligent, Web-based algorithms. One of the main objectives is to 'make reasoning explicit': statements that enable conclusions to be drawn are to be represented explicitly on the Web, for machine use. The W3C has standardized the XML-based Semantic Web languages RDF and OWL. RDF allows statements to be expressed, in the form of subject-predicate-object triples [RDF]. OWL makes it possible to express *ontologies* that give machine-usable accounts of the meaning of terms used in RDF statements [OWL]. RDF and OWL knowledge bases allow algorithms to derive *entailments*, i.e. to derive conclusions implied by semantic information on the Web.

Wim F.J. Verhaegh et al. (Eds.), Intelligent Algorithms in Ambient and Biomedical Computing, 215-242.
© 2006 *Springer. Printed in the Netherlands.*

This chapter considers the computational complexity of reasoning with RDF and OWL knowledge bases. Before giving an overview of this chapter, we describe briefly some of the background relating to RDF and OWL.

13.1.1 RDF

The RDF (Resource Description Framework) language enables content to be described with metadata [RDF]. The basic entities described are called resources. RDF can be used to record statements, which are triples that describe the values of properties for resources. For example, an RDF statement could be that the book Iliad (a resource) has as author (property) Homer (a value). Using an abbreviated form of the N-Triples syntax for RDF [Grant & Beckett, 2004], this RDF statement can simply be written as

```
Iliad hasAuthor Homer .
```

RDF statements consist of a subject (a resource), a predicate (property) and an object (a resource or a data value such as a string or integer). Properties therefore essentially stand for binary relations. RDF statements can also include variables which are, implicitly, existentially quantified; such variables are called blank nodes. For example, an RDF statement might use a blank node b to express that a certain movie has a director, without stating the identity of the director:

```
b isDirectorOf TheSoundOfMusic .
```

There is a standard XML syntax for RDF, which is more verbose than the N-Triples syntax used in the examples. However, the meaning of an RDF document abstracts from this XML serialization. An (unordered) set of RDF statements is formalized as an RDF graph: the property of an RDF statement is viewed as a 'link' between the subject and object 'nodes' of the statement.

13.1.2 RDFS (RDF Schema)

The RDF Schema (RDFS) language is added to RDF to allow the specification of simple ontologies for describing the meaning of metadata for a particular application domain [RDF]. RDFS can be used to specify classes of resources. Moreover, RDFS enables the definition of domain and range classes for properties, which can be used to derive information about subjects and objects of RDF statements. For example, the standard properties rdfs:domain and rdfs:range can be used to state that the 'hasAuthor' property has domain class 'Book' and range class 'Person':

```
hasAuthor rdfs:domain Book .
hasAuthor rdfs:range Person .
```

RDFS also allows subclass hierarchies and subproperty hierarchies to be specified by means of the standard properties `rdfs:subClassOf` and `rdfs:subPropertyOf`.

13.1.3 OWL

Building further on RDF and RDF Schema, the W3C has standardized the Web Ontology Language, OWL. This is a richer language for specifying ontologies on the Web [OWL]. Ontologies define concepts (i.e. classes) and relationships between concepts (i.e. properties). We mention briefly the part of OWL that will be considered in this chapter. OWL offers the possibility, for example, to specify that certain properties are functional (i.e. have unique values), transitive or symmetric, or that one property is the inverse of another. OWL can also be used to state that a resource denotes an entity that is the same as or different from the entity denoted by another resource, or that a class is disjoint with another class. OWL also offers the possibility to define classes in terms of constraints on properties. Such classes are called restrictions. As an example, the `PersonParentsUK` class of persons whose parents are both British can be defined as an OWL restriction by means of the following two RDF statements:

```
PersonParentsUK owl:allValuesFrom PersonUK .
PersonParentsUK owl:onProperty hasParent .
```

In addition to `allValuesFrom` restrictions, it is also possible to define `someValuesFrom` restrictions and `hasValue` restrictions, in a similar way.

13.1.4 Semantics and entailment

As was already mentioned, information specified by means of RDF or OWL entails (i.e. implies) other information. For example, the RDF triples given as examples RDFS-entail the following triples, which express that Iliad is a book and Homer a person:

```
Iliad rdf:type Book .
Homer rdf:type Person .
```

The valid entailments for RDF and OWL knowledge bases are determined by the formal semantics of these languages [Hayes, 2004] [Patel-Schneider et al., 2004], which are defined in terms of a model theory. RDF reasoners and OWL reasoners are based on these formal semantics. For RDF and RDFS, a proof theory has been developed, leading to completeness results which provide graph-based characterizations of entailment [Hayes, 2004]. These results use axiomatic triples and entailment rules to form deductive closures of RDF graphs. These completeness results are used by all RDF and RDFS reasoners. For OWL, two semantics have been defined in standard form: OWL DL and OWL Full. OWL Full entailment is known to be undecidable [Horrocks & Patel-Schneider, 2004]. For OWL DL, restrictions have been imposed on the use of the language in order to obtain decidability. For example, OWL DL does not allow the use of classes as instances. OWL DL entailment and OWL DL reasoners are based on description logics [Baader et al., 2003]. With regard to computational complexity, OWL DL entailment is NEXPTIME-complete; a subset of OWL DL has been defined, OWL Lite, for which entailment is EXPTIME-complete [Horrocks & Patel-Schneider, 2004].

The completeness results of [Hayes, 2004] do not show decidability of RDFS entailment because the closure graphs used in the proof are infinite for finite RDF graphs. In [Ter Horst, 2004a] a completeness result for RDFS was proved that makes use of partial closure graphs that can be taken to be finite for finite RDF graphs and that can be computed in polynomial time. This result was used to prove that RDFS entailment is decidable, in NP, and in P if the target RDF graph is assumed to contain no blank nodes. These results were also extended to two stronger entailment relations, $D*$ entailment and $pD*$ entailment, which extend RDFS entailment to reasoning with datatypes and to reasoning with a subset of OWL, respectively. The $pD*$ entailment relation uses a weaker semantics than the standard OWL semantics: just like RDFS, the $pD*$ semantics is largely defined by if conditions, rather than OWL's if-and-only-if conditions. The $pD*$ semantics seems to be sufficient for many applications where an ontology is combined with data relating to instances. Moreover, $pD*$ entailment does not require restrictions on the use of the language, such as those imposed by OWL DL, and has improved computational complexity. The entailment relations considered in [Ter Horst, 2004a] were proved to be NP-complete in [Ter Horst, 2004b]. In [Ter Horst, 2005a], a revised and extended version of [Ter Horst, 2004a] is provided, which extends the subset of OWL considered by the $pD*$ entailment relation, removes an assumption from the results (the assumption that datatype maps are discriminating) and includes complete proofs of the results just mentioned. This chapter is a revised and extended version of [Ter Horst, 2004b], which includes all the extensions to the $pD*$ semantics made in [Ter Horst, 2005a].

13.1.5 Overview

This chapter gives an overview of complexity results for RDFS, $D*$ and $pD*$ entailment and includes statements of the completeness results that form the basis for these complexity results. These completeness results are stated in complete detail but without proof; we refer to [Ter Horst, 2005a] for the completeness proofs and for the underlying model theory. This chapter includes new bounds on the size of partial closures as well as a new proof of the central lemma that partial closures can be computed in polynomial time. It also includes an informal discussion of why the latter result is true. Moreover, several complexity-related corollaries of the completeness results in [Ter Horst, 2005a] are proved here in more detail than in [Ter Horst, 2005a]. The tables included in this chapter provide sufficient information for development of algorithms for RDFS, $D*$ and $pD*$ entailment. Section 13.2 considers the basic entailment notion that underlies RDFS, called simple entailment. Section 13.3 proves that RDFS entailment and $D*$ entailment are NP-complete, and in P if the target graph does not contain blank nodes. In Section 13.4 we prove the same results for $pD*$ entailment, compare $pD*$ entailment and OWL entailment, and discuss briefly the combination of ontologies with rules. Finally, Section 13.5 summarizes the conclusions.

13.2 RDF graphs and simple entailment

This section summarizes basic terminology relating to RDF graphs [Klyne & Carroll, 2004] [Hayes, 2004] and introduces notation. We also consider the basic notion of entailment for RDF graphs, called simple entailment [Hayes, 2004], and state a graph-based characterization of simple entailment (the interpolation lemma). We use this characterization to prove that simple entailment is NP-complete.

13.2.1 URI references, blank nodes, literals

RDF distinguishes three sets of syntactic entities [Klyne & Carroll, 2004]. Let U denote the set of *URI references* used to describe resources. URI stands for Uniform Resource Identifier; URIs include the familiar URLs. Let B denote the set of *blank nodes*, i.e. variables. The set B is assumed to be infinite. The set L of *literals* is used to describe data values, such as strings and integers; L is the union of the set L_p of *plain literals* and the set L_t of *typed literals*. A typed literal l consists of a lexical form s and a datatype URI t: we shall write l as a pair, $l = (s, t)$. Examples of datatype URIs are the XML Schema datatypes xsd:string and xsd:integer. The sets U, B, L_p and L_t are pairwise disjoint. A *vocabulary* is a subset of $U \cup L$. RDF has a special datatype URI, called rdf:XMLLiteral, which is also written as XMLLiteral. An *XML literal* is a typed literal of the form $(s, \text{XMLLiteral})$. XML literals enable pieces of XML content to be used as data values. The phrase *RDF term* is used to denote either a URI reference, blank node or literal. The set of RDF terms is denoted by T: $T = U \cup B \cup L$.

13.2.2 RDF graphs and generalized RDF graphs

An RDF or OWL knowledge base is formalized as an *RDF graph G*, defined
to be a subset of the set

$$U \cup B \times U \times U \cup B \cup L. \tag{13.1}$$

A *generalized RDF graph G* is defined to be a subset of the set

$$U \cup B \times U \cup B \times U \cup B \cup L. \tag{13.2}$$

The elements (s, p, o) of a (generalized) RDF graph are called *(generalized)
RDF statements* or *(generalized) RDF triples* and consist of a *subject*, a *predicate* (or *property*) and an *object*, respectively. We write triples (s, p, o) simply
as $s\,p\,o$. The notation can be viewed as an abbreviation of the N-Triples notation [Grant & Beckett, 2004] used in the examples in the introduction.[1] RDF
graphs require properties to be URI references; generalized RDF graphs, which
allow properties to be blank nodes, were introduced in [Ter Horst, 2005a] to
solve the problem that the standard set of entailment rules for RDFS is incomplete if only RDF graphs are used.

We denote the projection mappings on the three factor sets of the product
sets given in (13.1) and (13.2) by π_1, π_2 and π_3. The *set of RDF terms* of a
generalized RDF graph G is denoted by

$$T(G) = \pi_1(G) \cup \pi_2(G) \cup \pi_3(G),$$

which is a subset of $U \cup B \cup L$. The set of *blank nodes* of a generalized RDF
graph G is denoted by

$$bl(G) = T(G) \cap B.$$

The *vocabulary of a generalized RDF graph G*, which is denoted by $V(G)$, is
the set of URIs or literals that occur as subject, predicate or object of a triple
in G:

$$V(G) = T(G) \cap (U \cup L).$$

The *set of nodes* of a generalized RDF graph G is

$$nd(G) = \pi_1(G) \cup \pi_3(G).$$

A blank node in an RDF statement is viewed as a variable that is, implicitly,
existentially quantified. A generalized graph is *ground* if it has no blank nodes.

Given a generalized RDF graph G and a partial function $h : B \rightharpoonup T$, another
generalized RDF graph is defined, the *instance* G_h of G, which is obtained by
replacing the blank nodes v that are in both G and the domain of h by $h(v)$.

---

[1] As in the expression $s\,p\,o \in G$, where G is an RDF graph, the context will always make clear what the triple
is.

Two generalized RDF graphs G and G' are *equivalent* if there is a bijection $f : T(G) \rightarrow T(G')$ such that $f(bl(G)) \subseteq bl(G')$, such that $f(v) = v$ for each $v \in V(G)$, and such that $s\,p\,o \in G$ if and only if $f(s)\,f(p)\,f(o) \in G'$. Given a set S of generalized RDF graphs, a *merge* of S is a graph that is obtained by replacing the graphs G in S by equivalent graphs G' that do not share blank nodes, and by taking the union of these graphs G'. The merge of a set of generalized RDF graphs S is uniquely defined up to equivalence. A merge of S is denoted by $M(S)$.

13.2.3   Simple entailment

An entailment relation for RDF specifies conditions for a set S of RDF graphs to imply (i.e. to entail) an RDF graph G. In this chapter, we consider four entailment relations of increasing strength: simple entailment, handling blank nodes; RDFS entailment, handling the RDF Schema vocabulary; D^* entailment, handling datatypes; pD^* entailment, handling a subset of OWL (the Web Ontology Language). These entailment relations are defined in terms of interpretations in accordance with a general pattern derived from mathematical logic [Cori & Lascar, 2000]. For example, simple entailment is defined in terms of simple interpretations in the following way. A set S of generalized RDF graphs *simply entails* a generalized RDF graph G if each simple interpretation that satisfies S also satisfies G. In this case we write

$$S \models G.$$

In this chapter we do not go into model-theoretic details: see [Hayes, 2004] or [Ter Horst, 2005a] for the definition of simple interpretation. The intuition behind the simple entailment relation $S \models G$ is that replacements can be found for the blank nodes of the RDF graph G in a way that is consistent with the given set of graphs S. This intuition is formalized in the interpolation lemma [Hayes, 2004], which provides a graph-based characterization of simple entailment. In a form that applies to generalized RDF graphs [Ter Horst, 2005a], this lemma is formulated as follows.

Lemma 13.1 (Generalized interpolation lemma). *If S is a set of generalized RDF graphs and G a generalized RDF graph, then S simply entails G if and only if a subset of $M(S)$ is an instance of G.*

This lemma shows that the simple entailment relation $S \models G$ between finite sets S of finite generalized RDF graphs and finite generalized RDF graphs G is decidable. It also shows that this problem is in NP: guess an instance function $h : bl(G) \rightarrow T(M(S))$ and check that the instance G_h of G defined by means of this function is a subset of $M(S)$. It is clear that this problem, the simple entailment relation $S \models G$ between finite sets S of finite generalized

RDF graphs and finite generalized RDF graphs G, is in P if G is assumed to be ground, for in this case it needs only to be checked that G is a subset of $M(S)$.

According to [Hayes, 2004], the full simple entailment problem (without restrictive assumptions) is NP-complete. The proof is outlined in one sentence [Hayes, 2004]: "This can be shown by encoding the problem of detecting a subgraph of an arbitrary directed graph as an RDF entailment, using only blank nodes to represent the graph (observation due to Jeremy Carroll)." Note that it is not trivial to work out the details of this proof sketch to obtain a complete proof, because the definition of instance (see Section 13.2.2) does not require the instance functions h to be injective. We give a full proof of the NP-completeness of simple entailment by reduction from another standard NP-complete problem [Garey & Johnson, 1979] - the clique problem. Although the NP-completeness results discussed in this chapter (see the following proposition and Propositions 13.6 and 13.11) are stated for generalized RDF graphs, they clearly also hold for RDF graphs, as their proofs use only RDF graphs.

Proposition 13.2. *The simple entailment relation $S \models G$ between finite sets S of finite generalized RDF graphs and finite generalized RDF graphs G is NP-complete.*

Proof. As shown above, this problem is in NP. We prove NP-completeness using a reduction from the clique problem. An instance of the clique problem consists of a finite undirected graph $G = (V, E)$, where E can be assumed to contain no loops $\{v, v\}$, and a positive integer $k \leq |V|$. The clique problem asks whether G has a clique of size $\geq k$, i.e. whether there exists a set of nodes $W \subseteq V$ of size $|W| \geq k$, such that each pair of distinct nodes $v, w \in W$ is linked by an edge in G: $\{v, w\} \in E$. An instance $G = (V, E), k$ of the clique problem is transformed to RDF graphs G' and H'. Each of the triples in G' and H' has the same predicate p, and all nodes of G' and H' are blank nodes. The RDF graph G' is formed by converting each pair $\{v, w\} \in E$ into two triples $v\,p\,w$ and $w\,p\,v$. The RDF graph H' consists of the triples $v\,p\,w$, where v and w are distinct elements of an arbitrary set of exactly k blank nodes. It is clear that the transformation from G, k to G', H' can be done in polynomial time. We need to prove that the finite undirected graph without loops G has a clique of size $\geq k$ if and only if G' simply entails H'. In view of the interpolation lemma, it is sufficient to prove that the finite undirected graph without loops G has a clique of size $\geq k$ if and only if there is a function $h : bl(H') \rightarrow bl(G')$ that satisfies $H'_h \subseteq G'$. The only if direction is clear. Furthermore, if there is such a function h, then h needs to be injective. Note that H' otherwise has distinct blank nodes v and w such that $h(v) = h(w)$. However, then it would follow that $h(v)\,p\,h(v) \in G'$, contradicting the assumption that G does not contain loops. Since h is injective, G has a clique of size $\geq k$. $\square$

Table 13.1.   RDF and RDFS URIs [Hayes, 2004].

rdf:type	rdf:Seq	rdfs:Datatype
rdf:Property	rdf:Bag	rdfs:Class
rdf:XMLLiteral	rdf:Alt	rdfs:subClassOf
rdf:nil	rdf:_1	rdfs:subPropertyOf
rdf:List	rdf:_2	rdfs:member
rdf:Statement	...	rdfs:Container
rdf:subject	rdf:value	rdfs:ContainerMembershipProperty
rdf:predicate	rdfs:domain	rdfs:comment
rdf:object	rdfs:range	rdfs:seeAlso
rdf:first	rdfs:Resource	rdfs:isDefinedBy
rdf:rest	rdfs:Literal	rdfs:label

13.3   RDFS entailment and D^* entailment

RDFS entailment [Hayes, 2004] provides a more powerful kind of entailment relation between RDF graphs than the simple entailment relation considered in the preceding section. RDFS entailment provides conclusions in terms of certain standard URI references for RDF and RDFS. Table 13.1 lists the URI references considered.   This table includes the three container types of RDF, for handling sequences, bags and alternatives (i.e. rdf:Seq, rdf:Bag and rdf:Alt, respectively) as well as the container membership properties rdf:_i which can be used to indicate that a certain resource is the i-th element of such a container. The URI rdfs:ContainerMemberShipProperty stands for the class of all the properties rdf:_i. With the exception of the container membership properties rdf:_i, we often omit the prefixes rdf: and rdfs: from the URIs in Table 13.1. RDFS entailment includes reasoning with the standard datatype rdf:XMLLiteral (see Section 13.2.1). The semantic conditions imposed by RDFS on this datatype can be extended to other datatypes, such as the XML Schema datatypes xsd:string and xsd:integer. This leads to a generalization of RDFS entailment, called D^* entailment [Ter Horst, 2004a]. D^* entailment is defined in terms of datatype maps [Hayes, 2004]. There are completeness results providing graph-based characterizations of RDFS entailment and D^* entailment [Hayes, 2004] [Ter Horst, 2004a] which, unlike the interpolation lemma from the preceding section, work by using certain axiomatic triples and entailment rules to compute a deductive closure of an RDF graph. In this section we summarize the definitions involved, state a completeness result for D^* entailment and use this result to prove that D^* entailment (and therefore also RDFS entailment) is NP-complete.

13.3.1 Datatype maps

We first summarize some terminology and introduce some notation relating to datatype maps [Klyne & Carroll, 2004] [Hayes, 2004]. Intuitively, a datatype is characterized by certain strings (e.g. "11") and corresponding values (e.g. the natural number 11). Formally, a *datatype d* is defined by a non-empty set of strings $L(d)$, the *lexical space*, a nonempty set of values $V(d)$, the *value space*, and a mapping $L2V(d) : L(d) \rightarrow V(d)$, which is the *lexical-to-value mapping*. A *datatype map* is a partial function D from the set U of URI references to the class of datatypes; each URI in the datatype map denotes the corresponding datatype. Each datatype map is required [Hayes, 2004] to contain the pair (X,x), where X is `rdf:XMLLiteral` and x is the standard XML literal datatype [Klyne & Carroll, 2004].

Suppose that a datatype map D is given. The *D-vocabulary* consists of the datatype URIs in D: it is the *domain* dom(D) of D, i.e. the set of URI references $a \in U$ such that $(a,d) \in D$ for some datatype d. The set of *D-literals* is the set of typed literals $(s,a) \in L_t$ with $a \in$ dom(D). The set of all *well-typed D-literals* is denoted by L_D^+:

$$L_D^+ = \{(s,a) \in L_t : a \in \text{dom}(D), s \in L(D(a))\}.$$

We assume that a few basic operations with respect to datatype maps can be executed in polynomial time. These operations include, in particular, the operation that determines whether a given D-literal is well typed, the computation of values of well-typed D-literals, and the operation that determines whether the values of two well-typed D-literals are equal.

13.3.2 *D** entailment

The definition of D^* entailment is patterned like the definition of simple entailment (see Section 13.2.3): the only change is that simple interpretations are replaced by D^* interpretations [Ter Horst, 2004a], which are not described in detail here. With regard to D^* interpretations, we note only that each D^* interpretation is a simple interpretation, so that each simple entailment is also a D^* entailment. RDFS entailment coincides with D^* entailment if the datatype map D is assumed to consist of only the type `rdf:XMLLiteral`. We use the notation

$$S \models_s G$$

for D^* entailment (and also for the special case of RDFS entailment). We proceed with the definitions needed to describe a completeness result that gives a graph-based characterization of D^* entailment.

Definition 13.1 (RDF, RDFS and *D*-axiomatic triples). See Tables 13.2 and 13.3 for the RDF and RDFS axiomatic triples. If D is a datatype map, then

Table 13.2. RDF axiomatic triples [Hayes, 2004].

type type Property .	value type Property .
subject type Property .	rdf:_1 type Property .
predicate type Property .	rdf:_2 type Property .
object type Property .	...
first type Property .	nil type List .
rest type Property .	

Table 13.3. RDFS axiomatic triples [Hayes, 2004].

type domain Resource .	first range Resource .
domain domain Property .	rest range List .
range domain Property .	seeAlso range Resource .
subPropertyOf domain Property .	isDefinedBy range Resource .
subClassOf domain Class .	comment range Literal .
subject domain Statement .	label range Literal .
predicate domain Statement .	value range Resource .
object domain Statement .	Alt subClassOf Container .
member domain Resource .	Bag subClassOf Container .
first domain List .	Seq subClassOf Container .
rest domain List .	ContainerMembershipProperty
seeAlso domain Resource .	subClassOf Property .
isDefinedBy domain Resource .	isDefinedBy subPropertyOf seeAlso .
comment domain Resource .	XMLLiteral type Datatype .
label domain Resource .	XMLLiteral subClassOf Literal .
value domain Resource .	Datatype subClassOf Class .
type range Class .	rdf:_1 type
domain range Class .	ContainerMembershipProperty .
range range Class .	rdf:_1 domain Resource .
subPropertyOf range Property .	rdf:_1 range Resource .
subClassOf range Class .	rdf:_2 type
subject range Resource .	ContainerMembershipProperty .
predicate range Resource .	rdf:_2 domain Resource .
object range Resource .	rdf:_2 range Resource .
member range Resource .	...

the D-axiomatic triples are the triples a type Datatype and a subClassOf Literal where $a \in \text{dom}(D)$.

Definition 13.2 (D^* entailment rules). Given a datatype map D, the D^* entailment rules are defined in Table 13.4. These rules consist of the 18

Table 13.4. D^* entailment rules [Ter Horst, 2005a].

	If G contains	where	then add to G
lg	$v\,p\,l$	$l \in L$	$v\,p\,b_l$
gl	$v\,p\,b_l$	$l \in L$	$v\,p\,l$
rdf1	$v\,p\,w$		p type Property
rdf2-D	$v\,p\,l$	$l = (s,a) \in L_D^+$	b_l type a
rdfs1	$v\,p\,l$	$l \in L_p$	b_l type Literal
rdfs2	p domain u		
	$v\,p\,w$		v type u
rdfs3	p range u		
	$v\,p\,w$	$w \in U \cup B$	w type u
rdfs4a	$v\,p\,w$		v type Resource
rdfs4b	$v\,p\,w$	$w \in U \cup B$	w type Resource
rdfs5	v subPropertyOf w		
	w subPropertyOf u		v subPropertyOf u
rdfs6	v type Property		v subPropertyOf v
rdfs7x	p subPropertyOf q		
	$v\,p\,w$	$q \in U \cup B$	$v\,q\,w$
rdfs8	v type Class		v subClassOf Resource
rdfs9	v subClassOf w		
	u type v		u type w
rdfs10	v type Class		v subClassOf v
rdfs11	v subClassOf w		
	w subClassOf u		v subClassOf u
rdfs12	v type Container-		
	MembershipProperty		v subPropertyOf member
rdfs13	v type Datatype		v subClassOf Literal

entailment rules defined in [Hayes, 2004] for RDFS, with two differences that involve rules rdf 2 and rdfs 7. Rule rdfs7x differs from rule rdfs 7 in that it can produce generalized RDF triples with blank nodes in predicate position when applied to ordinary RDF triples; the condition $q \in U$ is replaced by $q \in U \cup B$. In order to deal with datatypes, rule rdf 2 is replaced by the more general rule rdf 2-D. The first rule lg ('literal generalization') prescribes that if G contains $v\,p\,l$, where l is a literal, then add $v\,p\,b_l$ to G, where b_l is a blank node allocated to l by this rule. Here *allocated to* means that the blank node b_l has been created by an application of rule lg on the same literal l, or, if there is no such blank node, b_l must be a new node which is not yet in the graph. In rule rdfs1, b_l is a blank node that is allocated by rule lg to the plain literal $l \in L_p$. In rule rdf 2-D, b_l is a blank node that is allocated by rule lg to the well-typed D-literal $l \in L_D^+$. Rule rdf 2-D is a direct generalization of entailment rule rdf 2 from [Hayes, 2004], which has the same effect only for well-typed XML literals l. If D contains only the datatype rdf:XMLLiteral, then rule rdf 2-D becomes exactly rule rdf 2.

Definition 13.3 (*D*-clash). The notion of XML clash [Hayes, 2004] is generalized in a straightforward way to any datatype map. Given a datatype map D, a *D-clash* is a triple b type Literal, where b is a blank node allocated by rule lg to an ill-typed D-literal.

Definition 13.4 (partial and full RDFS and *D** closures). The entailment relations considered in this paper are declarative. For example, the entailment rules of Table 13.4 can be applied in any order (see [Ter Horst, 2005a], Theorem 4.11). However, in order to prove decidability of entailment, we consider a special way of applying the entailment rules. Suppose that D is a datatype map and G an RDF graph. In the definitions that follow, the axiomatic triples that contain the URI references rdf:_i (i.e. the four triples rdf:_i type Property, rdf:_i type ContainerMembershipProperty, rdf:_i domain Resource and rdf:_i range Resource) are treated in a special way. Suppose that K is a non-empty subset of the positive integers $\{1, 2, ...\}$ chosen in such a way that for each rdf:_$i \in V(G)$ we have $i \in K$. The *partial D* closure* $G_{s,K}$ of G is defined in the following way. In the first step, all RDF and RDFS axiomatic triples and D-axiomatic triples are added to G, except for the axiomatic triples that include rdf:_i such that $i \notin K$. In the next step, rule lg is applied to each triple that contains a literal in such a way that distinct, well-typed D-literals l with the same value are associated with the same surrogate blank node b_l. Then, rules rdf 2-D and rdfs1 are applied to each triple containing a well-typed D-literal or a plain literal, respectively. Finally, arbitrary derivations are made using applications of the remaining D* entailment rules until the graph is unchanged. The steps used in the definition of partial D* closure are summarized in Figure 13.1. In addition to the partial D* closure $G_{s,K}$ defined in this way, the *full D* closure* G_s of G is defined by taking $G_s = G_{s,K}$, where K is the full set $\{1, 2, ...\}$. If the datatype map D consists of only the datatype rdf:XMLLiteral, a partial and a full D* closure of a generalized RDF graph G are called a *partial* and a *full RDFS closure* of G, respectively.

The following lemma shows that a partial closure can be computed in polynomial time. This lemma plays an important role in the results that follow. Before giving a formal proof, we discuss in an informal way the reasons 'why' partial closures can be computed in polynomial time. The fact that RDF graphs form a kind of graph seems to play a role here. There is an analogy with the graph connectivity problem [Homer & Selman, 2001]. Given a directed graph $G = (V, E)$ and a pair of nodes $v, w \in V$, this problem raises the question of whether there is a path in G from v to w. The algorithm depicted in Figure 13.2 solves the graph connectivity problem. This algorithm runs in quadratic time: the repeat step needs to be executed at most $|V|$ times, while the 'for all' step checks $|V|$ nodes.

---

input: generalized RDF graph G
add axiomatic triples to G;
apply rule lg;
apply rules rdf2-D and rdfs1;
repeat until graph unchanged:
      apply each entailment rule except lg, rdf 2-D, rdfs1;
return graph

Figure 13.1.   Algorithm for partial $D*$ closure.

---

input: directed graph $G = (V, E)$, nodes $v, w \in V$
mark v;
repeat until no new nodes are marked:
    for all nodes $y \in V$
      if there is a marked node $x \in V$ such that $(x, y) \in E$
      then mark y;
if w is marked then accept else reject

Figure 13.2.   Algorithm for graph connectivity problem [Homer & Selman, 2001].

There is a similarity between this algorithm for the graph connectivity problem and the algorithm described in Definition 13.4 for computing partial closures (see Figure 13.1). Marking of nodes corresponds to adding new triples to the partial closure under construction in the final, recursive step described in Definition 13.4. In this final step the entailment rules do not introduce new terms. In other words, 'marking' (adding RDF triples by applying entailment rules) is done within a fixed, finite RDF graph that is a superset of the given RDF graph.

Lemma 13.3. *Let D be a finite datatype map. If G is a finite generalized RDF graph, then each partial $D*$ closure $G_{s,K}$ of G is finite for K finite, and a partial $D*$ closure of G can be computed in polynomial time. If $g = |G|$, $k = |K - \{i : \text{rdf}:\_i \in V(G)\}|$ and $d = |D|$, then $|G_{s,K}| \leq (4g + k + d + 30)^3$ and a partial $D*$ closure $G_{s,K}$ can be computed in polynomial time.*

Proof. The graph obtained from a finite generalized RDF graph G in the first step of the definition of partial closure is clearly finite if K is finite. Then, rule lg adds only finitely many new triples, leading to a finite graph G' containing G. In the remaining steps, no new URIs, literals or blank nodes are added to G', so it follows that there exist finite sets $U_0 \subseteq U$, $B_0 \subseteq B$ and $L_0 \subseteq L$ such that

$$G_{s,K} \subseteq U_0 \cup B_0 \times U_0 \cup B_0 \times U_0 \cup B_0 \cup L_0. \tag{13.3}$$

Hence, $G_{s,K}$ is a finite graph.

To prove the polynomial bound on the size of partial D^* closures, choose a finite, nonempty set $K \subseteq \{1, 2, ...\}$ such that $i \in K$ for each $\text{rdf}:\_i \in V(G)$. Let $g = |G|$, $d = |D|$ and $k = |K - \{i : \text{rdf}:\_i \in V(G)\}|$. Note that d is a constant. We first estimate $|T(G_{s,K})|$. We have $|T(G)| \leq 3g$. With regard to the first step of the closure construction process, the RDF, RDFS and D-axiomatic triples include at most $30 + k + d$ new terms not in $T(G)$: there are 15 URIs for RDF, 15 URIs for RDFS (see Table 13.1), k container membership properties $\text{rdf}:\_i$, and d datatype URIs. In the next step, rule lg adds at most g new blank nodes. In the final, recursive step, no new terms are added. It follows that $|T(G_{s,K})| \leq 4g + k + d + 30$. Therefore, $|G_{s,K}| \leq (4g + k + d + 30)^3$.

There are $4k$ axiomatic triples that include a container membership property, 8 additional RDF axiomatic triples, 40 additional RDFS axiomatic triples (see Tables 13.2 and 13.3) and $2d$ D-axiomatic triples (see Definition 13.1). Therefore, the first step of the closure construction process adds at most $48 + 4k + 2d$ axiomatic triples. In the next steps, rule lg adds at most g new triples, while rules rdf2-D and rdfs1 also add at most g new triples. Therefore, the total number of triples added in these steps to G is at most $2g + 4k + 2d + 48$. The computations needed for the applications of rule lg, which relate to the allocation of surrogate blank nodes, can be done in polynomial time, since it was assumed that the basic operations involved that are connected to datatype maps can be done in polynomial time, in particular the operation to determine whether two well-typed D-literals have the same value (see Section 13.3.1). Since K is required to be non-empty, $k = |K - \{i : \text{rdf}:\_i \in V(G)\}|$ can be taken to be 1. It follows that the preliminary part of the computation of a partial D^* closure $G_{s,K}$ (i.e. the part before the final, recursive step) can be done in polynomial time.

Since $|G_{s,K}| \leq (4g + k + d + 30)^3$, it follows that in the final, recursive step at most $(4g + k + d + 30)^3$ rule applications can add a new triple to the partial closure graph under construction. For each of the entailment rules used it can be determined whether a successful rule application exists in at most linear or quadratic time as a function of the size of the partial closure graph under construction (cf. Table 13.4). For example, for rule rdfs2 quadratic time is sufficient, while linear time is sufficient for rule rdf1. It follows that the final, recursive part of the computation of $G_{s,K}$ can be done in time $O((4g + k + d + 30)^5)$. This shows that a partial D^* closure $G_{s,K}$ can be computed in polynomial time. □

The following completeness result provides a graph-based characterization of D^* entailment. It is a generalization of the RDFS entailment lemma [Hayes, 2004] to D^* entailment and generalized RDF graphs. The RDFS entailment lemma is proved in [Hayes, 2004] in terms of full RDFS closures, which are infinite graphs. The following result (see [Ter Horst, 2005a], Theorem 4.12) is stated and proved in terms of partial closures, which can be taken to be finite for finite RDF graphs, thus allowing the proof of decidability of $D*$ entailment and thereby also of RDFS entailment.

Theorem 13.4 ($D*$ entailment lemma). *Let D be a datatype map, S a set of generalized RDF graphs and G a generalized RDF graph. Let H be a partial $D*$ closure $M(S)_{s,K}$ of $M(S)$ and suppose that $i \in K$ for each $\text{rdf}:\_i \in V(G)$. Then, $S \models_s G$ if and only if either H contains an instance of G as a subset or H contains a D-clash.*

Corollary 13.5. *Let D be a finite datatype map. The $D*$ entailment relation $S \models_s G$ between finite sets S of finite generalized RDF graphs and finite generalized RDF graphs G is decidable. This problem is in NP, and in P if G is ground.*

Proof of corollary. If D is a finite datatype map, then the $D*$ entailment relation $S \models_s G$ between finite sets S of finite generalized RDF graphs and finite generalized RDF graphs G can be decided by first computing a partial $D*$ closure $H = M(S)_{s,K}$, such that $i \in K$ for each $\text{rdf}:\_i \in V(G)$, following the steps described in Definition 13.4. Theorem 13.4 shows that it is sufficient to then check if H has an instance of G as a subset or if H contains a D-clash. A non-deterministic guess is used of an instance function $h : bl(G) \to T(H)$ that satisfies $G_h \subseteq H$. If G is ground, then a non-deterministic guess is not necessary. The proof of Lemma 13.3 shows that the preliminary part of the computation of H adds at most $2s + 4k + 2d + 48$ new triples to G and can be done in polynomial time in s and k, while the final, recursive part of the computation of H can be done in time $O((4s+k+d+30)^5)$, where $s = |M(S)|$ and $g = |G|$, and where $d = |D|$ is a constant. Since K is required to be non-empty and since K is required to contain the values of i that satisfy $\text{rdf}:\_i \in G$, it follows that $k = |K - \{i : \text{rdf}:\_i \in V(M(S))\}|$ can be taken to be $3g + 1$. It can be checked in polynomial time whether a generalized RDF graph contains a D-clash (see Definition 13.3), because we have assumed that it is possible to determine in polynomial time whether a D-literal is well typed (see Section 13.3.1). It follows that the entire computation can be done in non-deterministic polynomial time, and in polynomial time if G is ground.   □

Proposition 13.6. *Let D be a finite datatype map. The $D*$ entailment relation $S \models_s G$ between finite sets S of finite generalized RDF graphs and finite generalized RDF graphs G is NP-complete.*

Proof. Membership of NP has already been shown above. As in the proof of Proposition 13.2, we prove NP-completeness using a reduction from the clique problem. An instance $G = (V, E), k$ of the clique problem (see the proof of Proposition 13.2) is transformed to RDF graphs G' and H'. The RDF graph H' consists, again, of the triples $v p w$, where p is a fixed property and where

v and w are distinct elements of an arbitrary set of exactly k blank nodes. It is now also assumed that the URI reference p is not in the RDF and RDFS vocabulary or in the D-vocabulary. The RDF graph G' is formed in two steps. In the first step, each pair $\{v, w\} \in E$ is converted into two triples $v\,p\,w$ and $w\,p\,v$, where v and w are viewed as blank nodes. This leads to an RDF graph G''. In the second step, the graph G' is taken to be a partial D^* closure of G''. In view of Lemma 13.3, it is clear that the transformation from $G = (V, E), k$ to G', H' can be done in polynomial time. We need to prove that G has a clique of size $\geq k$ if and only if $G' \models_s H'$. It follows in the same way as in the proof of Proposition 13.2 that G has a clique of size $\geq k$ if and only if there is a function $h : bl(H') \to bl(G'')$ that satisfies $H'_h \subseteq G''$. We still have to prove that there is a function $h : bl(H') \to bl(G'')$ that satisfies $H'_h \subseteq G''$ if and only if $G' \models_s H'$. If there is a function $h : bl(H') \to bl(G'')$ that satisfies $H'_h \subseteq G''$, then the interpolation lemma (see Lemma 13.1) shows that $G'' \models H'$, so that by $G'' \subseteq G'$ we get $G' \models_s H'$. To prove the converse implication, suppose that $G' \models_s H'$. It is clear that G' itself is a partial D^* closure of G', so the D^* entailment lemma (Theorem 13.4) shows that there is a function $h : bl(H') \to nd(G')$ that satisfies $H'_h \subseteq G'$. The proof can be concluded by showing that none of the triples in $G' - G''$ has p as predicate, for then it follows that $H'_h \subseteq G''$. The graph G'' has a simple structure: each of the triples in G'' has the same predicate p, which is not in the RDF and RDFS vocabulary or in the D-vocabulary, and all nodes in G'' are blank nodes. By considering the axiomatic triples and entailment rules, it is not difficult to see that G' contains only three further triples containing p: rules rdf1, rdfs4a and rdfs6 produce the triples p type Property, p type Resource and p subPropertyOf p, respectively. Therefore, none of the triples in $G' - G''$ has p as predicate. □

Given a fixed generalized RDF graph G, it is natural to choose a cutoff number $k > 1$ and to compute the partial closure $G_{s,k} := G_{s,\{1,\dots,k\}}$ of G. Theorem 13.4 shows that this partial closure $G_{s,k}$ can be used to decide if G D^* entails any other RDF graph H that satisfies $i \leq k$ for each $\mathtt{rdf:\_}i \in V(H)$.

A generalized RDF graph is *inconsistent* with respect to the D^* semantics if it is not satisfied by any D^* interpretation (see [Ter Horst, 2005a] for the precise definition). The following result characterizes consistency with respect to the D^* semantics and shows that D^* consistency is in P.

Theorem 13.7. *Let D be a datatype map, S a set of generalized RDF graphs and H a partial D^* closure of $M(S)$. Then, S is D^* consistent if and only if H does not contain a D-clash. If D is a finite datatype map, the problem to determine whether a finite set of finite generalized RDF graphs is D^* consistent is in P.*

Proof. See [Ter Horst, 2005a], Theorem 4.15, for the proof of the first statement. The proof of the last statement is similar to and simpler than the proof of Corollary 13.5. If D is a finite datatype map, then the problem of determining whether a finite set S of finite generalized RDF graphs is D^* consistent can be decided by first computing a partial D^* closure $H = M(S)_{s,K}$, following the steps described in Definition 13.4. The first part of the theorem shows that it is sufficient to determine next whether H contains a D-clash. The proof of Lemma 13.3 shows that the final, recursive part of the computation of H can be done in time $O((4s + k + d + 30)^5)$, where $s = |M(S)|$ and $g = |G|$, while $d = |D|$ is a constant. Since K is required to be non-empty, $k = |K - \{i : \mathtt{rdf:}\_i \in V(M(S))\}|$ can be taken to be 1. Since it can be checked in polynomial time whether a generalized RDF graph contains a D-clash, it follows that the entire computation can be done in polynomial time.   □

13.3.3 Datatyped reasoning

In [Hayes, 2004], the notion of D-entailment is defined for extending RDFS with datatyped reasoning. Unlike D^* entailment, for D-entailment no completeness or decidability result is known. The D^* entailment relation considered in this section also extends RDFS with datatyped reasoning, but is weaker than D-entailment. Consider, for example, a datatype map D that contains the XML Schema datatypes string and integer. Theorem 13.7 shows that the following two triples, where b is a blank node, are D^* consistent:

> b `rdf:type xsd:string` .
> b `rdf:type xsd:int` .

On the other hand, these two triples are inconsistent under the D-semantics [Hayes, 2004], since the value spaces of the two datatypes string and integer are disjoint. To give another example, with regard to the XML Schema datatype `xsd:boolean`, the three triples a p `true`, a p `false` and b `type` `boolean` D-entail the triple a p b. However, this is not a D^* entailment, as can be seen with Theorem 13.4. It is possible to use meta-modeling to strengthen the datatyped reasoning capabilities obtained with D^* entailment, without increasing the computational complexity of entailment and consistency. The first example discussed above can be handled with a meta-modeling triple that uses the OWL vocabulary:

> `xsd:string owl:disjointWith xsd:int` .

The two triples of the first example, b `type` `string` and b `type` `int`, together with this meta-modeling triple, are pD^* inconsistent, as can be seen with Theorem 13.12: the three triples form a P-clash (see Definition 13.6). The second example discussed above can be handled by using rules [Ter Horst, 2005b] to model finite datatypes (cf. 13.4.5).

13.4 *pD** entailment

In this section, we extend the above results on RDFS entailment and *D** entailment to a subset of OWL. See Table 13.5 for the URI references from the OWL vocabulary considered. In [Ter Horst, 2004a] [Ter Horst, 2005a] a semantics was defined for this subset of OWL. The induced entailment relation is called *pD* entailment*. The *pD** semantics is weaker than the standard OWL semantics, but seems to be sufficient for many applications of this subset of OWL and leads to improved computational complexity of entailment.

13.4.1 *pD** entailment and OWL: 'if-semantics' versus 'iff-semantics'

The definition of the standard OWL semantics uses many if-and-only-if conditions, while the RDFS, *D** and *pD** semantics are defined by means of if conditions. For example, the *pD** semantics assumes, like RDFS, that if *c* is a subclass of *d*, then each instance of *c* is an instance of *d*, but does not assume, as OWL does, that the converse condition also holds. This means that fewer entailments are supported that relate to entire classes or properties. For example, an iff-semantics may lead to entailments that are not supported under an if-semantics, stating that a class is a subclass of another class. We briefly compare these types of reasoning for RDFS. See Sections 4.2 and 7.3.1 of [Hayes, 2004] for a description of a (non-standard) iff-semantics for RDFS. With respect to the iff-semantics for RDFS, the two triples *p* domain *v* and *v* subClassOf *w* entail the triple *p* domain *w*. As can be seen from Theorem 13.4, this entailment is not an RDFS entailment. However, triples entailed when the iff-entailed triple is added are already entailed under the standard RDFS semantics when only the original two triples are given; for example, with or without the triple *p* domain *w*, the three triples *p* domain *v*, *v* subClassOf *w* and *a p b* RDFS-entail the triple *a* type *w*. This can be shown using the entailment rules (see Table 13.4). It turns out that entailment rules for the stronger iff-semantics become more complex; no complete set of entailment rules has been described. The standard RDFS semantics is "deliberately chosen to be the weakest 'reasonable' interpretation of the RDFS vocabulary" [Hayes, 2004]. The *pD** semantics is designed along similar lines. We continue the comparison of *pD** entailment and OWL entailment in Sections 13.4.3 and 13.4.4.

13.4.2 *pD** entailment

In the treatment of *pD** entailment in this section we follow the same pattern as in the preceding section for *D** entailment.

Table 13.5.   OWL URIs used in the *pD** semantics.

owl:FunctionalProperty	owl:Restriction
owl:InverseFunctionalProperty	owl:onProperty
owl:SymmetricProperty	owl:hasValue
owl:TransitiveProperty	owl:someValuesFrom
owl:sameAs	owl:allValuesFrom
owl:inverseOf	owl:differentFrom
owl:equivalentClass	owl:disjointWith
owl:equivalentProperty	

Table 13.6.   P-axiomatic triples [Ter Horst, 2005a].

```
FunctionalProperty subClassOf Property .
InverseFunctionalProperty subClassOf Property .
SymmetricProperty subClassOf Property .
TransitiveProperty subClassOf Property .
sameAs type Property .
inverseOf type Property .
inverseOf domain Property .
inverseOf range Property .
equivalentClass type Property .
equivalentProperty type Property .
equivalentClass domain Class .
equivalentClass range Class .
equivalentProperty domain Property .
equivalentProperty range Property .
Restriction subClassOf Class .
onProperty domain Restriction .
onProperty range Property .
hasValue domain Restriction .
someValuesFrom domain Restriction .
someValuesFrom range Class .
allValuesFrom domain Restriction .
allValuesFrom range Class .
differentFrom type Property .
disjointWith domain Class .
disjointWith range Class .
```

Definition 13.5 (P-axiomatic triples, P-entailment rules). See Table 13.6 for the P-axiomatic triples. See Table 13.7 for the P-entailment rules.

Table 13.7. P-entailment rules [Ter Horst, 2005a].

	If G contains	where	then add to G
rdfp1	p type FunctionalProperty $u\,p\,v$ $u\,p\,w$	$v \in U \cup B$	v sameAs w
rdfp2	p type Inverse- FunctionalProperty $u\,p\,w$ $v\,p\,w$		u sameAs v
rdfp3	p type SymmetricProperty $v\,p\,w$	$w \in U \cup B$	$w\,p\,v$
rdfp4	p type TransitiveProperty $u\,p\,v$ $v\,p\,w$		$u\,p\,w$
rdfp5a	$v\,p\,w$		v sameAs v
rdfp5b	$v\,p\,w$	$w \in U \cup B$	w sameAs w
rdfp6	v sameAs w	$w \in U \cup B$	w sameAs v
rdfp7	u sameAs v v sameAs w		u sameAs w
rdfp8ax	p inverseOf q $v\,p\,w$	$w, q \in U \cup B$	$w\,q\,v$
rdfp8bx	p inverseOf q $v\,q\,w$	$w \in U \cup B$	$w\,p\,v$
rdfp9	v type Class v sameAs w		v subClassOf w
rdfp10	p type Property p sameAs q		p subPropertyOf q
rdfp11	$u\,p\,v$ u sameAs u' v sameAs v'	$u' \in U \cup B$	$u'\,p\,v'$
rdfp12a	v equivalentClass w		v subClassOf w
rdfp12b	v equivalentClass w	$w \in U \cup B$	w subClassOf v
rdfp12c	v subClassOf w w subClassOf v		v equivalentClass w
rdfp13a	v equivalentProperty w		v subPropertyOf w
rdfp13b	v equivalentProperty w	$w \in U \cup B$	w subPropertyOf v
rdfp13c	v subPropertyOf w w subPropertyOf v		v equivalentProperty w
rdfp14a	v hasValue w v onProperty p $u\,p\,w$		u type v
rdfp14bx	v hasValue w v onProperty p u type v	$p \in U \cup B$	$u\,p\,w$
rdfp15	v someValuesFrom w v onProperty p $u\,p\,x$ x type w		u type v
rdfp16	v allValuesFrom w v onProperty p u type v $u\,p\,x$	$x \in U \cup B$	x type w

Definition 13.6 (P-clash). In addition to D-clashes (see Definition 13.3), the $pD*$ semantics also leads to possible inconsistencies in connection with distinctFrom and disjointWith. A *P-clash* is either a combination of two triples of the form v differentFrom w, v sameAs w, or a combination of three triples of the form v disjointWith w, u type v, u type w.

Definition 13.7 (partial and full $pD*$ closures). Suppose that G is a generalized RDF graph and D a datatype map. The *partial $pD*$ closure* of G with respect to a set K, denoted by $G_{p,K}$, and the *full $pD*$ closure* of G, denoted by G_p, are defined in a way similar to the definition of the partial and full $D*$ closures $G_{s,K}$ and G_s of G (see Definition 13.4). The only differences are in the first and last steps. In the first step, the P-axiomatic triples are also added to G. In the last step, the P-entailment rules are used as well. Just like a partial $D*$ closure, a partial $pD*$ closure is, in general, a generalized RDF graph, even if the given graph is an ordinary RDF graph.

Lemma 13.8. *Let D be a finite datatype map. If G is a finite generalized RDF graph, then each partial $pD*$ closure $G_{p,K}$ of G is finite for K finite, and a partial $pD*$ closure of G can be computed in polynomial time. If $g = |G|$, $k = |K - \{i : \text{rdf:\_}i \in V(G)\}|$ and $d = |D|$, then $|G_{p,K}| \leq (4g + k + d + 45)^3$ and a partial $pD*$ closure $G_{p,K}$ can be computed in polynomial time.*

Proof. This is proved as for partial $D*$ closures (see Lemma 13.3), noting in addition that 15 URIs and 25 P-axiomatic triples are used for the $pD*$ semantics (see Tables 13.5 and 13.6). For the last part of the proof, note that for rules rdfp1, rdfp2, rdfp4, rdfp11, rdfp14a and rdfp14bx the existence of a successful rule application can be detected in time $O(n^3)$, where n is the size of the partial closure graph under construction. Since two triples are required to define a someValuesFrom or allValuesFrom restriction, the complexity is higher for rules rdfp15 and rdfp16: these rules can be handled in time $O(n^4)$. The other P-entailment rules can be handled in linear or quadratic time, just like the $D*$ entailment rules. It follows that the final, recursive part of the computation of $G_{p,K}$ can be done in time $O((4g + k + d + 45)^7)$ and that a partial $pD*$ closure $G_{p,K}$ can be computed in polynomial time. $\square$

Theorem 13.9 ($pD*$ entailment lemma). *Let D be a datatype map, S a set of generalized RDF graphs and G a generalized RDF graph. Let H be a partial $pD*$ closure $M(S)_{p,K}$ of $M(S)$ and suppose that $i \in K$ for each $\text{rdf:\_}i \in V(G)$. Then, $S \models_p G$ if and only if either H contains an instance of G as a subset or H contains a P-clash or a D-clash.*

Corollary 13.10. *Let D be a finite datatype map. The $pD*$ entailment relation $S \models_p G$ between finite sets S of finite generalized RDF graphs and finite generalized RDF graphs G is decidable. This problem is in NP, and in P if G is ground.*

Proof. See [Ter Horst, 2005a], Theorem 5.12, for the proof of Theorem 13.9. The corollary follows in the same way as in the case of $D*$ entailment (see Corollary 13.5). If D is a finite datatype map, the $pD*$ entailment relation $S \models_p G$ between finite sets S of finite generalized RDF graphs and finite generalized RDF graphs G can be decided by first computing a partial $pD*$ closure $H = M(S)_{p,K}$, such that $i \in K$ for each $\mathtt{rdf:\_} i \in V(G)$, following the steps described in Definitions 13.4 and 13.7. Theorem 13.9 shows that it is sufficient to check next whether H has an instance of G as a subset or whether H contains a P-clash or a D-clash. A non-deterministic guess is used of an instance function $h : bl(G) \rightarrow T(H)$ that satisfies $G_h \subseteq H$. If G is ground, then a non-deterministic guess is not necessary. The proof of Lemma 13.8 shows that the final, recursive step in the computation of H can be done in time $O((4s + k + d + 45)^7)$, where $s = |M(S)|$ and $g = |G|$, while $d = |D|$ is a constant and $k = |K - \{i : \mathtt{rdf:\_} i \in V(M(S))\}|$ can be taken to be $3g + 1$. It can be determined in time $O(n^3)$ whether a generalized RDF graph of size n contains a P-clash (see Definition 13.6). It can be checked in polynomial time whether a generalized RDF graph contains a D-clash (see Definition 13.3), given that we have assumed that it can be determined in polynomial time whether a D-literal is well typed (see Section 13.3.1). It follows that the entire computation can be done in non-deterministic polynomial time, and in polynomial time if G is ground. □

Proposition 13.11. *Let D be a finite datatype map. The $pD*$ entailment relation $S \models_p G$ between finite sets S of finite generalized RDF graphs and finite generalized RDF graphs G is NP-complete.*

Proof. Two additions need to be made to the proof of Proposition 13.6. In the first part of the proof, it is also assumed, of course, that the property p of the triples of G'' and H' is not in the P-vocabulary. As to the last part of the proof, it should be noted that $G' - G''$ contains a fourth triple containing p, and no further triples containing p: rule rdfp5a produces the triple p sameAs p. Therefore, it still holds that none of the triples in $G' - G''$ has p as predicate. □

Theorem 13.12. *Let D be a datatype map, S a set of generalized RDF graphs and H a partial $pD*$ closure of $M(S)$. Then, S is $pD*$ consistent if and only if H does not contain a P-clash or a D-clash. If D is a finite datatype map, the problem to determine whether a finite set of finite generalized RDF graphs is $pD*$ consistent is in P.*

Proof. See [Ter Horst, 2005a], Theorem 5.15, for the proof of the first statement. The last statement follows in a similar way to that in Theorem 13.7; compare also the proof of Corollary 13.10. If D is a finite datatype map, then the problem to determine whether a finite set S of finite generalized RDF graphs is $pD*$ consistent can be decided by first computing a partial

$pD*$ closure $H = M(S)_{p,K}$, following the steps described in Definition 13.4 and 13.7. The first part of the theorem shows that it is sufficient to determine next whether H contains a P-clash or a D-clash. The proof of Lemmas 13.3 and 13.8 shows that the final, recursive part of the computation of H can be done in time $O((4s+k+d+45)^5)$, where $s = |M(S)|$ and $g = |G|$, while $d = |D|$ is a constant. Since K is required to be non-empty, $k = |K - \{i : \mathtt{rdf:\_i} \in V(M(S))\}|$ can be taken to be 1. Since it can be checked in polynomial time whether a generalized RDF graph contains a P-clash or a D-clash, it follows that the entire computation can be done in polynomial time. □

13.4.3 Entailment rule rdf-svx for someValuesFrom

Note that entailment rules such as rdfs2, rdfs9, rdfp1 and rdfp16 for domain, subClassOf, FunctionalProperty, allValuesFrom (see Tables 13.4 and 13.7) share a common pattern, which can be described as follows. If certain instances satisfy certain conditions involving certain classes and/or properties, and if there is information about these classes and/or properties phrased in terms of the RDFS and/or OWL vocabulary, then it can be concluded that these instances satisfy certain other conditions. Entailment rule rdfp15 for someValuesFrom is formulated in a way that is different from the other entailment rules just mentioned. OWL's complete semantic condition for someValuesFrom can be captured by adding a second entailment rule for someValuesFrom, called rdf-svx and analogous to entailment rule rdfp16 for allValuesFrom. Some applications may need this extension to $pD*$ entailment. Entailment rule *rdf-svx* is defined as follows: if a generalized RDF graph G contains the three triples v someValuesFrom w, v onProperty p and u type v, where $p \in U \cup B$, then add to G the two triples $u\,p\,b$ and b type w, where b is a new blank node. Theorem 13.9 has been extended to include this stronger semantics for someValuesFrom (see [Ter Horst, 2005a], Theorem 6.8). However, Lemma 13.8 has not been extended to include the use of entailment rule rdf-svx. The issue is that, unlike the other entailment rules used in the final, recursive step in the computation of partial closures, rule rdf-svx introduces new blank nodes. Even though applications of rule rdf-svx may be restricted to situations where there is no x such that the triples $u\,p\,x$ and x type w are included in the partial closure graph under construction (see [Ter Horst, 2005a], Section 6), it has not been proved that these partial closures that make use of rule rdf-svx are always finite for finite RDF graphs. With regard to someValuesFrom, the $pD*$ semantics seems to be sufficient in a number of cases, even though entailment rule rdf-svx is not supported by the $pD*$ semantics; applications using someValuesFrom may introduce new blank nodes in a more controlled way (cf. [Ter Horst, 2005a], Section 1.8): if a person or an application intends to add the statement that a certain item u belongs to the someValuesFrom restriction class v with respect to the class w and property p and if there is no item x such that the triples $u\ p\ x$ and x type w are available, then, instead of adding the triple u type v, the two triples $u\ p\ b$ and b type w can be added, where b is a new blank node (this allows entailment rule rdfp15 to derive the triple u type v).

13.4.4 *pD** entailment and OWL

To continue the discussion started in Sections 13.4.1 and 13.4.3, we briefly compare *pD** entailment with the standard OWL entailment relations, OWL DL and OWL Full [Patel-Schneider et al., 2004]. See [Ter Horst, 2005a], Sections 1.8 and 5.1, for a more extensive discussion. Just as for RDFS, certain iff-entailments are missing from the *pD** semantics. We give several examples: with respect to the OWL Full semantics, the triple p inverseOf q entails q inverseOf p; the two triples p type FunctionalProperty and p inverseOf q entail q type InverseFunctionalProperty; the two triples p domain u and p inverseOf q entail q range u. These iff-entailments are not *pD** entailments, as can be seen with Theorem 13.9. However, just as for the corresponding example for RDFS given in Section 13.4.1, if the combination is made with data relating to instances, the consequences of the entailed triples are also entailed under the *pD** semantics from the triples originally given. This can be shown with the entailment rules (see Tables 13.4 and 13.7). For example, the two triples p type FunctionalProperty and p inverseOf q plus the two triples $u\ q\ w$ and $v\ q\ w$ *pD** entail, with or without the triple q type InverseFunctionalProperty, the triple u sameAs v. In view of examples such as these, it seems that *pD** entailment is sufficient for many applications that combine ontologies with data relating to instances.

The *pD** semantics does not apply to all of the OWL vocabulary; in particular, the *pD** semantics does not deal with the owl:oneOf class constructor, which allows enumeration classes to be defined, or with OWL's possibilities for defining restriction classes in terms of cardinalities. Although the *pD** semantics does not include OWL's Boolean class constructors owl:unionOf and owl:intersectionOf either, part of these constructs is available in an alternative way, by means of rdfs:subClassOf. It can be expressed that the union of the classes $c_1, \ldots, c_n$ is contained in a class c by saying that each class c_j is a subclass of c. Moreover, it can be expressed that a class c is contained in the intersection of the classes $c_1, \ldots, c_n$ by saying that c is a subclass of each class c_j.

OWL Full entailment is undecidable, while OWL DL entailment is decidable and NEXPTIME-complete [Horrocks & Patel-Schneider, 2004]. OWL DL imposes certain restrictions on the use of the language to obtain decidability. For example, classes cannot be used as instances, while the use of the properties `FunctionalProperty` and `TransitiveProperty` is restricted [Patel-Schneider et al., 2004]. The *pD** semantics imposes no restrictions on the use of the language and supports meta-modeling expressivity. Just as for RDFS, an if-semantics also leads to computational advantages for OWL: there is a complete set of simple entailment rules for *pD** entailment, while *pD** entailment is in NP, and in P if the target graph does not contain blank nodes. The result that RDFS entailment and *pD** entailment are in P if the target graph does not have blank nodes is relevant in practice. The target graph *G* typically forms a relatively small part of the combined problem data *S, G* and may contain relatively few blank nodes. The results clearly indicate the computational cost of using blank nodes in a target graph.

13.4.5   Ontologies and rules

There is much interest in extending the Semantic Web languages RDF and OWL with facilities for representation of and reasoning with rules. For example, OWL cannot express the notion of 'uncle' by means of a rule such as the following:

```
IF      ?a hasParent ?b .
        ?b hasBrother ?c .
THEN    ?a hasUncle ?c .
```

The IF and THEN sides of such rules consist of 'rule graphs' that consist of 'triple patterns' which extend RDF triples with the possibility to include 'rule variables' (such as ?*a* in the example) in subject, predicate and object positions. If the IF side of such a rule can be matched with certain triples in an RDF graph, then the rule prescribes that the THEN side can be added to the RDF graph, with the rule variables being substituted with the matching terms in the given RDF graph. The entailment rules used in this chapter can be viewed as arising from such rules. For example, entailment rule rdfp1 for `owl:FunctionalProperty` (see Table 13.7) arises from the following rule which uses rule variables that correspond to the meta-variables *p*, *u*, *v* and *w* used in Table 13.7:

```
IF      ?p rdf:type owl:FunctionalProperty .
        ?u ?p ?v .
        ?u ?p ?w .
THEN    ?v owl:sameAs ?w .
```

There is not yet a standard Semantic Web language for rules. There is a proposal to extend OWL DL with rules in the language SWRL [Horrocks et al., 2005]. SWRL extends OWL DL with a restricted form of Horn rules. Entailment for SWRL is undecidable [Horrocks et al., 2005].

In [Ter Horst, 2005b] the model theory of RDF is extended to apply to rules. For an arbitrary set of rules *R* of the kind just discussed, a general notion of

R-entailment is defined. *R*-entailment extends RDFS and its meta-modeling capabilities. *R*-entailment also extends *pD** entailment. In [Ter Horst, 2005b] it is shown that the decidability and NP (P) complexity results for RDFS and *pD** entailment extend to include a large class of rules.

13.5 Conclusion

In this chapter we provided an overview of results relating to complexity of entailment with respect to Semantic Web ontologies. We also included an overview of the graph-based entailment characterizations that form the basis for these complexity results. In summary, the complexity results proved in this chapter are: Simple entailment and entailment for RDF Schema (RDFS) are NP-complete, and in P if the target graph is assumed to contain no blank nodes. These results also apply to the *D** entailment and *pD** entailment relations, which extend RDFS entailment to reasoning with datatypes and to reasoning with a property-related subset of OWL, respectively. We proved new bounds on the size of the partial closure graphs used in these results. The results discussed support the conclusion that an if-semantics applied to (a subset of) the OWL vocabulary leads to computational advantages.

Acknowledgements

I am grateful to Jan Korst for his suggestion with regard to the proof of Proposition 13.2 and to Warner ten Kate and Jan Korst for their useful comments on earlier versions of this chapter.

References

Baader, F., et al. [2003]. (Eds.), *The Description Logic Handbook.* Cambridge University Press, Cambridge.

Berners-Lee, T., J. Hendler, and O. Lassila [2001]. The Semantic Web. *Scientific American,* May issue, pages 34–43.

Cori, R., and D. Lascar [2000]. *Mathematical Logic,* Part I. Oxford University Press.

Garey, M.R., and D.S. Johnson [1979]. *Computers and Intractability: A Guide to the Theory of NP-Completeness.* Freeman, New York.

Grant, J., and D. Beckett (eds.) [2004]. *RDF Test Cases.* W3C Recommendation, http://www.w3.org/TR/2004/REC-rdf-testcases-20040210/

Hayes, P. (ed.) [2004]. *RDF Semantics.* W3C Recommendation, http://www.w3.org/TR/2004/REC-rdf-mt-20040210/

Homer, S., and A.L. Selman [2001]. *Computability and Complexity Theory.* Springer, New York.

Horrocks, I., and P.F. Patel-Schneider [2004]. Reducing OWL entailment to description logic satisfiability. *Journal of Web Semantics,* 1:345–357.

Horrocks, I., P.F. Patel-Schneider, S. Bechhofer, and D. Tsarkov [2005]. OWL rules: A proposal and prototype implementation. *Journal of Web Semantics,* 3: 23–40.

Klyne, G., and J. Carroll (eds.) [2004]. *Resource Description Framework (RDF): Concepts and Abstract Syntax.* W3C Recommendation, http://www.w3.org/TR/2004/REC-rdf-concepts-20040210/

OWL (Web Ontology Language). W3C, http://www.w3.org/2004/OWL/

Patel-Schneider, P.F., P. Hayes, and I. Horrocks (eds.) [2004]. *OWL Web Ontology Language Semantics and Abstract Syntax.* W3C Recommendation, http://www.w3.org/TR/2004/REC-owl-semantics-20040210/

RDF (Resource Description Framework). W3C, http://www.w3.org/RDF/

ter Horst, H.J. [2004a]. Extending the RDFS entailment lemma. *Proceedings of the Third International Semantic Web Conference (ISWC2004), Lecture Notes in Computer Science,* 3298:77–91.

ter Horst, H.J. [2004b]. Semantic web ontologies and entailment: Some complexity results. *Proceedings of the 2nd Philips Symposium on Intelligent Algorithms (SOIA'04),* W.F.J. Verhaegh et al. (eds.), Philips Research, Eindhoven, The Netherlands, pages 59–71.

ter Horst, H.J. [2005a]. Completeness, decidability and complexity of entailment for RDF schema and a semantic extension involving the OWL vocabulary. *Journal of Web Semantics,* 3:79–115.

ter Horst, H.J. [2005b]. Combining RDF and part of OWL with rules: Semantics, decidability, complexity. *Proceedings of the Fourth International Semantic Web Conference, Lecture Notes in Computer Science,* 3729:668–684.

Chapter 14

BAYESIAN METHODS FOR TRACKING AND LOCALIZATION

Wojciech Zajdel, Ben J.A. Kröse, and Nikos Vlassis

Abstract   This chapter presents a tutorial-type introduction to dynamic Bayesian networks (DBNs), which provide a computational framework for the analysis of stochastic dynamic systems. The first part of the chapter provides a short overview of the work on probabilistic state estimation and system identification. The second part presents two example applications where the DBNs lead to elegant algorithms: robot localization and tracking of multiple persons with multiple cameras.

Keywords   Time-series analysis, probabilistic inference, dynamic Bayesian networks, robot localization, multi-object tracking.

14.1   Introduction

Localization and tracking of moving objects are one of the central issues in research in intelligent environments. Typical examples are service or security applications, where the environment needs to localize moving persons. Alternatively, mobile embedded systems have to be localized and tracked, for example to deliver location dependent services. In all of these situations there is a need for sensors to give location information.

In typical problems, one can distinguish between two configurations: the sensors can either be mounted on the moving system and observe the environment, or the sensors may be fixed to the environment and observe the moving system. An instance of the first configuration is the GPS (global positioning system), which uses the position of satellites as reference points to calculate the position of the moving system. Other examples are systems which use radio beacons. If the environment is not artificially landmarked, natural features have to be used to localize the system. These features have to be derived from the sensory signal, for example from images of a camera mounted on a moving platform. However, in many cases the objects to be tracked are not equipped with sensors. In the second configuration, it is the environment that observes

Wim F.J. Verhaegh et al. (Eds.), Intelligent Algorithms in Ambient and Biomedical Computing, 243-258.
© 2006 *Springer. Printed in the Netherlands.*

the moving objects. Because in this case the sensors are not associated one-to-one with objects, the environment will not only have to *localize* the moving objects but also have to *identify* them. In the subsequent discussion we will assume color video cameras as sensing systems.

Irrespective of the sensor configuration, localization and tracking methods need to estimate the state of a dynamic system (i.e. object's identity and/or position) from observations. The estimation algorithm has to account for two problems. Firstly, the observations are affected by noise, which is typically caused by camera jitter, variations in illumination or viewing angle, occlusions, and shadows. Secondly, cameras provide often high-dimensional observations, and modeling dynamics in a high-dimensional space is not trivial. To address these issues we consider a probabilistic framework to deal with the noise and furthermore use the fact that the underlying system generating the observations has only a few degrees of freedom.

This chapter offers a tutorial on dynamic Bayesian networks and their applications to localization and tracking. Section 14.2 outlines basics forms of DBNs and presents suitable Bayesian inference algorithms. Section 14.3 presents a method for localizing a mobile robot from sensory information, and Section 14.4 — a method for tracking multiple people with multiple cameras.

14.2 Bayesian networks for dynamic systems analysis

Throughout this tutorial we discuss stochastic discrete-time dynamic systems. We denote the subsequent time steps with $k = 1, 2, \ldots$, and the state of the system at the kth step with s_k, and a time-sequence of states $s_n, s_{n+1}, \ldots, s_m$ as $s_{n:m}$. Our key assumption is that the state of the system in question cannot be directly observed. Instead, at every step we have access to a noisy measurement or observation y_k provided by the sensor(s).

The Bayesian framework [Gelman et al., 1995; Murphy, 2002; Jordan, 1998] embeds the relation between states and observations into a probability distribution $p(s_{1:k}, y_{1:k})$. Further, the distribution $p(s_{1:k}, y_{1:k}) = p(s_{1:k})p(y_{1:k}|s_{1:k})$ is decoupled into conceptually simpler parts: a model of system internal dynamics $p(s_{1:k})$ and a sensor noise model $p(y_{1:k}|s_{1:k})$. Given the measurements we can compute (often only approximately) posterior distributions, in the form $p(s_k|y_{1:k})$, which convey useful information about the system states. Such a framework places the designer's emphasis on accurate modeling of system dynamics and sensor noise, while computing posterior state densities is left to Bayesian numerical inference methods.

14.2.1 Representations

Practical representation of time-series models, like $p(s_{1:k}, y_{1:k})$, relies on the notion of causal dependency [Jensen, 2001]. Given a series of variables

$x_{1:n}$, causal dependency states that some variable x_i is assumed to be generated from – or caused by – a limited subset $\text{Pa}(x_i)$ of variables from $x_{1:n}$. We refer to variables in $\text{Pa}(x_i)$ as the causes or parents for x_i. One can intuitively represent causal dependencies as directed graphs, where nodes correspond to variables, and directed edges lead from causes to the resulting variables. For time-series models such graphs are called dynamic Bayesian networks (DBNs). In DBNs the edges usually point forward in time reflecting a natural assumption that variables future in time are caused by (a subset of) past variables. Formally, a DBN defines a distribution as a product

$$p(x_{1:n}) = \prod_i p(x_i | \text{Pa}(x_i)),$$

where i enumerates the nodes, and $p(x_i | \text{Pa}(x_i))$ are generally simple probability density functions. One usually refers to distributions defined by this framework as *directed graphical* or *generative* models [Murphy, 2002].

Below we briefly review and motivate the most common structures of graphical models for dynamic systems with noisy measurements. Our outline includes only a limited selection of models; more comprehensive surveys can be found in [Murphy, 2002; Jordan, 1998].

System dynamics. A simple design for system dynamics follows from the Markov assumption, which says that the state s_k is generated exclusively by s_{k-1}. Typically, the causal dependency $p(s_k | s_{k-1})$ is time-invariant (i.e., the same for all time-steps). This leads to the following model

$$p(s_{1:k}) = p(s_1) \prod_{\tau=2}^{k} p(s_\tau | s_{\tau-1}), \tag{14.1}$$

where $p(s_1)$ is the prior state distribution. The most common instances of such a design are hidden Markov models (HMMs) [Rabiner, 1990] or Kalman filter models (KFMs) [Rowies & Ghahramani, 1999].

Sometimes however one may need so called non-Markovian models, where the causal dependency $p(s_k | s_{1:k-1})$ incorporates all (or a subset) of past states [Neal, 2000]. An interesting class of systems takes $p(s_k | s_{1:k-1})$ as a weighted sum of simpler functions, each depending on a single past state

$$p(s_{1:k}) = p(s_1) \prod_{\tau=2}^{k} p(s_\tau | s_{1:\tau-1}) \quad \text{with} \quad p(s_\tau | s_{1:\tau-1}) = \sum_{\kappa=1}^{\tau-1} \pi(\tau - \kappa) p(s_\tau | s_{\tau-\kappa}), \tag{14.2}$$

where π is a vector of weights. The right panel of Figure 14.1 shows the corresponding graphical structure. Examples of such systems include Dirichlet processes [Neal, 2000] or its variations, like the model for multi-object tracking, as described in Section 14.4.

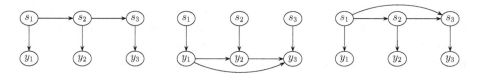

Figure 14.1. Dynamic Bayesian networks representing standard dynamic systems. (Left) System with Markovian dynamics. (Middle) Mixed memory Markov model, where y_k depends on $\{y_{1:k-1}, s_k\}$ and states are independent from each other. (Right) System with non-Markovian dynamics, where s_k depends on $s_{1:k-1}$, and y_k depends on s_k.

Sensor models. In the simplest setup, sensor models assume that the observation y_k is generated exclusively from the underlying state s_k. Therefore

$$p(y_{1:k}|s_{1:k}) = \prod_{\tau=1}^{k} p(y_\tau|s_\tau).$$

Another class of models assumes that the current observation y_k depends on s_k and also on the past observations $y_{1:k-1}$. An example is a mixed memory Markov model [Saul & Jordan, 1999], where $y_{1:k-1}$ become a "memory", and s_k is a "switch" that selects a single memory item that generates y_k:

$$p(y_{1:k}|s_{1:k}) = \prod_{\tau=1}^{k} p(y_\tau|y_{1:\tau-1}, s_\tau) \quad \text{with} \quad p(y_\tau|y_{1:\tau-1}, s_\tau) = p(y_\tau|y_{f(s_\tau)}),$$

where $1 < f(s_\tau) < \tau$ is an index function. From the inference perspective, the models $p(y_k|s_k)$ and $p(y_k|y_{1:k-1}, s_k)$ do not differ substantially, since in both cases s_k is the only hidden variable at every time-step (the variables $y_{1:k-1}$ are fixed). Therefore, without loss of generality, we will discuss the case $p(y_k|s_k)$.

Example. A popular probabilistic time-series models are linear dynamical systems, also known as Kalman filter models [Rowies & Ghahramani, 1999]. These models assume a Normal prior $p(s_1) = N(s_1|\mathbf{m}_0, \mathbf{V}_0)$, and linear state transitions with additive Gaussian noise $p(s_k|s_{k-1}) = N(s_k|\mathbf{A}s_{k-1}, \mathbf{Q})$. The observation y_k is a linear projection of s_k with additive Gaussian noise; $p(y_k|s_k) = N(y_k|\mathbf{C}s_k, \mathbf{R})$. Here, $\mathbf{A}$ is the transition matrix, $\mathbf{C}$ the observation or generative matrix, and $\mathbf{R}, \mathbf{Q}$ are noise covariances, and $N(s|\mathbf{m}, \mathbf{V})$ denotes a Gaussian (Normal) probability distribution on variable s with mean $\mathbf{m}$ and covariance $\mathbf{V}$.

14.2.2 Inference

Bayesian inference refers to a collection of techniques for reasoning about hidden states on the basis of available data and the assumed model that ties the hidden and the observed quantities. In the Bayesian framework, any information about the state(s) follows from posterior state distribution(s) conditioned on the data. Given such a distribution — or a *belief* for short — one can compute quantities like the expected or most likely state or confidence intervals.

In particular for DBNs, the following inference problems are the most important. The first involves computing $\arg\max_{s_{1:k}} p(s_{1:k}|y_{1:k})$ to find out the most likely sequence of states. The other type of problems involve computing marginal posterior distributions in the form $p(s_k|y_{1:k+\tau})$, which provide information about individual states from a sequence of measurements. When $\tau < 0$ such distributions are referred to as *predictive* distributions, when $\tau = 0$ — as *filtering* distributions, and when $\tau > 0$ — as *smoothing* distributions. Below we focus on computing filtering distributions. More details on various exact and approximate methods for solving the above problems are discussed in [Jordan, 1998; Murphy, 2002; Lerner & Parr, 2001].

Filtering. In on-line problems, including tracking and robot localization, one wishes to estimate the current state of the system s_k given the data available so far $y_{1:k}$, therefore the filtering distribution $p(s_k|y_{1:k})$ is of particular interest.

For systems with Markovian dynamics (see (14.1)), the filtering distribution can be computed recursively using the distribution from previous time step, denoted as $p(s_{k-1}|y_{1:k-1})$. First, we find the *predictive* distribution

$$p(s_k|y_{1:k-1}) = \int p(s_k|s_{k-1})p(s_{k-1}|y_{1:k-1})\,ds_{k-1}. \qquad (14.3)$$

The predictive distribution summarizes our knowledge about state s_k given the past data. When the observation y_k arrives we incorporate it to the filtering distribution, using Bayes' theorem

$$p(s_k|y_{1:k}) = \frac{1}{L_k}p(y_k|s_k)p(s_k|y_{1:k-1}), \qquad (14.4)$$

where $L_k = p(y_k|y_{1:k-1})$ is a normalization constant. At $k = 1$ we set the predictive distribution equal to the prior $p(s_1)$. The procedure (14.3)–(14.4) is usually called Bayesian filtering [Murphy, 2002; Rowies & Ghahramani, 1999].

For systems with non-Markovian dynamics (see (14.2)) the procedure is more complicated. Although, one is still interested in estimation of the current state, the recursive scheme now requires propagation of the complicated density $p(s_{1:k}|y_{1:k})$, because the current state depends on all past states. Theoretically, we can follow a similar derivation as for Markovian systems

$$p(s_{1:k}|y_{1:k-1}) = p(s_k|s_{1:k-1})p(s_{1:k-1}|y_{1:k-1}) \qquad (14.5)$$

$$p(s_{1:k}|y_{1:k}) = \frac{1}{L_k}p(y_k|s_k)p(s_{1:k}|y_{1:k-1}), \qquad (14.6)$$

but in practice the resulting expression can only be approximated.

In either type of systems, feasibility of Bayesian filtering relies on compact representation of the filtering distribution. Ideally, this distribution falls into some parametric family $p(s_k|y_{1:k}) = f(s_k, \lambda_k)$, where λ_k is a fixed-size set of parameters. In this case, filtering simply recomputes parameters λ_k from λ_{k-1} and y_k. The most common examples of such systems are HMMs and KFMs. For HMMs, s_k is a discrete variable, λ_k is a vector representing a discrete distribution. For KFMs, s_k is a continuous variable, and $f(s_k, \lambda_k) = \mathcal{N}(s_k|\mathbf{m}_k, \mathbf{V}_k)$ is a Normal density function, where $\mathbf{m}_k$ is the mean vector, $\mathbf{V}_k$ is covariance matrix; and $\lambda_k = \{\mathbf{m}_k, \mathbf{V}_k\}$.

Unfortunately, for many problems the filtering distribution cannot be compactly represented [Murphy, 2002]. In some cases, the integral (14.3) does not have a closed-form solution. In some other cases, including non-Markovian systems, the size of parameter vector λ_k grows linearly, or even exponentially with time, rendering filtering intractable. In the rest of this section, we present two popular techniques for efficient approximating the filtering distribution.

Particle filtering. The particle filter is a simulation-based technique for approximating intractable filtering distributions in Markovian models [Doucet et al., 2001]. It is particularly useful for continuous-state systems, where the integral (14.3) is complicated. The idea is to represent the continuous density $p(s_k|y_{1:k})$ at each time step k by a random sample of I particles s_k^i with corresponding probability masses (weights) π_k^i. The filtering density at step $k-1$ is approximated by

$$p(s_{k-1}|y_{1:k-1}) \approx \sum_{i=1}^{I} \pi_{k-1}^i \, \delta(s_{k-1} - s_{k-1}^i), \qquad (14.7)$$

where $\delta(s_{k-1} - s_{k-1}^i)$ is a delta function centered on the particle s_{k-1}^i. Using (14.7), the integration for computing the predictive density in (14.3) is now replaced by the much easier summation

$$p(s_k|y_{1:k-1}) = \sum_{i=1}^{I} \pi_{k-1}^i \, p(s_k|s_{k-1}^i) \qquad (14.8)$$

The filtered distribution evaluates to

$$p(s_k|y_{1:k}) \approx \frac{1}{L_k} p(y_k|s_k) \sum_{i=1}^{I} \pi_{k-1}^i \, p(s_k|s_{k-1}^i). \qquad (14.9)$$

Since all integrals are replaced by sums and the continuous densities by discrete ones, the normalization term L_k of the filtered distribution is trivial, namely, the sum of the weights.

Assuming a set of particles that approximate the posterior density $p(s_{k-1}|y_{1:k-1})$ sufficiently well, the problem is how to project the new posterior (14.9) to the form required by (14.7). In other words, how to sample a set of particles from the new posterior $p(s_k|y_{1:k})$. Efficient sampling from the posterior is the central theme of most methods in the particle filters literature [Doucet et al., 2001].

Assumed-density filtering. Unlike a particle filter, assumed-density filtering (ADF) is a deterministic technique [Boyen & Koller, 1998; Murphy, 2002]. It approximates the filtering distribution with a parametric analytical family $p(s_k|y_{1:k}) \approx q(s_k, \lambda_k)$. Importantly, ADF is applicable to both Markovian and non-Markovian systems.

Below, we show how ADF applies to non-Markovian systems, where $p(s_{1:k}|y_{1:k})$ can be compactly represented as a product of simpler density functions. Assume at step $k-1$

$$p(s_{1:k-1}|y_{1:k-1}) \approx q(s_{1:k-1}, \lambda_{k-1}) = \prod_{\tau=1}^{k-1} f(s_\tau|\lambda_{k-1,\tau}).$$

If states are discrete, than f has to be a multinomial (discrete) distribution. When states are continuous, we are free to choose any function f with parameters $\lambda_{k,\tau}$ that will represent the density. After executing (14.5) and (14.6), the next-step filtering density becomes

$$p(s_{1:k}|y_{1:k}) \approx \tilde{q}(s_{1:k}) = \frac{1}{L_k} p(y_k|s_k) p(s_k|s_{1:k}) \prod_{\tau=1}^{k-1} f(s_\tau, \lambda_{k-1,\tau}) \qquad (14.10)$$

The expression $\tilde{q}(s_{1:k})$ generally does not belong to the assumed factorial family $q(s_{1:k})$. We will approximate it with such a function form the family that minimizes the Kullback-Leibler (KL) divergence. KL divergence measures the distance between two distributions $\tilde{q}(x)$ and $q(x)$

$$\text{KL}\left(\tilde{q}(x)||q(x)\right) = \int \tilde{q}(x) \log\left(\frac{\tilde{q}(x)}{q(x)}\right) dx.$$

According to a standard result [Cover & Thomas, 1991], the closest factorial distribution to any $\tilde{q}(s_{1:k})$ is a product of its marginals $\prod_\tau^k \tilde{q}(s_\tau)$. If the states are discrete, the marginals will already be multinomials and $f(s_\tau, \lambda_{k,\tau}) = \tilde{q}(s_\tau)$. For continuous states, the marginals $\tilde{q}(s_\tau)$ need to be further approximated by the KL-closest density function $f(s_\tau, \lambda_{k,\tau})$. The parameters of this function follow from

$$\lambda_{k,\tau} = \text{argmin}_\lambda \text{KL}\left(\tilde{q}(s_\tau)||f(s_\tau, \lambda)\right).$$

Efficiency of ADF methods crucially depends on the feasibility of this minimization problem. Examples and extensions of this technique are provided in [Murphy, 2002].

14.3 Localization of a mobile platform

A problem that can be addressed using the above techniques is *robot localization*. The term refers to the ability of a mobile robot to predict and maintain at any time step its state (position and orientation) within its environment. As in object tracking, robot localization can be regarded as an on-line filtering problem: estimate the current state of the robot given an initial state estimate and a sequence of observations.

Formally, we assume that at time step k the state of the robot is a random variable $s_k \in \mathbb{R}^3$ that involves the position (x, y) and orientation (θ) of the robot. Moreover, we assume a given stochastic transition (motion) model $p(s_{k+1}|s_k, u_k)$ for a robot action u_k that is issued at time step k, and which changes the state of the robot stochastically from s_k to s_{k+1}. The transition model is assumed Gaussian, with mean computed from the issued action of the robot (translation-rotation), and standard deviation given by the odometry noise characteristics (which are known for the particular robot or have been computed in advance from a training set).[1] We also assume that in each time step k the robot observes a high-dimensional sensor vector $y_k \in \mathbb{R}^d$, which is related to the robot state through a stochastic observation model $p(y_k|s_k)$. The observations $\{y_k\}$ are assumed conditionally independent given the states $\{s_k\}$. Robot localization amounts to estimating in each time step k a posterior density $p(s_k|y_{1:k})$ over the state space, that characterizes the *belief* of the robot about its current state at time k given its initial belief $p(s_0)$ and the sequence of observations $y_1, \ldots, y_k$ up to time step k.

Many techniques have been developed for robot localization in the last couple of years, most of which rely on the use of a Kalman filter: this estimates the state of the robot by means of a Gaussian distribution with a certain mean and covariance matrix. Such an approach essentially relies on the assumption that the state vector is always Gaussian distributed, which can be a restrictive assumption in case the environment exhibits *perceptual aliasing*: two perceptually distinct locations may look identical to the sensor of the robot.

To deal with perceptually aliased environments, an alternative representation involves the use of a particle filter. As explained above, a particle filter represents the distribution of the robot state using a set of 'particles' scattered over the state space. Each particle can be viewed as a hypothesis about the true location of the robot at any time step, and the complete set of particles defines

---

[1] In the following, we assume the existence of an action u_k in the transition model and write $p(s_k|s_{k-1})$.

the set of all possible hypotheses where the robot could be. Such an (approximate) filter has been employed in several mobile robot applications recently, with reported success [Thrun et al., 2001].

14.3.1 Particle filter implementation

In this section we provide more details about the use of a particle filter in robot localization. As explained above, a particle filter represents the filtering density at time step k by a weighted set of I particles scattered over the robot's state space, given by (14.7). Consequently, the predictive density (14.8) is a mixture of I components (transition kernels), one for each particle s_{k-1}^i. A way to sample a new set of particles for the next step filtering density, referred to in the literature as Sampling/Importance Resampling (SIR) [Gordon et al., 1993; Isard & Blake, 1998; Delaert et al., 1999], involves first sampling from the above predictive density: select the i-th mixture component $p(s_k|s_{k-1}^i)$ with probability π_{k-1}^i, and then draw a sample from it (which is trivial if the transition model is Gaussian). Each sampled particle s_k^j is then assigned weight π_k^j proportional to the likelihood $p(y_k|s_k^j)$. Finally a resampling step is taking place in order to make all particle weights equal.

A problem with the SIR filter is that it requires very many particles to converge when the likelihood function $p(y|s)$ is too peaked or is situated in one of the prior's tails [Pitt & Shephard, 1999]. The latter is much more severe in case of *outliers*, model-implausible observations that occur when there is image occlusion or other unexpected effects in the environment. An alternative sampling method has been proposed in [Pitt & Shephard, 1999] under the name *auxiliary particle filter*. The main idea is to sample from the posterior in (14.9) after inserting the likelihood inside the mixture:

$$p(s_k|y_{1:k}) \propto \sum_{i=1}^{I} \pi_{k-1}^i p(y_k|s_k) p(s_k|s_{k-1}^i) \qquad (14.11)$$

and treat the products $\pi_{k-1}^i p(y_k|s_k)$ as component probabilities in order to sample from the respective mixture. Because the likelihood $p(y_k|s_k)$ in the above product involves the unobserved state vector s_k, an approximation of the mixture (14.11) can be used as

$$\tilde{p}(s_k|y_{1:k}) \propto \sum_{i=1}^{I} \pi_{k-1}^i p(y_k|\mu_k^i) p(s_k|s_{k-1}^i) \qquad (14.12)$$

where μ_k^i is any value associated with the i-th component transition density $p(s_k|s_{k-1}^i)$, for example its mean. After a set of $j = 1, \ldots, I$ particles have been sampled[2] from the mixture (14.12), with locations s_k^j, their weights are set proportional to

$$\pi_k^j \propto \frac{p(y_k|s_k^j)}{p(y_k|\mu_k^{ij})} \tag{14.13}$$

where μ_k^{ij} is the associated value of the mixture component $p(s_k|s_{k-1}^{ij})$ in (14.12) from which the particle j was sampled. Setting the weights as in (14.13) has the additional benefit of creating particles with much less variable weights than for the original SIR method, which has advantages in case of outliers.

The auxiliary particle filter can be regarded as a one-step look-ahead procedure, where a particle s_{k-1}^i is propagated to μ_k^i in the next time step in order to assist the sampling from the posterior. The resulting method is particularly efficient since it requires only the ability to sample from the transition model and evaluate the likelihood function $p(y_k|s_k)$. This makes it very attractive compared to alternative methods that require specialized data structures for sampling from the posterior.

14.3.2 The sensor model

Since the sensor observations y_k are typically high-dimensional (e.g., camera images), computing a good observation model $p(y_k|s_k)$ is a computationally expensive problem, and one has to resort to simpler, lower-dimensional models. A solution we have adopted in our system is to linearly project (using principal component analysis (PCA) or some other linear projection method) a raw observation y_k to a low-dimensional *feature* vector z_k, and define an observation model for the low-dimensional features z_k. The sensor model $p(z_k|s_k)$ in the reduced space is computed non-parametrically from a training set of state-feature pairs $\{s_n^*, z_n^*\}$ (with the z_n^* computed by projecting from corresponding y_n^*) using nearest neighbor density estimation:[3]

$$p(y_k|s_k) = p(z_k|s_k) = \alpha \sum_{j=1}^{J} \lambda_j(z_k)\,\phi(s_k|s_j^*), \tag{14.14}$$

hence, it is a mixture of J components $\phi(s|s_j^*)$, each weighted by $\lambda_j(z)$, which is computed as follows: We first find the J nearest neighbors z_j^* of z among the $\{z_n^*\}$ training data. This can be done efficiently, with average cost $O(J\log K)$,

---

[2]This involves a multinomial sampling on the weights, and there is an $O(I)$ procedure for doing this [Pitt & Shephard, 1999].

[3]Note that this is much easier than modeling the density $p(y|s)$ directly in the y space because the dimensionality of the state s is low (three).

using methods from computational geometry (e.g., *kd*-trees). We then sort these neighbors z_j^* by increasing distance to z, and for each nearest neighbor z_j^* we extract from the training set the corresponding state s_j^*. Each s_j^* defines a respective component $\phi(s|s_j^*)$ in the mixture (14.14), where $\phi(s|s_j^*)$ is a Gaussian kernel centered on s_j^* with bandwidth equal to half the bin size of the grid of the $\{s_n^*\}$ points. Finally, the mixing weights $\lambda_j(z)$ are positive and sum to one, and decrease linearly with j:

$$\lambda_j(z) = \frac{2(J - j + 1)}{J(J + 1)}.$$

Our sensor model also detects outliers (e.g., occlusions in the image), by using a simple threshold test of the distance of an observation z to its first nearest neighbor $z_{j=1}$. If occlusion is detected, the auxiliary particle filter sampling is not used and the filter just propagates the particles from the previous time step according to the transition model. Further details and experiments are provided in [Vlassis et al., 2002].

14.4 Tracking with distributed cameras

In this section we demonstrate how inference in an appropriately defined DBN leads to a multi-camera multi-object tracking algorithm. The fundamental problem in multi-object tracking is association of incoming observations with trajectories. Typical cameras cannot directly observe the identity of an object. Therefore, we design a probabilistic model that explicitly identifies objects with hidden labels, and provides a set of probabilistic dependencies that couple the readings from the cameras with the labels. Under such a framework, we resolve association ambiguities by finding the most likely label for each observation according to the filtering density.

14.4.1 Problem formulation

We consider tracking people in wide areas, such as airports, where the cameras observe relatively small, disjoint regions from the global area of interest [Pasula et al., 1999]. Moreover, we assume that every camera locally tracks a person within its field-of-view (FOV). When the person leaves the FOV, a camera reports a single observation y_k that *summarizes* the local trajectory of the person. In such a setup, we aim to (re-)identify people when they move between FOVs by association of the observations y_k with global trajectories.

We process observations from all cameras centrally, and treat $k = 1, 2, \ldots$, as a central index that preserves time-order of observations. It is also assumed, that each observation $y_k = \{o_k, d_k\}$ includes color features o_k, and spatio-temporal features $d_k = \{l_k, t_k^e, t_k^q, b_k^e, b_k^q\}$, where l_k is the camera location, t_k^e, b_k^e (t_k^q, b_k^q) are the time and frame border of entering (quitting) the FOV.

Appearance features. Typically, the color features o_k are noisy observations of some constant intrinsic properties of a person. When observed noiseless, these properties provide the key cues for distinguishing people from each other. The noise arises from camera jitter and variations in illumination or pose. To suppress the illumination-originated artifacts we will use channel-normalized color space [Zajdel et al., 2004]. Explicit compensation for the other noise sources requires complicated analysis. Instead, we assume that these sources introduce stochastic, Gaussian noise. For every observed r-vector o_k we introduce parameters $x_k = \{\mathbf{m}_k, \mathbf{V}_k\}$ of a Normal density function that generated o_k. The $r \times 1$ mean vector $\mathbf{m}$ describes the person-specific features. The $r \times r$ covariance matrix $\mathbf{V}$ models person-specific sensitivity to the jitter and changing pose. For instance, the appearance of somebody dressed uniformly is relatively independent of pose, so his/her covariance $\mathbf{V}$ has small eigenvalues. In contrast, the appearance of a person wearing non-uniform colors, is very sensitive to pose changes. This is modeled with a 'broad' density, i.e., $\mathbf{V}$ with large eigenvalues.

As we have seen, $x = \{\mathbf{m}, \mathbf{V}\}$ describes object-specific properties. These properties are not directly observed, therefore we treat x as a latent state of a person. The state is a hidden random variable with a prior distribution

$$\pi(x) = \phi(x|\theta_0), \tag{14.15}$$

where $\phi(x|\theta_i)$ is appropriate parametric density for mean and covariances and $\theta_i = \{\mathbf{a}_i, \kappa_i, \eta_i, \mathbf{C}_i\}$ are parameters [Gelman et al., 1995]. We will set θ_0 to define relatively vague prior distribution.

Spatio-temporal features. To understand the role of the spatio-temporal features d_k (time, location, etc.) consider a sequence $\{d_1^{(n)}, d_2^{(n)}, \ldots\}$ attributed to the nth person (denoted by the superscript). This sequence defines the path of the person in the global area of interest. Depending on the layout of camera locations, certain paths will be more likely than the other. We will define path probabilities using a simple, first-order Markov chain. This model assumes that a path starts by sampling $d_1^{(n)}$ from an initial distribution $p_{\delta_0}(d_1^{(n)})$, and is extended by sampling $d_{i+1}^{(n)}$ from a transition distribution $p_\delta(d_{i+1}^{(n)}|d_i^{(n)})$ that depends only on the last element in the path. Appropriately selected p_δ and p_{δ_0} will exclude physically impossible paths (e.g. with zero or negative travel times) and put high likelihood to the paths commonly followed in the considered area.

14.4.2   Graphical model

So far we have described our basic probabilistic assumptions about the process that yields observations of a single person. To make our model handle observations of multiple persons, we introduce additional hidden variables – *association* variables. Since the observations arrive from cameras one-by-one, our model will be organized into slices, each corresponding to a single y_k.

Association variables. For every y_k, the model maintains a discrete label ℓ_k that denotes the person represented by y_k. For convenience, at every slice k, the model maintains also a set of auxiliary variables: a counter c_k, and pointers $z_k^{(1)}, \ldots, z_k^{(k)}$. The counter, c indicates the number of objects present in the data $y_{1:k}$. The nth pointer, $z_k^{(n)}$, denotes the slice when the nth person was *last* observed *before* slice k. Value $z_k^{(n)} = 0$ indicates that person n has not yet been observed. At slice k there can be up to k persons, so we need $z_k^{(n)}$ for $n = 1, \ldots, k$. We will jointly denote association-related variables as $h_k \equiv \{\ell_k, c_k, z_k^{(1)}, \ldots, z_k^{(k)}\}$.

One-slice generation. Below we provide probabilistic dependencies that couple the hidden states, the hidden association variables and the observations. A simple (and almost mechanical) way to define these dependencies is by trying to reconstruct the process that generates observations.

We begin by initializing the counter, $c_0 = 0$, and generate the observations one-by-one. Generation of y_k starts by setting the label ℓ_k. We assume that people enter FOVs irregularly, so we choose uniformly between one of the known c_{k-1} persons or a new person who will receive the next available label (symbol "$\sim$" reads "distributed as");

$$\ell_k \sim \text{Uniform}(1, \ldots, c_{k-1}, c_{k-1} + 1). \tag{14.16}$$

Given the label we deterministically update the counter

$$c_k = c_{k-1} + [\ell_k > c_{k-1}] \tag{14.17}$$

where $[f]$ is an indicator function; $[f] \equiv 1(0)$ iff the binary proposition f is true (false). If the label indicates a new person, $\ell_k = c_{k-1} + 1$, then the counter increases, as in (14.17). A similar update rule can defined for the pointers which depend on past pointers values and the past label [Zajdel et al., 2004].Next, we generate the state x_k of the person indicated by ℓ_k. By our assumption the person's state does not change, so we set $x_k = x_j$, where $j = z_k^{(\ell_k)}$ points to the slice when the person was previously observed. If the person has not been yet observed, then we sample the state from prior;

$$x_k = x_j[j > 0] + x^{\text{new}}[j = 0] \qquad x^{\text{new}} \sim \pi(x), \tag{14.18}$$

Given the state $x_k = \{\mathbf{m}_k, \mathbf{V}_k\}$ (i.e. parameters of a Normal density) and the pointer to the last observation of the current person $j = z_k^{(\ell_k)}$; the model generates

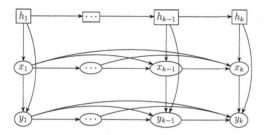

Figure 14.2. Graphical representation of the model for multi-camera multi-object tracking.

$$y_k = \{o_k, d_k\};$$

$$o_k \sim \mathcal{N}(\mathbf{m}_k, \mathbf{V}_k) \qquad\qquad d_k \sim p_\delta(d_k|d_j), \qquad (14.19)$$

Graphical representation. Figure 14.2 depicts the structure of proba-bilistic dependencies between the variables of our model. The state of our dynamic system $s_k = \{x_k, h_k\}$ includes the state of the current person and association variables (including the label of the current person). The sen-sor model $p(y_k|x_k, h_k, y_{1:k-1})$ is defined by (14.19). The transition model $p(h_k, x_k|x_{1:k-1}, h_{k-1})$ follows from (14.16)–(14.18). Depending on h_k, the state x_k may be a copy of any past state x_j, $j < k$, therefore the node x_k connects to all past nodes $x_{1:k-1}$. This shows that our model is a non-Markovian time-series model.

14.4.3 Assumed-density filtering

In the classical tracking scenario, one aims to associate an incoming ob-servation y_k on-line, that is, on the basis of the currently available data $y_{1:k}$. Under our model, this objective corresponds to inference on label ℓ_k from $y_{1:k}$. However, due to the coupling with the states and other hidden variables, the filtering distribution for our model is $p(x_{1:k}, h_k|y_{1:k})$. It is updated recursively:

$$p(x_{1:k}, h_k|y_{1:k}) = \frac{1}{L_k} p(y_k|x_k, h_k, y_{1:k-1}) p(x_{1:k}, h_k|y_{1:k-1}) \qquad (14.20)$$

$$p(x_{1:k}, h_k|y_{1:k-1}) = \sum_{h_{k-1}} p(h_k, x_k|x_{1:k-1}, h_{k-1}) p(x_{1:k-1}, h_{k-1}|y_{1:k-1}). \qquad (14.21)$$

Unfortunately, the continuous states depend on a joint label sequence $\ell_{1:k}$ that has $O(k!)$ possible instantiations. Thus, repeated summations over label ℓ_{k-1} in (14.21) result in a function that cannot be represented in a closed-form, ren-dering our model an intractable hybrid model [Lerner & Parr, 2001]. Note, that the intractability of multi-object tracking is a general problem. A popu-lar heuristic method to sidestep this problem is multiple hypothesis tracking, where one maintains only several hypothetical instantiations of the label sequence.

Bayesian framework allows to apply approximations with a stronger theoretical basis. Here, we follow the assumed-density filtering approach with the following approximating family

$$p(x_{1:k}, h_k|y_{1:k}) \approx q_k(\ell_k, c_k) \prod_{i=1}^{k} \phi(x_i|\theta_{i,k}) q_k(z_k^{(i)}), \qquad (14.22)$$

where q_k represents a probability table for appropriate discrete variable, and $\phi(x_i|\theta_{i,k})$ is a probability density function as defined in (14.15). At slice $k-1$ our algorithm maintains an approximation to the filtered distribution in the assumed factorial family (14.22). After one-slice update, the expression (14.20) will not admit the factorial representation. The nearest in the KL-sense factorial distribution is the product of marginals [Cover & Thomas, 1991], so ADF recovers the assumed family by computing the marginals of (14.20). The marginals on discrete variables immediately take the requested form. Each marginal on the continuous state evaluates to a mixture. This mixture has to be projected to the closest (in the KL-sense) assumed density function (14.15). For details and experiments see [Zajdel et al., 2004].

14.5   Conclusions and remaining issues

We have presented a probabilistic framework for solving complex problems that involve recovering meaningful low-dimensional state information from a series of noisy high-dimensional observations. The presented framework lays down a general pattern for the design of algorithms dealing with noisy data. It is based on establishing a generative – or – causal relation between the state of a latent, low-dimensional dynamic system and the observations. Given observations, this relation can be "inverted" using Bayesian inference techniques.

The discussion in this tutorial focused on on-line inference techniques (Bayesian filtering). However, the Bayesian framework provides also a class of methods for off-line improvement of state estimates using future observations. These techniques include Variational methods [Jordan, 1998], Expectation Propagation (EP) [Minka, 2001] and Markov Chain Monte Carlo (MCMC) sampling [Murphy, 2002].

Another aspect closely related to the presented methodology is learning of probabilistic models from data [Jordan, 1998]. In some problems, we have a set of supervised training examples, that includes observations and the underlying states. In some other, we want the learn the probabilistic model, despite the fact that the states are not available. This can be achieved with the powerful Expectation Maximization (EM) algorithm [Murphy, 2002].

References

Boyen, X., and D. Koller [1998]. Tractable inference for complex stochastic processes, *Proc. of Conf. on Uncertainty in Artificial Intelligence.*

Cover, T.M., and J.A. Thomas [1991]. *Elements of information theory*, John Wiley & Sons, New York.

Dellaert, F., D. Fox, W. Burgard, and S. Thrun [1999]. Monte carlo localization for mobile robots, *Proc. IEEE Int. Conf. on Robotics and Automation.*

Doucet, A., N. de Freitas, and N. Gordon (eds.) [2001]. *Sequential monte carlo methods in practice*, Springer-Verlag.

Gelman, A., J.B. Carlin, H.S. Stern, and D.D. Rubin [1995]. *Bayesian data analysis*, Chapman & Hall.

Gordon, N.J., D.J. Salmond, and A.F.M. Smith [1993]. Novel-approach to nonlinear non-Gaussian Bayesian state estimation. *IEE Proceedings-F Radar and signal processing*, 140 (2): 107–113.

Isard, M., and A. Blake [1998]. Condensation– conditional density propagation for visual tracking, *Int. Journal of Computer Vision*, 29(1): 5–28.

Jensen, F.V. [2001]. *Bayesian networks and decision graphs*, Springer-Verlag.

Jordan, M.I. (ed.) [1998]. *Learning in graphical models*, MIT Press.

Lerner, U., and R. Parr [2001]. Inference in hybrid networks: Theoretical limits and practical algorithms, *Proc. of Uncertainty in Artificial Intelligence.*

Minka, T. [2001]. Expectation propagation for approximate bayesian inference, *Proc. of Uncertainty in Artificial Intelligence.*

Murphy, K.P. [2002]. *Dynamic Bayesian networks: Representation, inference and learning*, Ph.D. thesis, University of California, Berkley.

Neal, R. [2000]. Markov chain sampling methods for Dirichlet process mixture models, *Journal of Computational and Graphical Statistics*, pages 249–265.

Pasula, H., S. Russell, M. Ostland, and Y. Ritov [1999]. Tracking many objects with many sensors, *Proc. of Int. Joint Conf. on Artificial Intelligence.*

Pitt, M.K., and N. Shephard [1999]. Filtering via simulation: Auxiliary particle filters, *J. Amer. Statist. Assoc.*, 94(446): 590–599.

Rabiner, L.R. [1990]. A tutorial on hidden Markov models and selected applications in speech recognition, in *Reading in Speech Recognition*, A. Weibel and K.-F. Lee (eds.), pages 267–296.

Rowies, S., and Z. Ghahramani [1999]. A unifying review of linear gaussian models, *Neural Computation*, 11: 305–345.

Saul, L.K., and M. I. Jordan [1999]. Mixed memory Markov models: Decomposing complex stochastic processes as mixtures of simpler ones, *Machine Learning*, 37(1): 75–87.

Thrun, S., D. Fox, W. Burgard, and F. Dellaert [2001]. Robust Monte Carlo localization for mobile robots, *Artificial Intelligence*, 128(1-2): 99–141.

Vlassis, N., B. Terwijn, and B. Kröse [2002]. Auxiliary particle filter robot localization from high-dimensional sensor observations, *Proc. IEEE Int. Conf. on Robotics and Automation.*

Zajdel, W., A.T. Cemgil, and B. Kröse [2004]. Online multicamera tracking with a switching state-space model, *Proc. Int. Conference on Pattern Recognition.*

Chapter 15

PRIVATE PROFILE MATCHING

Berry Schoenmakers and Pim Tuyls

Abstract   In this chapter we present protocols for privately and securely matching two
profiles (modelled as bit vectors), where a match means that the Hamming distance between the profiles is sufficiently small. The protocol does not reveal any
information other than the fact whether the two profiles match.

We use the framework of secure computation based on threshold homomorphic cryptosystems, as put forth by Cramer, Damgård, and Nielsen at Eurocrypt'01. More in particular, we use the extension introduced by Schoenmakers
and Tuyls at Asiacrypt'04, which replaces the general multiplication gate by a
restricted multiplication gate, called the conditional gate. The advantage of the
conditional gate is that threshold homomorphic ElGamal suffices as the underlying cryptosystem, which allows for efficient discrete log based implementations,
whereas a general multiplication gate requires the use of more involved RSA-
like cryptosystems, such as Paillier's cryptosystem.

The computational complexity of the protocol is dominated by the number
of conditional gates needed. For profiles of length m, our protocol requires only
$O(m)$ conditional gates, where the hidden constant is small.

Keywords   Profile matching, secure computation, homomorphic encryption, threshold decryption, ElGamal cryptosystem.

15.1   Introduction

An increasing amount of information (address information, medical data,
financial data, biometric identifiers, preference data, ...) about human beings is
stored in electronic form. Originally these data were mainly passively used as
reference information (e.g., to give people access to certain services or build-
ings, for looking up information about people, etc.). Nowadays these data are
more and more also actively used by companies, service providers, and persons
themselves. Recommender systems, for instance, send their clients targeted ad-
vertisements for items corresponding to their profiles. Dating services on the
internet use personal data to find a partner for their clients. Currently there is a
growing concern about the way personal data are handled by service providers

Wim F.J. Verhaegh et al. (Eds.), Intelligent Algorithms in Ambient and Biomedical Computing, 259-272.

and companies. Clearly, active use of personal data makes them more vulnerable to leakage of sensitive information. In order to guarantee privacy a well-defined security system using privacy preserving protocols is needed.

In this chapter, we consider a simple case of the above sketched privacy problem called private profile matching. A profile is a list describing certain properties of a human being. As an example one can think of a file containing the preference list of a user rating a list of songs, restaurants, etc. according to his taste. A second example of a profile is a reference biometric identifier (fingerprint, iris scan, ...) of a user that is used for biometric authentication. Profile matching is defined as the process where two entities want to find out whether their profiles match; i.e., whether their profiles are similar with respect to some predefined similarity measure. *Private profile matching* is the problem of matching two profiles without revealing any other information about the entities' individual profiles than whether they match or not. We present two motivating examples. The first one gives a privacy problem in a case where data are actively used, while the second example provides an example for the passive case (access control).

Our first example concerns *shared interests*, where two users want to find out whether they have some shared music interest, for instance. In this case the profiles consist of a list of songs together with a rating expressing how much they like the song. The problem addressed here is that the two people do not want to reveal any information about their preferences unless the other person has a similar music interest. Therefore, they need a protocol to find out (in a secure way) whether they have a common music interest or not. In case they find a large similarity between their profiles, they only know that their interests are close but do not obtain any more details about each others profiles. Similarly, a low similarity between their profiles reveals only that their tastes are different.

As a second example we consider the setting of *private biometrics*. Biometric authentication is a special case of private profile matching in the following sense. During biometric authentication the biometric identifier of a user measured during the authentication phase has to be matched with her reference biometric identifier that was stored during the enrollment phase in some reference database. The comparison of these two biometric identifiers with respect to some similarity measure can be seen as a profile matching protocol. Since biometric identifiers contain sensitive information about people and can be used to track their behavior, there is a privacy problem with biometric authentication [Tuyls & Goseling, 2004]. This problem was studied in several papers [Linnartz & Tuyls, 2003; Dodis, Reyzin & Smith, 2004; Tuyls & Goseling, 2004], where solutions to this problem are presented in an information theoretic setting. It was however shown [Linnartz & Tuyls, 2003] that from an information theoretic point of view *full* privacy cannot be achieved.

The private profile matching protocol developed in this chapter can be used to implement biometric authentication without leaking any information (under a computational assumption) on the biometric identifier of the person involved.

We investigate the private profile matching problem within the setting of secure multi-party computation. More in particular we use the framework of [Cramer, Damgård & Nielsen, 2001] and its extension presented in [Schoenmakers & Tuyls, 2004]. The protocols that we present guarantee two properties to the users:

- The protocol does not reveal any other information on the inputs of the users than what leaks through the output of the protocol.

- The protocol guarantees that the two users performed their computations correctly. If one of the users (deliberately or not) does not follow the protocol, this is detected and the protocol is aborted. Moreover no additional information can be obtained by an adversary using such a strategy.

In this chapter we restrict ourselves to the case where the entries of the profiles are binary values. In the case of a preference list they indicate whether a user likes a certain item or not. We refer to [Freedman, Nissim & Pinkas, 2004] for a recent, more extensive overview of secure matching problems.

This chapter is organized as follows. In Section 15.2 we present a brief introduction to threshold homomorphic cryptosystems and their application in secure multi-party computation. First, we explain the basics of threshold homomorphic ElGamal encryption. Secondly, we give an overview of basic primitives that we need such as threshold ElGamal decryption and Σ-protocols. Finally, we describe two secure multiplication protocols (the private multiplier gate and the conditional gate) and explain how these can be used to evaluate any function consisting of multiplication and addition gates only in a secure way. The secure approximate matching problem is treated in Section 15.3. This computation is best described in terms of a secure adder and a secure comparison circuit which are presented in detail in this section. Finally, we present some variants to the posed problem.

15.2 Preliminaries

In this section we summarize the techniques used in our protocol for private matching of Section 15.3. We mainly recapitulate the results of [Schoenmakers & Tuyls, 2004].

15.2.1 Secure computation based on threshold homomorphic cryptosystems

Threshold homomorphic cryptosystems are a special type of public key cryptosystems with the distinguishing features that (i) key generation is a protocol between several parties for jointly generating a public key such that each party gets a share of the corresponding private key (but the private key is never known to a single party), (ii) encryption is a (probabilistic) algorithm behaving, for a fixed public key, as a group homomorphism mapping plaintexts to cipher-texts, and (iii) decryption is a protocol between sufficiently many of the parties holding shares of the private key, such that they are able to jointly recover the plaintext for a given ciphertext (again, without ever recovering the private key itself. The threshold homomorphic ElGamal cryptosystem, described below, is a primary example of such a cryptosytem.

Threshold homomorphic cryptosystems provide a basis for secure multi-party computation in the cryptographic model [Franklin & Haber, 1996; Juels & Jakobsson, 2000; Cramer, Damgård & Nielsen, 2001; Damgård & Nielsen, 2003; Schoenmakers & Tuyls, 2004]. For a given n-ary function f, one com-poses a circuit C of elementary gates that given encryptions of $x_1, \ldots, x_n$ on its input wires, produces an encryption of $f(x_1, \ldots, x_n)$ on its output wire. The elementary gates operate in the same fashion. The wires of the entire circuit C are all encrypted under the same public key; the corresponding private key is shared among a group of parties. It is customary to distinguish addition gates and multiplication gates. Addition gates can be evaluated without having to decrypt any value, taking full advantage of the homomorphic property of the cryptosystem. Multiplication gates, however, require at least one thresh-old decryption to succeed even for an honest-but-curious (passive) adversary. To deal with a malicious (active) adversary, multiplication gates additionally require the use of zero-knowledge proofs.

A major advantage of secure computation based on threshold homomor-phic cryptosystems is the fact that it results in particularly efficient solu-tions, even for active adversaries. The communication complexity, which is the dominating complexity measure, is $O(nk|C|)$ bits for [Juels & Jakobsson, 2000; Cramer, Damgård & Nielsen, 2001; Damgård & Nielsen, 2003; Schoen-makers & Tuyls, 2004], where n is the number of parties, k is a security para-meter, and $|C|$ is the number of gates of circuit C.

We apply the approach of [Schoenmakers & Tuyls, 2004], which allows for threshold homomorphic ElGamal to be used as the underlying cryptosystem. See [Schoenmakers & Tuyls, 2004] for a comparison with the approaches of [Juels & Jakobsson, 2000] and [Cramer, Damgård & Nielsen, 2001; Damgård & Nielsen, 2003].

15.2.2   Threshold homomorphic ElGamal

Discrete log setting.   Let $G = \langle g \rangle$ be a finite cyclic (multiplicative) group of prime order q, which means that the elements of G can be obtained as pow-ers of the generator g: $1, g, g^2, \ldots, g^{q-1} \in G$ and $g^q = 1$. As usual, we assume that the Decision Diffie-Hellman (DDH) problem is infeasible for G, which is the problem of distinguishing between the distribution $\{(g^\alpha, g^\beta, g^\gamma) \mid \alpha, \beta, \gamma \in_R \mathbb{Z}_q\}$ consisting of uniformly distributed random triples and the distribution $\{(g^\alpha, g^\beta, g^{\alpha\beta}) \mid \alpha, \beta \in_R \mathbb{Z}_q\}$ consisting of uniformly distributed Diffie-Hellman triples. This assumption implies that the Diffie-Hellman (DH) problem, which is to compute $g^{\alpha\beta}$ given $g^\alpha, g^\beta \in_R G$, is infeasible as well. In turn, this implies that the Discrete Log (DL) problem, which is to compute $\alpha = \log_g h$ given $h = g^\alpha \in_R G$, is infeasible.

Needless to say, we need to assume that multiplication in G is easy, so that exponentiation in G is easy as well. Using repeated squaring, one computes g^α, for any given integer α, $0 \le \alpha < q$, using $O(\log q)$ multiplications in G.

Homomorphic ElGamal encryption. For public key $h \in G$, a message $m \in \mathbb{Z}_q$ is encrypted as a pair $(a, b) = (g^r, g^m h^r)$, with $r \in_R \mathbb{Z}_q$. Encryption is *additively* homomorphic: given encryptions $(a, b), (a', b')$ of messages m, m', respectively, an encryption of $m + m'$ is obtained as $(a, b) \star (a', b') = (aa', bb') = (g^{r+r'}, g^{m+m'} h^{r+r'})$.

Given the private key $\alpha = \log_g h$, decryption of $(a, b) = (g^r, g^m h^r)$ is performed by first calculating $b/a^\alpha = g^m$, and then solving for $m \in \mathbb{Z}_q$. In general, this is exactly the DL problem, which we assume to be infeasible. The way out is to require that message m is constrained to a sufficiently small set $M \subseteq \mathbb{Z}_q$.[1] In our case, the cardinality of M will be very small, often $|M| = 2$.

Homomorphic ElGamal encryption is semantically secure assuming the infeasibility of the DDH problem. Throughout this chapter we use $[\![m]\!]$ to denote the set of all q possible ElGamal encryptions of m under some understood public key h, and, frequently, we also use $[\![m]\!]$ to denote one of its elements. More formally, using that $[\![0]\!]$ is a subgroup of $G \times G$, $[\![m]\!]$ is the coset of $[\![0]\!]$ in $G \times G$ containing encryption $(1, g^m)$. Hence, encryptions (a, b) and (a', b') belong to the same coset if and only if $\log_g(a/a') = \log_h(b/b')$. Lifting the operations on the direct product group $G \times G$ to the cosets, we thus have, for $x, y \in \mathbb{Z}_q$, that $[\![x]\!] \star [\![y]\!] = [\![x+y]\!]$, and $[\![x]\!]^y = [\![xy]\!]$, where $(a, b)^c = (a^c, b^c)$ for $c \in \mathbb{Z}_q$. Hence, $[\![x]\!] \star [\![y]\!]^{-1} = [\![x-y]\!]$. Addition and subtraction over $\mathbb{Z}_q$ and multiplication by a publicly known value in $\mathbb{Z}_q$ can thus be performed easily on encrypted values. These operations are deterministic. Another useful consequence is that any encryption in $[\![x]\!]$ can be transformed into a statistically independent encryption in $[\![x]\!]$ by multiplying it with a uniformly selected encryption in $[\![0]\!]$; this is often referred to as "random re-encryption."

---

[1] For intervals M, the Pollard-λ ("kangaroo") method runs in $O(\sqrt{|M|})$ time using $O(1)$ storage.

Σ-protocols. We need some well-known instances of Σ-protocols, which each take only a small, constant number of exponentiations in G for the prover and the verifier. The simplest case is Schnorr's protocol for proving knowledge of a discrete log α, on common input $a = g^\alpha$ [Schnorr, 1991], and Okamoto's variant for proving knowledge of α, β, on common input $a = g^\alpha h^\beta$ [Okamoto, 1993] . Another basic case is Chaum-Pedersen's protocol for proving knowledge of α, on common input $(a, b) = (g^\alpha, h^\alpha)$, which is a way to prove that $(a, b) \in [\![0]\!]$ without revealing any information on α [Chaum & Pedersen, 1993]. Applying OR-composition [Cramer, Damgård & Schoenmakers, 1994], these basic protocols can be combined into, for instance, a Σ-protocol for proving that $(a, b) \in [\![0]\!] \cup [\![1]\!]$, where the common input is an ElGamal encryption (a, b). The latter protocol thus proves that the message encrypted (which is an element of $\mathbb{Z}_q$) actually is a "bit", without divulging any further information on the message.

In general, a Σ-protocol for a relation $R = \{(v, w)\}$ is a three-move protocol between a prover and a verifier, where the prover does the first move. Both parties get a value v as common input, and the prover gets a "witness" w as private input, $(v, w) \in R$. A Σ-protocol is required to be a proof of knowledge for relation R satisfying special soundness and special honest-verifier zero-knowledge. See [Cramer, Damgård & Schoenmakers, 1994] for details.

For simplicity, we will only use the non-interactive versions of these Σ-protocols, which are obtained via the Fiat-Shamir heuristic, that is, by computing the challenge as a hash of the first message (and possibly other inputs).

Threshold ElGamal decryption. We use a $(t + 1, n)$-threshold ElGamal cryptosystem, $0 \le t < n$, in which encryptions are computed using a common public key h (as above) while decryptions are done using a joint protocol between n parties $P_1, \ldots, P_n$. Each party P_i holds a share $\alpha_i \in \mathbb{Z}_q$ of the private key $\alpha = \log_g h$, where the corresponding value $h_i = g^{\alpha_i}$ is public. As long as more than t parties take part, decryption will succeed, whereas t or fewer parties are not able to decrypt successfully.

The parties initially obtain their shares α_i by running a secure distributed key generation protocol; see [Pedersen, 1991; Gennaro et al., 1999] for details. We note that these protocols are very practical (the communication complexity is $O(n^2 k)$ bits for security parameter k, where the hidden constant is small).

For decryption of (a, b), party P_i, $i = 1, \ldots, n$, produces a decryption share $d_i = a^{\alpha_i}$ along with a proof that $\log_a d_i = \log_g h_i$. Assuming w.l.o.g. that parties $P_1, \ldots, P_{t+1}$ produce correct decryption shares, the message can be recovered from $g^m = b/a^\alpha$, where a^α is obtained from $d_1, \ldots, d_{t+1}$ by Lagrange interpolation. Assuming homomorphic ElGamal, $m \in M$ will hold for some small set M; if such m cannot be found decryption fails. Also, if fewer than $t + 1$ parties provide a correct decryption share, decryption fails.

15.2.3 Special multiplication protocols

The results of the previous section imply that a function f can be evaluated securely in a multiparty setting if f can be represented as a circuit over $\mathbb{Z}_q$ consisting only of addition gates and simple multiplication gates. Here, an addition gate takes encryptions $[\![x]\!]$ and $[\![y]\!]$ as input and produces $[\![x]\!] \star [\![y]\!] = [\![x+y]\!]$ as output, and a simple multiplication gate takes $[\![x]\!]$ as input and produces $[\![x]\!]^c = [\![cx]\!]$ as output, for a publicly known value $c \in \mathbb{Z}_q$. To be able to handle any function f, however, we need more general multiplication gates for which both inputs are encrypted.

In this section, we consider two special multiplication gates. If no restrictions are put on x or y, a multiplication gate, taking $[\![x]\!]$ and $[\![y]\!]$ as input and producing $[\![xy]\!]$ as output efficiently, cannot exist assuming that the DH problem is infeasible.[2] Therefore, two special multiplication gates were introduced in [Schoenmakers & Tuyls, 2004], putting some restrictions on the multiplier x. The first gate, referred to as the *private-multiplier gate*, requires that the multiplier x is *private*, which means that it is known by a single party. The second gate, referred to as the *conditional gate*, requires that the multiplier x is from a two-valued domain.

Private-multiplier gate. We describe a multiplication protocol where the multiplier x is a *private* input rather than a shared input. That is, the value of x is known by a single party P. No restriction is put on the multiplicand y. Multiplication with a private multiplier occurs as a sub-protocol in the protocol for the conditional gate and is of use in other protocols as well.

Given encryptions $[\![x]\!] = (a,b) = (g^r, g^x h^r)$ and $[\![y]\!] = (c,d)$, where party P knows r and x, party P computes on its own a randomized encryption $[\![xy]\!] = (e,f) = (g^s, h^s) \star [\![y]\!]^x$, with $s \in_R \mathbb{Z}_q$, using the homomorphic properties. Party P then broadcasts $[\![xy]\!]$ along with a proof showing that this is the correct output, which means that it proves knowledge of witnesses $r,s,x \in \mathbb{Z}_q$ satisfying $a = g^r$, $b = g^x h^r$, $e = g^s c^x$, $f = h^s d^x$.

Below, we will also use a variation of the above protocol, where the private multiplier x is multiplied with several multiplicands y_i at the same time.

Conditional gate. Next, we describe a multiplication gate for which the multiplier x is from a two-valued domain, whereas the multiplicand y is unrestricted, called the *conditional gate*, and show how to implement it by an

---

[2]Given g^x, g^y we form encryptions $[\![x]\!]$, $[\![y]\!]$ and feed these into the multiplication gate. The gate would return an encryption $[\![xy]\!]$, which would give g^{xy} upon decryption.

efficient protocol, using just homomorphic threshold ElGamal. We will formulate the conditional gate for the domain $\{-1,1\}$.[3]

Let $[\![x]\!], [\![y]\!]$ denote encryptions, with $x \in \{-1,1\} \subseteq \mathbb{Z}_q$ and $y \in \mathbb{Z}_q$. The following protocol enables parties $P_1, \ldots, P_n$, $n \geq 2$, to compute an encryption $[\![xy]\!]$ securely. For simplicity, we assume that these parties also share the private key of the $(t+1,n)$-threshold scheme $[\![\cdot]\!]$, where $t < n$. The protocol consists of two phases.

1. Let $x_0 = x$ and $y_0 = y$. For $i = 1, \ldots, n$, party P_i in turn takes $[\![x_{i-1}]\!]$ and $[\![y_{i-1}]\!]$ as input, and broadcasts an encryption $[\![s_i]\!]$, with $s_i \in_R \{-1,1\}$. Then P_i applies the private-multiplier multiplication protocol to multiplier $[\![s_i]\!]$ and multiplicands $[\![x_{i-1}]\!]$ and $[\![y_{i-1}]\!]$, yielding encryptions $[\![x_i]\!]$ and $[\![y_i]\!]$, where $x_i = s_i x_{i-1}$ and $y_i = s_i y_{i-1}$.

2. The parties jointly decrypt $[\![x_n]\!]$ to obtain x_n, and check that $x_n \in \{-1,1\}$. Given x_n and $[\![y_n]\!]$, an encryption $[\![x_n y_n]\!]$ is computed publicly.

The output of the protocol is $[\![x_n y_n]\!]$. Clearly, if all parties are honest, $x_n y_n = (\prod_{i=1}^{n} s_i)^2 xy = xy$.

The protocol requires a single threshold decryption only. Since $x_n \in \{-1,1\}$ is required to hold, decryption is feasible for the homomorphic ElGamal encryption scheme.

Note that we do not need to require that each s_i is in $\{-1,1\}$ in phase 1.. For instance, parties P_1 and P_2 may cheat by setting $s_1 = 2$ and $s_2 = 1/2$. Since $s_1 s_2 = 1$, this type of "cheating" will go unnoticed in phase 2. if all other parties are honest. However, the security of the protocol is not affected by such "cheating."

The fact that the conditional gate leaks no information on the inputs x and y follows from the fact that the encryption scheme is semantically secure and that the decrypted value x_n is a random element in $\{-1,1\}$ which is statistically independent from x. For a more formal proof of this statement, we refer the reader to [Schoenmakers & Tuyls, 2004].

15.2.4   Circuit evaluation

In this section, we briefly describe a protocol for evaluating a given circuit composed of elementary gates, following [Cramer, Damgård & Nielsen, 2001]. Recall that our elementary gates operate over $\mathbb{Z}_q$, except that the first input of a conditional gate is required to belong to a two-valued domain. It is clear that

---

[3]Domain $\{0,1\}$ or any other domain $\{a,b\}$, $a \neq b$, can be used instead, as these domains can be transformed into each other by linear transformations: $x \mapsto a' + (b'-a')(x-a)/(b-a)$ maps $\{a,b\}$ to $\{a',b'\}$. These transformations can be applied directly to homomorphic encryptions, transforming $[\![x]\!]$ with $x \in \{a,b\}$ into $[\![x']\!]$ with $x' \in \{a',b'\}$.

these elementary gates suffice to emulate any Boolean circuit. Specifically, any operator on two bits $x, y \in \{0,1\} \subseteq \mathbb{Z}_q$ can be expressed uniquely as a polynomial of the form $a_0 + a_1 x + a_2 y + a_3 xy$ with coefficients in $\mathbb{Z}_q$. Hence, any binary operator can be expressed using at most one conditional gate.

For convenience, we assume that parties $P_1, \ldots, P_n$ evaluating the circuit are exactly the same as the parties for which the $(t+1, n)$-threshold homomorphic cryptosystem has been set-up, where $t < n$. A circuit is evaluated in three phases:

1. The parties encrypt their inputs using the threshold homomorphic cryptosystem $[\![\cdot]\!]$, and the parties are required to provide a proof of knowledge for their inputs, and possibly that the inputs belong to a two-valued domain.

2. The parties then jointly evaluate the circuit gate-by-gate. Conditional gates at the same depth of the circuit are evaluated in parallel.

3. Finally, the parties jointly decrypt the outputs of the circuit.

Intuitively, this protocol is secure because the only values that are ever revealed by it are (i) the values decrypted during the evaluation of the conditional gates in phase 2 (but these values are statistically independent of the input values, assuming that at least one of the parties is honest) and (ii) the values decrypted in phase 3 (but these values are supposed to be revealed anyway). This intuition has been confirmed by a rigorous security proof given in [Cramer, Damgård & Nielsen, 2001], and in [Schoenmakers & Tuyls, 2004] we have shown that security is maintained if the general multiplication gate of [Cramer, Damgård & Nielsen, 2001] is replaced by our conditional gate.

When dealing with large circuits, it's common to divide up phase 2 in the evaluation of several *sub-circuits*. Each sub-circuit has a set of input wires and a set of output wires. At the start of the evaluation of a sub-circuit, an encrypted value is given for each input wire. Then the gates of the sub-circuit are evaluated until each of the output wires is assigned an encrypted value.

Thus, it makes sense to distinguish two basic cases of secure computations. In the first case, the inputs are *private*, i.e., each of the inputs is known by one of the parties. In the second case, the inputs are *shared*, meaning that the inputs are given as encrypted values, and no single party knows these values. It turns out that in the first case one may take advantage of the fact that (some of) the inputs are private.

15.3 Secure approximate matching w.r.t. Hamming distance

Using the tools presented in the previous section, we will now construct a protocol for privately matching two bit vectors. Two bit vectors (of equal length) are said to match approximately if the number of positions in which they differ does not exceed a certain threshold value.

15.3.1 Introduction

Let $X = (X_1, \ldots, X_m)$ and $Y = (Y_1, \ldots, Y_m)$ be bit vectors, $m \geq 1$, and let $d_H(X,Y)$ denote the Hamming distance between X and Y, defined by $d_H(X,Y) = \sum_{i=1}^{m} X_i \oplus Y_i$. Let T be a threshold value, $0 \leq T \leq m$. The problem of securely matching bit vectors X and Y is to compute the value $[d_H(X,Y) \leq T]$ securely, where $[B]$ denotes the characteristic function for a condition B (if B holds, then $[B] = 1$, and otherwise $[B] = 0$). We will at first assume that the value of the threshold T must remain hidden too, and later show how to optimize the protocol if T is a public value.

Thus, the problem is to compute the encrypted value $[\![[d_H(X,Y) \leq T]]\!]$, given encrypted bit vectors $([\![X_1]\!], \ldots, [\![X_m]\!])$ and $([\![Y_1]\!], \ldots, [\![Y_m]\!])$, and given encrypted bits $([\![T_{s-1}]\!], \ldots, [\![T_0]\!])$ representing threshold value $T = \sum_{j=0}^{s-1} T_j 2^j$, where $s = \lceil \log_2(m+2) \rceil$. We show how to solve this problem using a circuit with $4m$ conditional gates only, following the approach presented in [Schoenmakers & Tuyls, 2004].

As a first step, we note that $d_H(X,Y) = \sum_{i=1}^{m}(X_i - Y_i)^2 = \sum_{i=1}^{m} X_i + Y_i - 2X_iY_i$, which shows that $[\![d_H(X,Y)]\!]$ can easily be computed using m conditional gates and $O(m)$ addition/subtraction gates. However, in order to compare the value of $d_H(X,Y)$ securely with the value of T, we need to compute the binary representation of $d_H(X,Y)$ rather than as an integer between 0 and m.

The outline of our secure computation of $[d_H(X,Y) \leq T]$ is now as follows, see also Figure 15.1.

1. On input of $X = (X_1, \ldots, X_m)$ and $Y = (Y_1, \ldots, Y_m)$, compute the vector $(H_1, \ldots, H_m)$, where $H_i = X_i + Y_i - 2X_iY_i$.

2. Using an s-layer binary tree of adders, compute the sum $S = \sum_{i=1}^{m} H_i$, representing the values at the successive layers of the tree as bit vectors of increasing length. The value S is represented as a bit vector $(S_{s-1}, \ldots, S_0)$, $S = \sum_{j=0}^{s-1} S_j 2^j$.

3. Given bit vectors $(S_{s-1}, \ldots, S_0)$ and $(T_{s-1}, \ldots, T_0)$ representing S and T respectively, compute $[S \leq T]$ using a comparator circuit.

15.3.2 Secure adder

Given bit vectors $(x_{r-1}, \ldots, x_0)$ and $(y_{r-1}, \ldots, y_0)$ representing nonnegative integers x and y, respectively, we need to compute the bit vector $(z_r, \ldots, z_0)$ representing the integer $z = x + y$. This is a well-known problem, e.g. from electrical engineering. The common approach is based on the use of half-adders and full-adders, resulting in a circuit of about $5r$ non-trivial Boolean

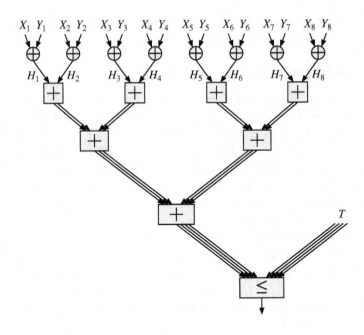

Figure 15.1. Secure matcher circuit, case $m = 8$.

gates (such as AND, OR, NAND, NOR, and XOR gates), leading to a secure computation using about $5r$ conditional gates.

The advantage of the conditional gate, however, is that one of its inputs is not restricted to be binary. Therefore, we get a much better solution by considering circuits defined over the integers, with the restriction that for each multiplication at least one of the inputs is binary. This way we are able to limit the number of conditional gates to about $2r$.

We propose the following circuit for the computation of $x + y$. The output of the circuit consists of the bits $(c_{r-1}, z_{r-1}, \dots, z_0)$, where

$$z_i = x_i + y_i + c_{i-1} - 2c_i,$$
$$c_{-1} = 0, \quad c_i = x_i y_i + x_i c_{i-1} + y_i c_{i-1} - 2x_i y_i c_{i-1}.$$

By writing $c_i = x_i(y_i + c_{i-1} - 2y_i c_{i-1}) + y_i c_{i-1}$ it follows that each c_i, $1 \leq i < r$, can be computed securely using two conditional gates only: first compute $w_i = y_i c_{i-1}$ and then compute $c_i = x_i(y_i + c_{i-1} - 2w_i) + w_i$. As $c_0 = x_0 y_0$ can be evaluated using just one conditional gate, a total of $2r - 1$ conditional gates suffices to compute $x + y$ securely. We conjecture that $2r - 1$ is actually the minimal number of conditional gates required, as one needs to compute in any case the terms $x_i y_i c_{i-1}$, each of which requiring two multiplications.

15.3.3 Secure comparator

Given bit vectors $(x_{r-1},\ldots,x_0)$ and $(y_{r-1},\ldots,y_0)$ representing nonnegative integers x and y, respectively, we need to compute $[x \leq y]$. For this we apply the circuit for computing $[x > y]$ of [Schoenmakers & Tuyls, 2004, Section 5.2], and we simply 'invert' the output of this circuit.

So, the output is given by $1 - t_{r-1}$, where

$$t_0 = x_0(1 - y_0), \qquad t_i = (1 - (x_i - y_i)^2)t_{i-1} + x_i(1 - y_i).$$

Note that this circuit inspects the bits of x and y, starting with the *least* significant bits—intuitively, one would start to compare the most significant bits first, but for an efficient circuit it actually seems better to start with the *least* significant ones (see [Schoenmakers & Tuyls, 2004, Section 5.1], where a circuit traversing the bits starting at the least significant one is given using about $3r$ conditional gates).

A secure computation based on this circuit requiring $2r - 1$ conditional gates is obtained by computing $v_i = y_i t_{i-1}$ and $t_i = t_{i-1} - v_i - x_i(t_{i-1} + y_i - 2v_i - 1)$, for $i = 1,\ldots,r-1$. Again, we conjecture $2r - 1$ to be the minimum number of conditional gates required to compute $x > y$.

15.3.4 Secure approximate matcher

As outlined above, we are now ready to present a secure computation for matching two bit vectors with respect to Hamming distance. We simply put together the circuits for the adders and the comparator, and we count the total number of conditional gates needed. Let m denote the length of the inputs X and Y and let s denote the bit length of $m + 1$, $s = \lceil \log_2(m + 2) \rceil$.

1. We need m conditional gates to compute the terms $H_i = X_i + Y_i - 2X_i Y_i$, $1 \leq i \leq m$. All of these conditional gates can be evaluated in parallel.

2. Next, we need a total of $m - 1$ adders, each adder taking inputs of length at most s, to compute the bit vector $(S_{s-1},\ldots,S_0)$. Assuming that $m = 2^k$, we note that the total number of conditional gates is bounded above by

$$\frac{m}{2}\cdot 1 + \frac{m}{4}\cdot 3 + \cdots + \frac{m}{2^k}\cdot(2k - 1) = 3m - 2k - 3 = 3m - 2s - 1.$$

3. The secure computation of $[S \leq T]$ requires $2s - 1$ conditional gates only.

Thus, the total number of conditional gates is bounded above by $4m - 2$, if m is a power of two.

The round complexity is as follows, assuming that each conditional gate takes n rounds. Our circuits for secure addition and secure comparison take $O(nr)$ rounds each for inputs of length r. The total number of rounds is therefore $O(n \log^2 m)$. We note that the round complexity can be improved at the expense of a larger number of conditional gates.

15.3.5 Variants

We describe briefly several variants of the above protocol for secure matching.

In the first variant, we assume that at least one of the input bit vectors is private. Let's say input X is private, whereas input Y is either private or shared. We may exploit this by replacing the conditional gates used for the evaluation of $X_i \oplus Y_i$ by private-multiplier gates. This reduces the total number of conditional gates by m. The total number of conditional gates is then bounded by $3m - 2$.

For a second variant, we assume that the value of the threshold T is not required to be hidden. As a consequence, the input $y = (y_0, \ldots, y_{r-1})$ to the secure comparator becomes public, and the computation of the intermediary values $w_i = y_i t_i$ can be done publicly. The total number of conditional gates for $[S \leq T]$ then drops from $2s - 1$ to $s - 1$, hence it is reduced by about $\log_2 m$.

Of course, both of these variants can be combined into a third variant that requires about $3m - \log_2 m$ conditional gates.

Finally, as a variant of a different nature, we mention that one may of course replace the ElGamal cryptosystem by any suitable threshold homomorphic cryptosystem to obtain solutions relying on different computational assumptions. For instance, Paillier's cryptosystem may be used instead, which allows one to replace the conditional gates in our circuits by the general multiplication gates of [Cramer, Damgård & Nielsen, 2001]. This will reduce the round complexity, since a conditional gate requires $O(n)$ rounds, whereas multiplication is done in $O(1)$ rounds in [Cramer, Damgård & Nielsen, 2001]. However, the computational cost of the multiplication gates will increase substantially, as computing with Paillier encryptions is much more expensive than computing with ElGamal encryptions, especially when elliptic curves are used to implement the latter.

15.4 Conclusion

In this chapter, we have shown how to securely match two profiles, given as bit vectors. A similar approach can be applied to related problems. For example, instead of bit vectors, one may consider vectors over $\{0, \ldots, 2^\ell - 1\}$ for some given ℓ, in which case the Hamming distance may be replaced by the Euclidean distance defined as $d_E(X, Y) = \sum_{i=1}^{m} (X_i - Y_i)^2$, for vectors X, Y of length m. Assuming that the elements of vectors X and Y are given by their encrypted bit representations, one may compute encryptions of the terms $(X_i - Y_i)^2$, also bitwise, and then add these values bitwise, in order to compare the sum with a given threshold.

References

Cramer, R., I. Damgård, and J.B. Nielsen [2001]. Multiparty computation from threshold homomorphic encryption. *Advances in Cryptology—EUROCRYPT'01, Lecture Notes in Computer Science*, 2045:280–300.

Cramer, R., I. Damgård, and B. Schoenmakers [1994]. Proofs of partial knowledge and simplified design of witness hiding protocols. *Advances in Cryptology—CRYPTO'94, Lecture Notes in Computer Science*, 839:174–187.

Chaum, D., and T.P. Pedersen [1993]. Wallet databases with observers. *Advances in Cryptology—CRYPTO'92, Lecture Notes in Computer Science*, 740:89–105.

Damgård, I., and J.B. Nielsen [2003]. Universally composable efficient multiparty computation from threshold homomorphic encryption. *Advances in Cryptology—CRYPTO'03, Lecture Notes in Computer Science*, 2729:247–264.

Dodis, Y., L. Reyzin, and A. Smith [2004]. Fuzzy extractors: How to generate strong secret keys from biometrics and other noisy data. *Advances in Cryptology–Eurocrypt'04, Lecture Notes in Computer Science*, 3027:523–540.

Franklin, M., and S. Haber [1996]. Joint encryption and message-efficient secure computation. *Journal of Cryptology*, 9(4):217–232.

Freedman, M.J., K. Nissim, and B. Pinkas [2004]. Efficient private matching and set intersection. *Advances in Cryptology—EUROCRYPT'04, Lecture Notes in Computer Science*, 3027:1–19.

Gennaro, R., S. Jarecki, H. Krawczyk, and T. Rabin [1999]. Secure distributed key generation for discrete-log based cryptosystems. *Advances in Cryptology—EUROCRYPT'99, Lecture Notes in Computer Science*, 1592: 295–310.

Juels, A., and M. Jakobsson [2000]. Mix and match: Secure function evaluation via ciphertexts. *Advances in Cryptology—ASIACRYPT'00, Lecture Notes in Computer Science*, 1976:162–177.

Okamoto, T. [1993]. Provably secure and practical identification schemes and corresponding signature schemes. *Advances in Cryptology—CRYPTO'92, Lecture Notes in Computer Science*, 740:31–53.

Linnartz, J.-P., and P. Tuyls [2003]. New shielding functions to enhance privacy and prevent misuse of biometric templates, *4th International Conference on Audio- and Video-Based Biometric Person Authentication*.

Pedersen, T. [1991]. A threshold cryptosystem without a trusted party. *Advances in Cryptology—EUROCRYPT'91, Lecture Notes in Computer Science*, 547:522–526.

Schnorr, C.P. [1991]. Efficient signature generation by smart cards. *Journal of Cryptology*, 4(3):161–174.

Schoenmakers, B., and P. Tuyls [2004]. Practical two-party computation based on the conditional gate. *Advances in Cryptology—ASIACRYPT'04, Lecture Notes in Computer Science*, 3329:119–136.

Tuyls, P., and J. Goseling [2004]. Capacity and examples of template protecting biometric authentication systems. *Biometric Authentication Workshop (BioAW 2004), Lecture Notes in Computer Science*, 3087:158–170.

Chapter 16

AIR FAIR SCHEDULING
FOR MULTIMEDIA TRANSMISSION
OVER MULTI-RATE WIRELESS LANS

Sai Shankar N., Richard Y. Chen, Ruediger Schmitt, Chun-Ting Chou, and Kang G. Shin

Abstract   As wireless local area networks (WLANs) are becoming ubiquitous, there is an increasing demand for their application in multimedia transmission for both professional and personal uses. Multimedia applications are delay-sensitive and require specific Quality of Service (QoS) support. Scheduling is very crucial to support multimedia transmissions with performance guarantees in the IEEE 802.11e WLAN. Most of the work on scheduling are based on bandwidth sharing techniques which are not suitable for wireless systems where the physical transmission rate varies as a function of distance or signal to noise ratio as in IEEE 802.11/GPRS/3G networks. In such a network, we show that the widely-adopted concept of throughput fairness forces low-transmission rate stations to overuse air time, thus causing significant unfairness in air time usage among the different stations. In this chapter, we first present a scheme called *Air Fair Scheduling* (AFS), which (1) always guarantees fairness in terms of received air time, regardless of the underlying stations' transmission rates; (2) keeps service for individual stations unaffected by other stations' transmission rate fluctuations or transmission errors; and (3) also achieves a higher overall system throughput than generalized processor sharing (GPS). We provide an algorithm for AFS implementation in IEEE 802.11a/e system and show how the performance is enhanced through analysis and experiments.

Keywords   IEEE 802.11 wireless LANs, air fair scheduling, time fairness, HCCA, per flow QoS.

16.1   Introduction

Wireless local area networks (WLANs) have gained a prevailing position, as they provide simple and low-cost wireless connectivity and data delivery. Together with the wide deployment of WLANs both in professional and

273

Wim F.J. Verhaegh et al. (Eds.), Intelligent Algorithms in Ambient and Biomedical Computing, 273-297.
© *2006 Springer. Printed in the Netherlands.*

residential markets, there is increasing demand for QoS support for multimedia applications such as voice-telephony, video-on-demand, and wireless audio streaming. WLANs are also being considered as low-cost replacement for 3G broadband services in public hot-spots. However, the lack of QoS support in legacy WLAN has prevented multimedia applications from taking off rapidly. The emerging 802.11e standard provides mechanisms for QoS to support a variety of multimedia applications and to provide Diffserv/Intserv-like QoS at the data link or MAC layer. However, the challenge remains in designing an efficient algorithm for scheduling that is simple to implement and more efficient in achieving the goal of satisfying the multimedia QoS. Fair scheduling of data transmissions in wired networks has long been studied as a means of resource allocation to contending traffic. In general, the objective of fair scheduling is to allocate contending traffic flows the system resource proportional to their weights, which are translated functions of the applications QoS requirements. Fair scheduling prevents greedy or misbehaving users from starving others, thus achieving fairness among the users. It can also be used to allocate more resources to preferred users for achieving service differentiation. Fair scheduling is primarily based on Generalized Processor Sharing (GPS) [Parekh & Gallager, 1992], and numerous scheduling algorithms have been proposed to approximate its performance [Parekh & Gallager, 1992; Golestani, 1994; Bennett & Cheng, 1996]. These scheduling algorithms, originally designed for wired networks, are adapted to deal with the rapid growth of wireless communication services and applications. However, there are several challenges unseen in wired networks that need to be resolved in order to achieve fairness in wireless/mobile networks.

16.1.1 Distributed environment

As an example, let us consider the IEEE 802.11e wireless LAN shown in Figure 16.1. A station may have several concurrent traffic flows to schedule for transmission, and may also contend with other stations for access to the shared radio/channel. Even though a wireless station can schedule its local traffic flows as a wired node does to achieve the fairness among them, the absence of coordination between stations makes system-wide fairness very difficult, if not impossible. To solve this problem, the QoS Access Point[1] (QAP) usually takes charge of scheduling the frame transmissions from/to all stations. Since the QAP does not have fine information about traffic flows of individual stations (when a frame exactly arrived in the station's queue), these stations need to provide their queue/traffic status information to the QAP such that the scheduler at the QAP can function properly. For example, the scheduler at

---

[1]The term QAP is used in IEEE 802.11e instead of just AP

least needs to know whether or not a station has packets[2] to send; most fair scheduling algorithms require more detailed information than that, such as the arrival time of individual packets, for correct computation of transmission order [Golestani, 1994; Bennett & Cheng, 1996]. The self-clocked fair queueing (SCFQ) [Golestani, 1994] originally designed for wired networks is modified for use in wireless networks such that it can work in a distributed environment [Kautz & Garcia, 1997]. However, the information required for SCFQ, such as the service tag (or the arrival time) of each packet and a 'system-wide' virtual clock, is still necessary and could incur extra scheduling overhead. Moreover, the delay in relaying this status information could make the information obsolete when it is to be used by the QAP scheduler. More frequent transmission of this information could alleviate the problem but it will incur substantial control overhead. Thus, a better and more robust way to achieve system-wide fairness is to use a distributed or a centralized scheduling algorithm that does not rely on fine information about traffic or channel.

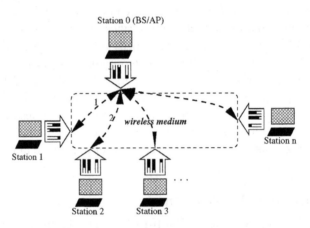

Figure 16.1. A generic wireless/mobile network.

16.1.2 Location-dependent errors

Another well-known property in wireless networks is the high probability of transmission errors and its intrinsic location-dependency. Each wireless station, due to multi-path fading or electromagnetic interference, may experience different probabilities of transmission errors. For example, flow 1 of station 0 in Figure 16.1 may suffer from high transmission errors while flow 2 is error-free. Thus, a higher error-rate flow may end up with receiving a

---

[2]Although the terms 'packet' and 'frame' are used interchangeably here, one should note that a packet is usually referred to as the transmission unit at the IP layer.

smaller share of effective system throughput than an error-free flow, simply because many of its transmissions get corrupted. If strict throughput fairness is mandatory, the flows with higher error rates will use more airtime to compensate for their losses of throughput caused by transmission errors. This overuse of radio resource by the error-prone flows, in turn, reduces the total system throughput. The most common solution to these problems is to defer the transmission of error-prone flows and compensate them after the channel condition gets better. Wireless packet scheduling (WPS) [Lu, Bharghavan & Srikant; Lu, Nandagopal & Bharghavan, 1998], channel condition independent packet fair queueing (CIF-Q) [Ng, Stoica & Zhang, 1998] and channel state dependent packet scheduling (CSDPS) [Fragouli, Sivaraman & Srivastava, 1998] are examples of this approach. A long-term fairness server is proposed such that the impact of a compensation mechanism on error-free flows can be reduced [Ramanathan & Agarwal, 1998]. The concept of adaptive weights is also used such that the scheduler can dynamically adjust the weights of error-prone flows to compensate for their throughput loss. The *power factor* of Eckhardt & Steenkiste [2000] and the *compensation index* of Jeong, Morikawa & Aoyama [2001] are the main control parameters to adjust the weights such that compensation can be made without degrading error-free flows too much.

16.1.3   Location-dependent transmission rate

Many existing/emerging wireless networks, such as the general packet radio service (GPRS), third generation (3G) mobile communication, wideband code division multiple access (W-CDMA) and IEEE 802.11 WLAN standard, support more than one physical transmission rate. Consider the IEEE 802.11a standard as an example. It can support 8 different rates, namely, 54, 48, 36, 24, 18, 12, 9, and 6 Mbps and IEEE 802.11b can support 4 different rates, namely, 11, 5.5, 2 and 1 Mbps. Depending on the channel conditions, especially their distance from the QAP, wireless stations may choose different transmission rates (i.e., so-called "link adaptation" [Qiao & Sunghyun, 2001]) in order to increase the probability of successful transmission. As also shown in Figure 16.1, for example, station n may choose 2 Mbps to transmit/receive data frames to/from the QAP while station 1 chooses 11 Mbps. It is very tricky to define a 'fair share of resource' among the stations in such a network, because serving an equal amount of traffic from individual stations with different transmission rates requires allocation of different amounts of airtime. That is, a fair share of system throughput is no longer synonymous with a fair share of airtime in a system that supports multi-transmission rates. Since no such location-dependent transmission rates exist in wired networks, applying the existing scheduling algorithms without considering this property may result in a station's misuse of radio resource.

In this chapter, we first investigate some potential disadvantages of using throughput fairness in a multi-rate wireless LAN. We then give a formal definition of airtime fairness, and discuss its advantage over throughput fairness. Sadeghi, Kanodia, Sabharwal & Knightly [2002] propose an Opportunistic Auto Rate (OAR) protocol to increase the system throughput in the presence of multi-transmission rates while still maintaining the same temporal fairness as in an unirate IEEE 802.11 wireless LAN. Our work is different from OAR [Sadeghi, Kanodia, Sabharwal & Knightly, 2002] in that (1) we want to achieve the weighted airtime fairness in order to support service differentiation/prioritization, but OAR manages to improve the system throughput while still maintaining the same (egalitarian) temporal fairness achievable by the unirate wireless LAN. Some weighted temporal fairness schemes were proposed for cellular wireless networks [Liu, Chong & Shroff, 2001]. Moreover, their scheme requires estimation of stations' channel conditions in order to choose the 'best' station to transmit. In our scheme, the only information needed — the physical transmission rates of individual stations, can be found in the preamble of the IEEE 802.11a/e frame header. It will not incur any control overhead or introduce errors. It should be noted that the unfairness caused by location-dependent errors (LDEs) can be solved by the above-cited proposals and thus is omitted. In fact, these schemes can be combined with our scheme to achieve the system-wide airtime fairness in the presence of LDEs.

16.2 Fairness in wireless/mobile networks

Fairness is defined as meeting all of the agreed quality of service (QoS) requirements by controlling the order of service for all active connections. This could be allocation of bandwidth, buffer space, network time, etc., proportional to a connection's requirements. The objective of fair scheduling is to provide different flows[3] with different amounts of 'work' proportional to their assigned weights. Usually, 'work' is measured by the amount of data transmitted (either in number of bytes or packets/frames) during a certain period of time. Let $S_i(t_1, t_2)$ be the amount of flow i's traffic served in a time interval (t_1, t_2), and ϕ_i be the weight assigned to flow i. ϕ_i represents the weight proportionality factor of the network resources that a particular flow i would require in order to satisfy its QoS requirements. Then, an ideal fair scheduler (i.e., the GPS server in [Parekh & Gallager, 1992]) with N flows must satisfy

$$\frac{S_i(t_1, t_2)}{S_j(t_1, t_2)} \geq \frac{\phi_i}{\phi_j}, j = 1, 2 \cdots, N, \qquad (16.1)$$

---

[3] In this chapter, a flow can be a traffic flow or an aggregation of traffic flows from a station.

for any flow i that is continuously backlogged during (t_1, t_2). A flow i is said to be continuously backlogged if it has packets to transmit in the time period (t_1, t_2) of observation. If the flow j is also continuously backlogged, then the above equation would be equal. On the other hand if the flow j was not continuously backlogged, the resources used by flow j is given to flow i. Whenever the flow j has packets to transmit, the fair scheduler would assign its rightful proportion, ϕ_j, of the network resource. If all flows' data are transmitted at a fixed rate, we can obtain from Eq. (16.1)

$$\frac{S_i(t_1, t_2)}{t_2 - t_1} \geq \frac{\phi_i}{\sum_j \phi_j} C, \qquad (16.2)$$

where C is the physical transmission rate or the channel capacity. Thus, flow i is guaranteed to have the throughput given by Eq. (16.2) regardless of the states of the queues and frame arrivals of the other flows. However, the advantages of using GPS, such as the guaranteed throughput and independent service, cannot be preserved if the flows are given different processing/transmission rates, which is the case in an IEEE 802.11b wireless LAN. Let us consider two stations, WSTA1 and WSTA2, in such a wireless LAN. Each station may transmit/receive frames to/from the QAP at any of the 4 rates mentioned above, and may dynamically adapt the transmission rate to its location. For simplicity, we assume both stations have an equal weight (i.e., $\phi_1 = \phi_2$). We further assume that both WSTA1 and WSTA2 transmit at 11 Mbps before $t = 4$. After $t = 4$, WSTA2 moves away from the QAP and adapts its transmission rate to 1 Mbps at the new location. This is shown in Figure 16.2(a)-(1). From Figure 16.2(a)-(1), the amount of traffic served in $(0, t)$ for WSTA1 and WSTA2 are always the same as required by the GPS, and so are their throughput as shown in Figure 16.2(a)-(3). However, the QAP has to allocate more transmission time to WSTA2 after $t \geq 4$ as shown in Figure 16.2(a)-(2) in order to compensate WSTA2 for its change of transmission rate. That is, Eq. (16.1) can only be achieved at the expense of degrading the throughput of both the system and the high-transmission rate flow WSTA1. Multimedia applications suffer most from airtime unfairness due to the fact that most multimedia content is encoded at a certain quality, which rules out congestion control that will result in dramatic quality reduction. For instance, assume that both WSTAs are running broadcast-quality standard-definition video streaming sessions. WSTA2 has dropped its PHY rate to 1 Mbps due to moving away from the QAP. As a result WSTA2 cannot sustain its video streaming application anymore. However, the bandwidth fairness causes the throughput of WSTA1 to drop to the same rate as WSTA2 even though nothing has changed for WSTA1 including location and link quality. It is absolutely unfair to WSTA1 in terms of resource usage and it destroys both videos instead of one video. This observation has led us to the use of AFS in which each flow may use a different transmission rate.

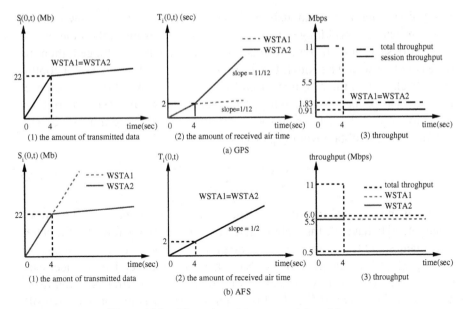

Figure 16.2. Throughput fairness vs. airtime fairness.

16.2.1 Air fair scheduling

As shown in Figures 16.2(a)-(1) and 16.2(a)-(2), there exists some inconsistency between fair sharing of system throughput and that of transmission time if stations have varying transmission rates (for $t \geq 4$). In a conventional wired network, such an inconsistency does not exist, but this inconsistency in wireless networks implies the misallocation (in the sense of fairness) of scarce radio resource. We argue that fairness in a wireless network supporting multiple transmission rates, should be measured by the transmission times used by individual stations, not their resultant throughput. Specifically, we propose *Air Fair Scheduling* (AFS) for such networks. Let $T_i(t_1, t_2)$ be the amount of time flow i receives to transmit its traffic during (t_1, t_2) and ϕ_i be its assigned weight. Then, a AFS server is defined as

$$\frac{T_i(t_1, t_2)}{T_j(t_1, t_2)} \geq \frac{\phi_i}{\phi_j}, \quad j = 1, 2 \cdots, N, \quad (16.3)$$

for flow i that is continuously backlogged in the interval (t_1, t_2). We consider the transmission time allocated to each flow, not the amount of traffic served during that interval, as we want to eliminate the effect of different transmission rates (i.e., the potential unfairness it may cause). From Eq. (16.3), we can obtain

$$T_i(t_1, t_2) \geq \frac{\phi_i}{\sum_j \phi_j}(t_2 - t_1). \quad (16.4)$$

That is, a continuously-backlogged flow is guaranteed to receive a certain portion of transmission time within any arbitrary interval of time, instead of a guaranteed throughput given by Eq. (16.2). Now, let us revisit the previous example but use the proposed AFS instead. The results are shown in Figure 16.2(b). For $t < 4$, using Eq. (16.1) or Eq. (16.3) does not make any difference with respect to all performance metrics. That is, AFS is the same as GPS if all of the flows use the same transmission rate; otherwise, using AFS is more advantageous as follows.

1. Under GPS, the amount of traffic transmitted by WSTA1 varies with the transmission rate of WSTA2, and so does WSTA1's throughput. However, under AFS, as shown in Figure 16.2(b)-(1), the amount of WSTA2's transmitted data will remain unaffected. That is, the service for individual flows can still be kept independent of each other under AFS.

2. As shown in Figure 16.2(b)-(2), the fair usage of radio resource by stations can always be guaranteed under AFS regardless of their transmission rates. In stark contrast, low-transmission-rate stations may abuse the radio resource under GPS.

3. Because of (1) and (2) above, the overall system throughput will be much higher under AFS (6 Mbps) than under GPS (1.83 Mbps). (See Figure 16.2.)

One can interpret the conventional GPS-based fairness as 'strict fairness' because each station is guaranteed with an share of system throughput without considering the corresponding transmission time required, and hence may result in degradation of the overall throughput in certain conditions. However, our fairness tries to maintain a higher system throughput while still providing fairness in terms of the transmission time received by the stations.

It should be noted that under AFS, it is the low-rate station, WSTA2, instead of both WSTA1 and WSTA2, that absorbs the throughput reduction. At a first glance, AFS may appear unsuitable for those applications (either originated from, or destined for, the low transmission-rate station) with strict bandwidth requirements. However, this problem can be solved in two ways: (1) the scheduler of that WSTA2 may adjust the weights of its applications such that the reduction of its throughput will not affect the applications with strict bandwidth requirements but affect only best-effort traffic [Eckhardt & Steenkiste, 2000]; (2) graceful degradation can be applied to these applications such that the resultant quality is still within their acceptable range [Chou & Shin, 2002].

16.2.2 Packetized AFS

As PGPS is used in packet networks to approximate GPS, we also need a packetized version of AFS to implement AFS. Time-Division-Multiplexing (TDM) may appear to be a good realization of AFS. Even though TDM does approximate Eq.(16.3), the 'time slots' allocated to the non-backlogged flows in TDM are simply wasted (i.e., non-work-conserving). In fact, we can generate a work-conserving, *packetized AFS* (P-AFS) just like PGPS for GPS [Parekh & Gallager, 1992]. The P-AFS server will pick the packet that would complete service first in the AFS simulation if no additional packets were to arrive thereafter. The difference between the service times of a flow received under AFS and P-AFS is bounded by

$$T_i(0,t) - T_i^p(0,t) \le T_{\max}, \qquad (16.5)$$

where $T_i(0,t)$ and $T_i^p(0,t)$ are the service times that flow i receives during $(0,t)$ under AFS and P-AFS, respectively, and $T_{\max}$ is the maximum amount of time a server may need to transmit a single packet. The proof of this bound follows the derivation in [Parekh & Gallager, 1992], and thus, we only present the modification needed for P-AFS.

Lemma 1: Let p and p' be packets in a AFS system at time t. Suppose packet p completes its service before p' if there were no packet arrivals after time t. Then, p will also complete its service before p' even if there are packet arrivals after time t.

Proof: The flows to which packets p and p' belong are both backlogged from t until packet p or p' completes its transmission. By Eq. (16.3), the ratio between the service times received by these backlogged flows is independent of future arrivals. Thus, the lemma follows[4]. □

Next, we need to prove the difference between packet finish times under AFS and P-AFS is bounded. Let $L_{\max}^i$ be the maximum packet length of flow i, r_i the transmission rate of flow i, and $T_p(T_p')$ the packet p's finish time under AFS (P-AFS). Then, for any packet p

$$T_p' - T_p \le T_{\max}, \qquad (16.6)$$

where $T_{\max} = \max_i(L_{\max}^i/r_i)$. Eq. (16.6) can be proven in the same way as in [Parekh & Gallager, 1992] by using Lemma 1 and replacing the fixed processing rate r with the flow-dependent processing rate r_i. So, the details of the proof are omitted here. Now, we can prove Eq. (16.5) as follows.

Proof: The slope of $T_i^p(0,t)$ alternates between 1 if the packet of flow i is being transmitted, and 0 otherwise. Since the slope of $T_i(0,t)$ is between

---

[4]Here we assume a flow can not change its transmission rate during the transmission of a frame.

$\phi_i / \sum_j \phi_j$ and 1, the difference, $T_i(0,t) - T_i^P(0,t)$, attains its maximum when flow i is transmitted under P-AFS. Let τ_p be the time when this difference attains its maximum, and T_t be the packet-transmission time. Then, the packet completes its transmission at $\tau_p + T_t$ under P-AFS. Let τ be the time at which the given packet completes its transmission under AFS. Then,

$$T_i(0,\tau) = T_i^P(0,\tau_p + T_t) = T_i^P(0,\tau_p) + T_t. \tag{16.7}$$

The first equality holds because the packets of flow i are transmitted in the same order under AFS and P-AFS, and the second equality holds because the slope of T_i^P in $(\tau_p, \tau_p + T_t)$ is 1. From Eq. (16.6), we have $\tau_p + T_t - T_{max} \leq \tau$. So,

$$T_i(0, \tau_p + T_t - T_{max}) \leq T_i(0,\tau). \tag{16.8}$$

Since $T_t - T_{max} \leq 0$ and the slope of T_i is always less than, or equal to 1, we get

$$T_i(0,\tau_p) + T_t - T_{max} \leq T_i(0, \tau_p + T_t - T_{max}). \tag{16.9}$$

By Eqs. (16.7)–(16.9), we get

$$T_i(0,\tau_p) \leq T_i^P(0,\tau_p) + T_{max}, \tag{16.10}$$

and thus, Eq. (16.5) follows.   Various algorithms exist to implement PGPS [Golestani, 1994; Bennett & Cheng, 1996] and their adaptation to P-AFS requires little modification. As we discussed in the Introduction, the only problem is that these algorithms require a scheduler to compute the transmission order based on all flows's traffic and queue status (e.g., packet length, packet arrival time). This information is very difficult for a scheduler to acquire in a distributed environment (e.g., a wireless network), or will at least incur a considerable amount of overhead to collect it. Thus, we will only try to achieve the proposed PAFS based on a coarser time granularity (i.e., $t_2 - t_1 \gg T_{max}$).

16.3   AFS in an IEEE 802.11e wireless LAN

The centrally controlled access mechanism of the IEEE 802.11e medium access control (MAC) [IEEE 802.11e, 2002], called the HCF[5] controlled channel access (HCCA), adopts a poll and response protocol to control the access to the wireless medium and eliminate contention among wireless STAs. It makes use of the PCF[6] Inter frame Space (PIFS), which is the shortest arbitration inter frame space (AIFS) value, to seize and maintain control of the medium. Once the hybrid coordinator (HC) has control of the medium, it starts to deliver parameterized downlink traffic to stations (STAs) or issue QoS

---

[5]HCF = hybrid coordination function.
[6]Point coordination function.

contention-free polls (QoS CF-Polls) frames to those STAs that have requested uplink or sidelink parameterized services. The QoS CF-Poll frames include the TXOP duration granted to the STA. If the STA being polled has traffic to send, it may transmit several frames for each QoS CF-Poll received, respecting the TXOP limit specified in the poll frame. Besides, in order to utilize the medium more efficiently, the STAs/APs are allowed to piggyback both the contention free acknowledgment (CF-ACK) and the CF-Poll onto data frames. Differently from the PCF of the IEEE 802.11-99 Std. [IEEE 802.11b, 1999], HCCA operates during both the contention free period (CFP) and contention period (CP). During the CFP, the STAs cannot contend for the medium since their Network Allocation Vector (NAV), also known as virtual carrier sensing, is set, and therefore the HC enjoys free access to the medium. During the CP, the HC can also use free access to the medium once it becomes idle, in order to deliver downlink parameterized traffic or issue QoS CF-Polls. This is achieved by using the highest EDCA priority, i.e., AIFS = PIFS and $CW_{min} = CW_{max} = 0$. Note that the minimum time for any access category (AC) to access the medium is DIFS, which is longer than PIFS. For details the reader is referred to [IEEE 802.11e, 2002].

16.3.1   How does the HC allocate TXOP?

An STA can request parameterized services using the Traffic Specification (TSPEC) element [IEEE 802.11e, 2002]. The TSPEC element contains the set of parameters that characterize the traffic stream that the STA wishes to establish with the HC. Once the TSPEC request is received by the HC, it analyzes the TSPEC parameters and decides whether to admit the stream into the network. This is also known as the admission control process. If the stream is admitted, the HC schedules the delivery of downlink traffic and/or QoS CF-Polls (for uplink/sidelink traffic) in order to satisfy the QoS requirements of the stream as specified in the TSPEC. Several scheduling disciplines can be used in the HC. The performance of the HCCA is dependent on the choice of the admission control and the scheduling algorithm. Admission control algorithm is out of scope in this chapter. We assume that there exists an admission control algorithm and the bandwidth allocated by the admission control algorithm determines the weight each flow gets the overall system. The fields of a TSPEC element are shown in the Figure 16.3. These fields are used by the admission control algorithm to admit the flows and the scheduler translates it into TXOP per flow as well as per station. We need to note that all the parameters in the TSPEC element are negotiable between the QAP and wireless station (WSTA). Of the many parameters in TSPEC we consider the minimum physical transmission rate that is relevant for further understanding of this chapter. The minimum PHY rate indicates the minimum physical transmission rate that

Octets 1	1	2	2	4	4	4	4
Element ID	Length	TS Info	Nominal MSDU Size	Max. MSDU Size	Min. Service Interval	Max. Service Interval	Inactivity Interval

4	4	4	4	4	4	4	4	4
Service Start Time	Min. Data Rate	Mean Data Rate	Peak Data Rate	Max. Burst Size	Delay bound	Min. PHY Rate	Surplus Bandwidth Allowance	Medium Time

Figure 16.3. TSPEC element as defined in IEEE 802.11e standard.

is required for providing QoS guarantees. Once the *TXOP* and the service interval (*SI*) are calculated, the schedule element is transmitted by the HC to a non-AP QSTA[7] to announce the schedule that the HC/QAP follows for admitted streams originating from or destined to that non-AP QSTA in future. The information in this element may be used by the non-AP QSTA for power management, internal scheduling or for any other purpose. The structure of the schedule element is shown in Figure 16.4. We consider two fields, namely,

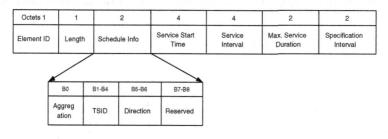

Figure 16.4. Structure of schedule element.

service interval and maximum service duration. The Service Interval field represents the measured time from the start of one service period to the start of the next service period. The Maximum Service Duration specifies the maximum transmission opportunities (TXOP) allocated to this flow. These fields along with the minimum PHY rate will be used by the air fair scheduling (AFS) scheduler.

16.3.2 Operation of the air fair scheduler

One of the most important problems in any wireless LAN systems is distributed nature of the location of non-AP QSTAs and absence of information at the QAP of the status of different queues at the non-AP QSTAs. Additionally each non-AP QSTA may have more than one streams originating from it. This

---

[7]QoS Station.

forces for a design of a two-stage scheduler. They are *station scheduler* (implemented at the QAP only) and the *flow scheduler* (implemented at the non-AP QSTA's and QAP). The scheduler architecture is shown in Figure 16.5. In this approach, the station scheduler at the QAP serves each non-AP QSTA (having upstreams and/or sidestreams) and the QAP itself (having downstreams) based on the relative weights. For this scheduler, residing at the QAP, the QAP also appears as a non-AP QSTA requiring channel time that is proportional to weight. In the second stage, called flow scheduler or *local scheduler*, the scheduler decides which packet among competing parameterized streams need to be selected at a single non-AP QSTA/QAP.

16.4 Station scheduler

This station scheduler is co-located with the HC. The purpose of this scheduler is to select a particular QSTA including the QAP. For this scheduler the QAP is also a QSTA requesting TXOP. The proposed scheduling scheme in Section 16.3 and the local scheduler (LS) form a two-stage fair scheduling system as shown in Figure 16.5. If there are several flows destined to the same

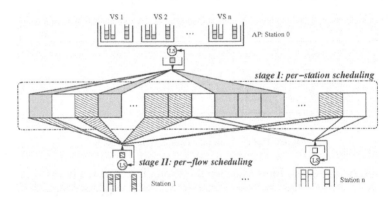

Figure 16.5. Two-stage scheduling.

station from QAP, the weights of those flows are aggregated and treated as one virtual flow to that particular station by the station scheduler. Since the flows within the QAP, which are destined for the same wireless station, will use the same transmission rate and experience the same transmission errors, they are aggregated as if they are a "virtual" stream (VS) for the purpose of scheduling. Suppose, for example, if flows 1 to 3 in the QAP are destined for wireless station 1, these flows are said to belong to virtual stream 1. Similarly if there are several flows originating from one QSTA to other QSTAs (side-link) or to QAP (uplink), they can be aggregated to form one virtual stream as the station scheduler can allocate $TXOP$ to this QSTA as if there is only one flow. In general, let VS_i be the virtual stream associated with the flows destined for the same wireless station i. At Stage I, the per-station fairness is achieved by Eq. (16.11) if the wireless LAN operates in the HCCA mode. The weight of each station will be the sum of all its flows' weights for that particular station:

$$W_i = \sum_{j=VS_1}^{VS_n} \phi_{ij} \qquad (16.11)$$

where ϕ_{ij} is the weight of station i's j-th flow. Since the virtual streams are aggregates of several streams, the station scheduler in the QAP will schedule transmissions for these virtual streams by treating them as a single stream. Indirectly this will select the station and use a QoS CF Poll to provide TXOP or if it is the opportunity of the QAP itself, it will use PIFS access to send downstreams. Thus W_i will determine the fraction of time a station or QAP will get.

By doing so, we can only guarantee fairness among stations in a wireless LAN, not among the individual flows in a station. Therefore, at Stage II, station i's LS allocates and schedules the transmission time acquired at Stage I to all its local traffic as shown in Figure 16.5 such that per-flow fairness can also be achieved.

As explained in section 16.3.1, each stream is quantified by two variables, $(TXOP_{i,j}, D_{i,j})$, where $TXOP_{i,j}$ and $D_{i,j}$ are the requested $TXOP$ and the delay bound of the flow j belonging to QSTA i. i takes the values from 1 to total number of QSTAs and j takes the values of total number of streams admitted. Now the following equation holds:

$$\sum_{i=1}^{n} \sum_{j=1}^{m_i} \frac{TXOP_{i,j}}{D_{i,j}} \le 1 \qquad (16.12)$$

Now the scheduler residing at the HC should decide the value of SI. Since the tuple $(TXOP_{i,j}, D_{i,j})$ represent the air time requirement and delay requirement (or distance constraint of the scheduler), we can infer that SI is given by

$$SI = \frac{1}{2}\min\{D_{i,j} \mid i = 0,\ldots,n \wedge j = 0,\ldots,m_i\} \qquad (16.13)$$

This value of SI is necessary to satisfy the QoS requirements on delay for each stream. This SI also satisfies any order of schedule for all streams within SI and is optimal. Once the SI is computed, we need to compute the new $TXOP'_{i,j}$. This is given by

$$TXOP'_{i,j} = \frac{TXOP_{i,j}}{D_{i,j}} SI. \qquad (16.14)$$

Then the $TXOP_i'$ for a QSTA is given by

$$TXOP_i = \sum_{j=1}^{m_i} TXOP_{i,j}' \qquad (16.15)$$

Based on the above equation the station scheduler will poll any of the stations in any order since the distance constraint is satisfied such that the delay requirements of all the flows are satisfied by properly choosing *SI*.

16.5 Local scheduler (LS)

Once the $TXOP$ is granted to a QSTA or to the QAP itself by the station scheduler, it is objective of the local scheduler (LS) to schedule a particular frame among competing flows originating from this QSTA or QAP. Any fair scheduling algorithm can be implemented so that the QoS constraints are satisfied for the particular flow. However we have two problems for the local scheduler. They are, (1) Location dependent errors and (2) Location dependent transmission rates. We will look into each of them.

16.5.1 Location dependent errors

So far, we have not considered any transmission error on the radio resource. However, this 'over-the-air' transmission error may cause unfairness as we mentioned in section 16.1. Blindly scheduling the transmissions without considering the location-dependent errors may not only reduce the overall system throughput, but also compromise fairness in terms of the radio resource usage. Various strategies have been proposed to alleviate these problems. The underlying principle is simple — The transmission opportunities scheduled to the flows that experience temporarily high errors should be reclaimed by the error-free flows, in order to increase the system throughput. After the channel condition is restored, the transmission opportunities relinquished by the high-error flows should be reimbursed to them, such that the fairness can be redeemed. Depending on their implementation, some use swapping of transmission opportunities between leading/laging flows [Lu, Bharghavan & Srikant; Ng, Stoica & Zhang, 1998], while in [Ramanathan & Agarwal, 1998], a certain portion of bandwidth is reserved such that the error-free flows will not be affected by the compensation mechanism.

An effort-limited fair scheduling is also proposed to limit the transmission attempts of flows that experiences high error rate (by a so-called 'power factor') such that the scheduler will not waste too much bandwidth on those invalid attempts [Eckhardt & Steenkiste, 2000]. In an IEEE 802.11 wireless LAN, the aforementioned problems may occur in a different way, due mainly to its MAC-layer retransmission mechanism. The purpose of retransmission at the MAC layer is to hide transmission errors from higher layers by means of controlling the value of *retry limit*. However, a scheduler without knowing

Table 16.1. Guarantee of soft fairness.

				scheduling order						
time	1	3			7	8	9		12	13
S1	v	v	–1	–2	v(–1)	v(0)	v	–1	v(0)	v
time	2	4	5	6				10	11	14
S2	v	x	x	v				x	v	v

v: transmission succeeds, x: transmission fails

the underlying retransmission process may over-schedule the transmission time for error-prone traffic flows.

For example, consider two traffic flows, S1 and S2, in the QAP with different destinations. Assume that both flows have the same weight and transmission rate, but the probability of transmission errors of S1 is 0 and that of S2 is 0.5. If S2 is scheduled to transmit, statistically it takes two MAC-layer transmission attempts to successfully transmit a frame while it takes only one for S1. Therefore, if these two flows are scheduled without taking into account the MAC-layer retransmissions, the AFS-based fairness cannot be achieved since S2 will acquire twice the transmission time of S1. The extra transmission time S2 acquires via the MAC-layer retransmission should be regarded as an overuse of resource as far as the air time is concerned. This could not only happen to the downlink flows (i.e., from the QAP to different wireless stations) but also to the uplink and side-link flows.

In order to solve this unfairness problem the LS considers the retransmission attempts of all its traffic. Since the LS only schedules the local traffic, any fair scheduling algorithm in the wired network can be applied directly. However, depending on the outcome of each frame transmission attempt, the LS should apply the algorithm shown in Figure 16.6, where the *fair_schedule*() can be any weighted fair scheduling algorithm. Consider round-robin scheduling as an example. As shown in Table 16.1, the first attempt of transmitting flow S2's second frame fails at $t = 4$. It is retransmitted by the MAC-layer retransmission mechanism, (not the LS) and succeeds by the second retransmission ($t = 6$). During this period, S1 should receive the same amount of transmission time as S2. Therefore, the LS will "reimburse" S1 at $t = 7$ and $t = 8$ for S2's overuse. This process repeats from $t = 9$ to $t = 12$ for another S2's unsuccessful transmission attempt, and then the LS returns to the normal scheduling at $t = 13$. This way the scheduler can offer two traffic flows with an equal share of transmission time, which is the desired soft fairness.

Note that the purpose of LS is to provide the AFS-based fairness at the MAC-layer. It dynamically adjusts the transmission order of traffic flows

```
/*  Sbacklog_in: Set of backlogged flows
/*  soveri: service overuse of flow i
/*  Reset to −TXOPi at the beginning of each SI
/*  TXOPremain: Remaining TXOP in SI after all flows have been allocated
/*  retry limit: retry limit in an IEEE 802.11 wireless station

/* local_ scheduler: schedule all backlogged flows*/
local_ scheduler(Sbacklog_in, retry limit)
{
    while (|Sbacklog| > 0) {
        /* flow i is scheduled based on any fair scheduling algorithm*/
        i=fair_ schedule(Sbacklog);
        Sbacklog = Sbacklog_in;
        /* if flow i overuses the radio, */
        /* scheduler should skip flow i */
        if (soveri > 0  && TXOPremain > 0) {
            if (i is only flow in overusing resource) {
                soveri = soveri − TXOPremain;
                TXOPremain == 0; }
            else {
                soveri = soveri − (Φi/Σj Φj) TXOPremain;
                TXOPremain− = (Φi/Σj Φj) TXOPremain } }
        if (soveri > 0  & TXOPremain = 0) {
            Sbacklog = Sbacklog − {i}; }
        /* else, transmit the frame of flow i, and continue */
        /* the scheduling for all backlogged flows */
        else {
            soveri=soveri+transmit(i, retry limit); }
    }
}

/* transmit: transmit a frame with flow ID=SID */
/* Return the time used for transmitting one frame. */
/* get_time() returns current real time */
int transmit(SID, retry limit)
{
    retry_count=0;
    time=get_time();
    /* send(): MAC-layer system call for sending a frame
    and will return -1 if transmission fails */
    status=send(SID);
    while (retry_count≤ retry limit & status< 0) {
        status=send(SID) ;
        retry_count=retry_count+1;
    }
    return(get_time()-time);
}
```

Figure 16.6. Pseudo code describing the implementation of the local scheduler (LS).

according to the retry count from the MAC-layer retransmission mechanism. Unlike many proactive schemes which require either estimation of error probability or prediction of an error burst, we provide fairness reactively. Its advantages are: (1) estimation of error probability or the start of an error burst can be eliminated because any inaccurate estimation itself may introduce unfairness (e.g., over-compensation or over-penalty), (2) probing the channel to search for the end of an error burst is unnecessary as we do not defer any flow's transmission for the error burst, (3) because of (2) and the fact that readjusting the transmission order is done at the end of every transmission,[8] the service lag of error-free flows is bounded since the number of MAC-layer transmission retries is limited. However, it should be noted that because of (2), the total system throughput under LS may be lower than that under the proactive schemes. But if we consider the potential inaccuracy of channel esimation/prediction and the resultant extra control overhead, this loss of throughput should be an acceptable compromise.

16.5.2 Location dependent transmission rates

As already mentioned, WSTA can operate in any one of the eight available rates in IEEE 802.11a/g WLAN. When the WSTA starts moving away from the QAP (in case of uplink) or moves away from another WSTA (in case of sidelink) it might have to do link adaptation by dropping its physical transmission rate or increasing its physical transmission rate. Increasing poses no problems as all frames will be transmitted in short time. On the contrary if the physical transmission rate was lowered because the WSTA moved away from its destination, then we might have overuse of airtime. The $TXOP'_{i,j}$ of a flow is based on a negotiated minimum physical transmission rate, $R_i^{\min}$. The LS first looks at flows whose current physical transmission rates $R_i < R_i^{\min}$. These flows are pooled together as lagging flows. The other flows are pooled together as normal flows. We implement a simple scheduling philosophy wherein the lagging flows are allocated their $TXOP$ and the MAC fragments frames or sends fewer frames than it would normally send if they had their $R_i \geq R_i^{\min}$. This provides fairness in terms of airtime usage as the lagging flows get their share of airtime proportional to their weights.

We can get a simple upper bound on delay for the normal flows whose $R_i \geq R_i^{\min}$. For the flow j belonging to station i, having a weight of ϕ_{ij}, we have the maximum delay between two $TXOP$ allocations as:

$$d_{\max} \leq 2SI - \frac{L_j}{\phi_{ij}R_{ij}^{\min}} \tag{16.16}$$

Here L_j is the frame size of the flow j.

---

[8]Here, it refers to the end of a successful transmission, or the end of *retry limit* consecutive unsuccessful transmission attempts for a frame.

16.6 Numerical and simulation results

16.6.1 Per-flow AFS-based fairness by LS

We consider the per-flow fairness provided by the proposed LS. In order to focus on the operation of LS, we assume that the QAP is the only station having MAC-frames to transmit. The destinations of its traffic flows are different, so different traffic flows have different transmission-error probabilities (i.e., the location-dependent error (LDE)). We compare the performance of the QAP's traffic flows with and without LS. Two performance metrics, the received transmission time (i.e., the air time) and scheduling delay, are considered for each traffic flow. The scheduling delay is defined as the difference between the time a given frame is transmitted in an error-free channel condition and that in the presence of location-dependent errors.

We consider 4 concurrent flows in the QAP with weights 4, 2, 1 and 1, respectively. All flows use IEEE 802.11b PHY of 11 Mbps transmission rate such that the computation of T_i in Figure 16.6 can be simplified. We also assume that the length of each frame is 1500 bytes, and thus, the transmission time of a single frame will be about 1.4 ms. Two different types of error patterns, sporadic and bursty, are considered. If it is sporadic, each transmission gets corrupted with a fixed probability, p. If it is bursty, transmission always succeeds during an ON-period (i.e, error-free period) but gets corrupted with probability 1 during an OFF-period (i.e., an error burst). The distribution of each ON/OFF duration is assumed to be *i.i.d.* Pareto distribution with different means. The parameters used for both error patterns are listed in Table 16.2. Note that we choose the parameters such that the average error probabilities under these two scenarios are equal.

Table 16.2. Parameters used for simulation in case of LDE.

		flow 1	flow 2	flow 3	flow 4
sporadic: p		0.25	0	0.2	0.1
bursty	off-period	$b = 800^\dagger$	NA	$b = 300$	$b = 400$
ab^a/x^{a+1}	on-period	$b = 2400$	NA	$b = 1200$	$b = 3600$

$\dagger\, a = 1.5$, mean $= 3b$ msecs, NA: not applicable

Since we use a Pareto distribution in the bursty-error case, we run that simulation for 1000 secs, instead of 100 secs in the case of sporadic error for the simulation accuracy. Table 16.3 shows the number of successful/unsuccessful transmissions for each flow. Under both error patterns, the error-free flow, flow 2, suffers a loss of its share of system transmission time due to the other stations' transmission errors. Even though the average error probability is small

Table 16.3. Successful/unsuccessful transmissions for each flow.

error pattern		*flow 1*	*flow 2*	*flow 3*	*flow 4*
sporadic	without LS	41244/13700	20622/0	10311/2655	10311/1157
	with LS	37567/12426	25667/0	9550/2388	11144/1168
actual ratio with LS		50.08%	25.66%	11.94%	12.32%
bursty	without LS	313649/225530	170898/0	73649/94730	80307/41237
	with LS	365918/134080	249997/0	96381/28622	112466/12525
actual ratio with LS		49.99%	24.99%	12.50%	12.49%
assigned ratio		50%	25%	12.5%	12.5%

(0.16), flow 2's loss can be up to 30% in case of bursty errors. Thanks to LS, however, each flow's received transmission time is consistent with its assigned ratio. For most of the existing scheduling algorithms used in wireless networks, it is difficult to maintain such fairness because it is almost impossible to predict/estimate the channel condition in case of sporadic errors (thus, transmission swapping/compensation will not work). In terms of station's throughput, the LS does not do much better in case of sporadic error because each transmission error is independent and the average error probability is relatively small. However, the LS can increase the station's throughput by $\frac{Succ. \ Trans. \ with \ LS - Succ. \ Trans. \ without \ LS}{Succ. \ Trans. \ without \ LS} = \frac{824762 - 638503}{638503} = 29\%$ in case of bursty errors. The improvement mainly comes from the error-prone flows. Take flow 3 as an example. If a frame gets corrupted during an error burst, it will very likely be retransmitted for *retry limit* consecutive times before being dropped. Thus, at the end of these retransmissions, the LS will re-adjust the transmission order such that the other flows can 'catch up' with flow 3. This implicitly prevents flow 3 from transmitting frames too quickly within an error burst and helps reduce the number of its invalid transmissions dramatically (from 94730 to 28622).

Finally, we consider the scheduling delays. Since the scheduling delays of error-prone flows are cumulative due to transmission error and MAC-layer retransmission mechanism, we are only interested in the scheduling delay of an error-free flow. Figure 16.7 plots the scheduling delays of flow 2's frames under bursty errors. The results under sporadic errors are similar and hence omitted here. Without LS, the error-prone flows will overuse the transmission time continuously, so flow 2 will be affected such that the scheduling delays of its frames keep increasing. In contrast, the LS will control the overuse of error-prone flows at the end of every 'round' of retransmissions. Thus, the

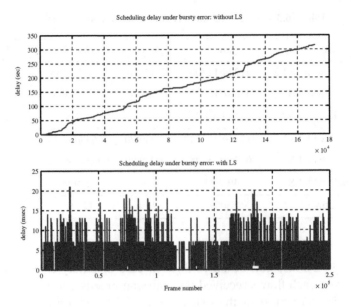

Figure 16.7.   Scheduling delays of flow 2's frames in case of bursty errors.

scheduling delays of flow 2 frames will be bounded. This complies with the results shown in Figure 16.7. We can draw a quick conclusion from this simulation: *the LS can always help maintain the AFS-based fairness among traffic flows within a station and bound the scheduling delay of error-free flows, regardless of the underlying error pattern. Moreover, it can improve the station's throughput in case of bursty errors.*

16.7   Experimental setup and results

In order to evaluate the performance of the scheduler, we use a simple network topology as shown in Figure 16.8 to conduct the experiments. Due to

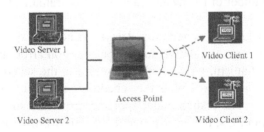

Figure 16.8.   Experimental setup to verify AFS against WFQ.

the lack of actual HCCA-capable 802.11e hardware at the client card, we only evaluate the performance of local scheduler with downstreams only using legacy 802.11a radio at the QAP with modified parameters of DCF channel access. The CW_{min} and CW_{max} was set to 0 so as to emulate the performance of HCCA. As seen in Figure 16.8, video server 1 and 2 are connected to a QAP through wired Ethernet connection and both are sending 4 Mbps standard-definition (720×480) video[9] to their wireless video client 1 and 2 respectively via the access point. Here the 4 Mbps throughput refers to video throughput measured just above the IP layer. Real Time Protocol (RTP) is used as the transport protocol to carry video from the source to destination. In wired network segment, 100 Mbps Ethernet was used to minimize queuing effect in wired network segment. In the wireless network segment, 18 Mpbs PHY rate is used with the maximum number of retransmissions[10] set to a default value . The reason for choosing 18 Mbps was to load the system when there are channel errors. If we were to set the physical transmission rate at 54 Mbps, we would need 7 to 8 streams to load the system. This would mandate lots of client displays and complicate the setup. Hence we fix the physical transmission rate as 18 Mbps.

Various local schedulers[11] are loaded on to the QAP at various sessions to benchmark the performance. We run the 2 experiments. In the first experiment, WFQ scheduler was loaded as the reference scheduler. In the second experiment the AFS was loaded to compare against WFQ. At the beginning of each session, both video players are located symmetrically with respect to the access point. 10 seconds later, video player 1 is moved away from the QAP, which results in more retransmissions. Another 10 seconds layer, video player 1 lowered its PHY rate to 6 Mbps in an attempt to achieve more robust transmission. The traces of throughput and loss for each session are captured in Table 16.4. We do not capture delay trace for synthetic data.

Each run in the experiment generates different statistics as each run sees different channel conditions. So the Table 16.4 gives a snapshot of the performance for that run. One can notice from the Table 16.4, that the air fair local scheduler: 1) achieved air fairness as expected; 2) keeps the flows throughput of S2 to be unaffected by S1's transmission rate or transmission errors in terms of both the throughput and delay; 3) achieved higher overall system throughput than WFQ.

---

[9]The video file name is Spacejam. Statistics for MPEG-4 video track 5 (computed from hint track 6): Mean Data Rate = 4.07616 Mbps, Peak Data Rate = 14.583 Mbps, Low Data Rate = 0.992256 Mbps, Maximum Burst Size = 193396 bytes, Average RTP packet size = 1003 bytes, Maximum RTP packet size = 1040 bytes, Minimum RTP packet size = 48 bytes, RTP packet size variance = 15953 bytes.

[10]10 on D-link DWL-AB650 airpro cardbus adaptor

[11]The local scheduler was implemented in Linux kernel and wireless device driver to take advantage of the Linux Traffic Control framework including packet classifier, queuing discipline, configuration utility and API.

Table 16.4. Comparison of the throughput and loss characteristics for AFS and WFQ by using a real video trace as well as generating synthetic data. Here R_1 and R_2 represents the physical transmission rates used by streams S1 and S2 Respectively. Both are down link flows. (NA: Not Available).

synthetic data	$R_1 = R_2 = 18$Mbps		$R_1 = R_2 = 18$Mbps R_1 has errors		$R_1 = 6$Mbps $R_2 = 18$Mbps	
scheme	AFS	WFQ	AFS	WFQ	AFS	WFQ
throughput (Mbps) of S1	4.169	4.15	2.157	3.1	2.61	NA
throughput (Mbps) of S2	4.169	4.10	4.169	3.0	4.19	NA
packet loss of S1 (no. lost/total sent)	0/4997	0/4997	2220/4997	858/4997	1869/4997	NA
packet loss of S2 (no. lost/total sent)	0/4996	0/4996	0/4996	1293/4996	0/4996	NA
real video	$R_1 = R_2 = 18$Mbps		$R_1 = R_2 = 18$Mbps R_1 has errors		$R_1 = 6$Mbps, $R_2 = 18$Mbps	
scheme	AFS	WFQ	AFS	WFQ	AFS	WFQ
throughput (Mbps) of S1	4.32	4.37	2.13	2.69	2.06	0.875
throughput (Mbps) of S2	4.31	4.3	4.3	2.7	4.3	0.83
packet loss of S1 (no. lost/total sent)	0/4966	0/4966	2396/4966	1881/4966	1558/4966	3987/4966
packet loss of S2 (no. lost/total sent)	0/5026	0/5026	0/5026	1455/5026	0/5026	4121/5026
delay (ms) of S1	NA	NA	718.61	330.41	543.08	1532.3
delay (ms) of S2	NA	NA	30.39	340.13	42.34	1525.9

Consider the flow S2. This flow does not experience a bad channel condition. Under AFS, this flow is guaranteed its required throughput as well as error rate. The maximum tolerable delay was fixed at 200 milliseconds taking into account the buffer at the encoder and decoder side. On the contrary, using WFQ, one sees the major degradation in the video quality because of loss of frames when there are links with errors. In our case, we fixed the physical transmission rate and moved the antenna of station 1 such that the signal to interference noise ratio (SINR) was lowered resulting in frame loss. This resulted in lots of retransmissions for flow S1. As the retry limit is reached or when the frames have exceeded the delay bounds, they are dropped. Each frame of S1 reaching the retry limit or exceeding the delay bound propagates its channel condition to S2 which also degrades. Looking at the Table 16.4, we find that the errors are localized to flow S1 in case of AFS and are distributed to flow S2 since lots of frames of S2 have to wait for a long time because of flow S1's error condition. For the synthetic traffic we find that the number of frames lost were 2220 and 0 for flows S1 and S2 respectively in case of AFS and was shared between S1 and S2 as 858 and 1293 respectively in case of WFQ. The sum of total frames lost is 2151(=858+1293). In the case of real video the errors are more in case of WFQ compared to AFS that localizes errors to S1 alone. We did not plot the delay as the delay bound was fixed to around 1 second and if the channel conditions were normal, both WFQ and AFS satisfied the delay QoS. Looking at the last column for throughput in case of real video data, it is inferred that differing link speeds as well as errors reduce the overall system throughput in case WFQ but not in case of AFS. This clearly shows that AFS achieves higher throughput than WFQ.

16.8   Conclusions

In this chapter, we proposed a new concept of fairness, Air Fairness and propose a scheduling scheme called AFS and showed its application to an IEEE 802.11e wireless LAN. To implement it in IEEE 802.11e WLAN we propose a 2 stage structure of scheduler with the stage 1 called the station scheduler and the second stage called the local scheduler. We present pseudo code for the local scheduler and show how it can achieve air fairness in case of downlink traffics through simulation and experiments. In order to provide different traffic flows a fair share of a station's transmission time, the LS may reschedule their transmissions based on the results of the MAC-layer retransmission in the IEEE 802.11e standard. Thus, it need not estimate or predict the channel condition, and the AFS-based per-flow fairness can be achieved easily. Based on AFS and our two-stage scheduling architecture, service for an individual station/flow will not be affected by the other stations/flows even in the presence of location-dependent transmission rates and location-dependent errors, and thus, the fairness can always be guaranteed. Moreover, AFS can always provide a higher system throughput than GPS.

References

Bennett, J.C.R., and H. Cheng [1996]. WF2Q: Worst-case fair weighted fair queueing. *IEEE INFOCOM'96*, pages 120–128.

Bianchi, G. [2000]. Performance analysis of the IEEE 802.11 distributed coordination function. *IEEE Journal on Selected Areas in Communications*, 3(3): 535–547.

Bianchi, G., L. Fratta, and M. Oliveri [1997]. Performance evaluation and enhancement of the CSMA/CA MAC protocol for 802.11 wireless LANs. *IEEE PIMRC'97*, pages 407–411.

Chou, C.T., and K.G. Shin [2002]. Analysis of combined adaptive bandwidth allocation and admission control in wireless networks. *IEEE INFOCOM'02*, pages 676–684.

Eckhardt, D., and P. Steenkiste [2000]. Effort-limited fair(ELF) scheduling for wireless networks. *IEEE INFOCOM'00*, pages 1097–1106.

Fragouli, C., V. Sivaraman, and M.B. Srivastava [1998]. Controlled multimedia wireless link sharing via enhanced class-based queuing with channel-state-dependent packet scheduling. *IEEE INFOCOM'98*, pages 572–580.

Golestani, S.J. [1994]. A self-clocked fair queueing scheme for broadband application. *IEEE INFOCOM'94*, pages 636–646.

Huang, H., D. Tsang, R. Sigle and P. Kuhn [2000]. Hierarchial scheduling with adaptive weights for ATM. *IEICE Transations on communication*, E83-B(2): 313–320.

IEEE 802.11b [1999]. Wireless LAN medium access control (MAC) and physical layer (PHY) specifications. *IEEE Standard*.

IEEE 802.11e [2002]. IEEE 802.11e/D3.0, draft supplement to part 11: Wireless medium access control (MAC) and physical layer (PHY) specifications: medium access control (MAC) enhancements for Quality of Service (QoS). *IEEE Standard*.

del Prado, J., Sai Shankar et al. [2002]. Mandatory TSPEC parameters and reference design of a simple scheduler. *IEEE 802.11-02/705r0*.

Jeong, M., H. Morikawa, and T. Aoyama [2001]. Fair scheduling algorithm for wireless packet networks. *IEICE Transactions on Fundamentals of Electronics, Communications and Computer Sciences*, E84-A(7): 1624–1635.

Kautz, R., and A.L. Garcia [1997]. Distributed self-clocked fair queueing architecture for wireless ATM networks. *IEEE PIMRC'97*, pages 189–193.

Liu, X., K.P. Chong, and N.B. Shroff [2001]. Transmission scheduling for efficient wireless utilization. *IEEE INFOCOM'01*, pages 776–785.

Lu, S., V. Bharghavan, and R. Srikant [1997]. Fair scheduling in wireless packet networks. *ACM SIGCOMM'97*, pages 63–74.

Lu, S., T. Nandagopal, and V. Bharghavan [1998]. A wireless fair service algorithm for packet cellualr networks. *ACM MOBICOM'98*, pages 10–20.

Ng, T.S.E., I. Stoica, and H. Zhang [1998]. Packet fair queueing algorithms for wireless networks with location-dependent errors. *IEEE INFOCOM'98*, pages 1103–1111.

Parekh, A.K., and R.G. Gallager [1992]. A generalized process sharing approach to flow control in integrated services networks: The single node case. *IEEE INFOCOM'92*, pages 915–924.

Qiao, D., and K.G. Shin [2002]. Achieving efficient channel utilization and weighter fairness for data communications in IEEE 802.11 WLAN under the DCF. *International Workshop on Quality of Service*.

Qiao, D., and S. Choi [2001]. Goodput enhancement of IEEE 802.11a wireless LAN via link adaptation. In *Proc. IEEE ICC'01*, Helsinki, Finland.

Ramanathan, P., and P. Agrawal [1998]. Adapting packet fair queueing algorithm to wireless networks. *ACM MobiCom'98*, pages 1–9.

Sadeghi, B., V. Kanodia, A. Sabharwal, and E. Knightly [2002]. Opportunistic media access for multirate ad hoc networks. *ACM Mobicom'02*, pages 24–35.

Vaidya, N.H., P. Bahl, and S. Gupta [2000]. Distributed fair scheduling in a wireless LAN. *ACM MobiCom'00*, pages 167–178.

Chapter 17

HIGH THROUGHPUT AND LOW POWER REED SOLOMON DECODER FOR ULTRA WIDE BAND

Akash Kumar and Sergei Sawitzki

Abstract   Reed Solomon (RS) codes are widely used in a variety of communication systems. Continual demand for ever higher data rates makes it necessary to devise very high-speed implementations of RS decoders. In this chapter, a uniform comparison is drawn for various algorithms and architectures proposed in the literature, which help in selecting the appropriate architecture for the intended application. Dual-line architecture of modified Berlekamp Massey algorithm is chosen for Ultra Wide Band (UWB). Using 0.12 μm technology the area of the design is 0.22 mm^2 and throughput is 1.6 Gbps. The design dissipates only 14 mW of power in the worst case, including memory, when operating at 1.0 Gbps data rate.

Keywords   Reed Solomon Decoder, low power, high throughput, ultra-wide band, dual-line Berlekamp Massey.

17.1   Motivation

Reed Solomon (RS) codes are widely used in a variety of communication systems such as space communication link, digital subscriber loops and wireless systems, as well as in networking communications and magnetic and data storage systems. Continual demand for ever higher data rates and storage capacity makes it necessary to devise very high-speed implementations of RS decoders. Newer and faster implementations of the decoder are being developed and implemented. A number of algorithms are available and this often makes it difficult to determine the best choice due to the number of variables and trade-offs available. Therefore, before making a good choice for the application a thorough research is needed into the options available.

For IEEE 802.15-03 standard proposal (commonly known as UWB) in particular, very high data rates for transmission are needed. According to the

Wim F.J. Verhaegh et al. (Eds.), Intelligent Algorithms in Ambient and Biomedical Computing, 299-316.
© 2006 *Springer. Printed in the Netherlands.*

current standard, the data rate for UWB will be as high as 480 Mbps. Since the standard is also meant for portable devices, power consumption is of prime concern, and at the same time the silicon area should be kept as low as possible. As such, a low power and high throughput codec is needed for UWB standard. Reed Solomon is seen as a promising codec for such a standard. It should be mentioned here that shortened code RS(23, 17) derived from RS(255, 249) is used in the Release 1.0 of the standard proposal and this is only used for the header, not for payload. This work looks ahead towards 1.0 Gbps throughput, where payload probably will be encoded as well and shortened code will not be good enough anymore.

17.2   Introduction to Reed Solomon

Reed Solomon codes are perhaps the most commonly used in all forms of transmission and data storage for forward error correction (FEC). The basic idea of FEC is to add redundancy at the end of the messages systematically so as to enable the retrieval of messages correctly despite errors in the received sequences. This eliminates the need of retransmission of messages over a noisy channel. RS codes are a subset of Bose-Chaudhuri-Hocquenghem (*BCH*) codes and are linear block codes. [Wicker & Bhargava, 1994] is one of the best references for RS Codes.

An $RS(n, k)$ code implies that the encoder takes in k symbols and adds $n - k$ parity symbols to make it an n-symbol code word. Each symbol is at least of m bits, where $2^m > n$. Conversely, the longest length of code word for a given bit-size m, is $2^m - 1$. For example, $RS(255, 239)$ code takes in 239 symbols and adds 16 parity symbols to make 255 symbols overall of 8 bits each. Figure 17.1 shows an example of a systematic RS code word. It is called systematic code word as the input symbols are left unchanged and the parity symbols are appended to it.

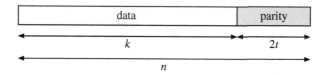

Figure 17.1.   A typical RS code word.

Reed Solomon codes are best for burst errors. If the code is not meant for erasures, the code can correct errors in up to t symbols where $2t = n - k$. A symbol has an error if at least one bit is wrong. Thus, $RS(255, 239)$ can correct errors in up to 8 symbols or 50 continuous bit errors. It is also interesting to see, that the hardware required is proportional to the error correction capability of the system and not the actual code word length as such.

When a code word is received at the receiver, it is often not the same as the one transmitted, since noise in the channel introduces errors in the system. Let us say if $r(x)$ is the received code word, we have

$$r(x) = c(x) + e(x) \qquad (17.1)$$

where $c(x)$ is the original codeword and $e(x)$ is the error introduced in the system. The aim of the decoder is to find the vector $e(x)$ and then subtract it from $r(x)$ to recover original code word transmitted. It should be added that there are two aspects of decoding - error detection and error correction. As mentioned before, the error can only be corrected if there are fewer than or equal to t errors. However, the Reed Solomon algorithm still allows one to detect if there are more than t errors. In such cases, the code word is declared as *uncorrectable*.

The basic decoder structure is shown in Figure 17.2. A detailed explanation on Reed Solomon decoders can be found in [Wicker & Bhargava, 1994] and [Blahut, 1983]. Decoder essentially consists of four modules. The first module computes the syndrome polynomial from the received sequence. This is used to solve a key equation in the second block, which generates two polynomials for determining the location and value of these errors in the received code word. The next block of Chien search uses the Error Locator Polynomial obtained from the second block to compute the error location, while the fourth block employs Forney algorithm to determine the value of error occurred. The correction block merely adds the values obtained from the output of the Forney block and the FIFO block.

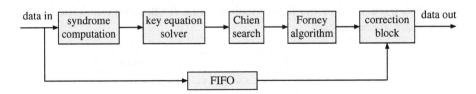

Figure 17.2. Decoder flow.

17.3 Channel model

Before we proceed to the actual decoder implementation, it is important to look at the channel model itself. Since UWB (Ultra Wide Band) is not very well explored yet, it is important to analyse how the channel would behave at the frequency and the data rate under consideration. One of the most common models used for modelling transmission over land mobile channels is the

Gilbert-Elliott model. In this model a channel can be either in a good state or a bad state depending on the signal-to-noise ratio (SNR) at the receiver. For different states, the probability of error is different. In [Ahlin, 1985], Ahlin presented a way to match the parameters of the GE model to the land mobile channel, an approach that was generalized in [Wang & Moayeri, 1995] to a Markov model with more than two states.

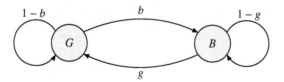

Figure 17.3.   The Gilbert-Elliott channel model.

Figure 17.3 shows the GE Channel Model. Two states are shown represented by G and B indicating the good and the bad state respectively. Further, the transition probability from the good state to the bad state is shown as b and from the bad to the good state as g. The probability for error in state G and B is denoted by $P(G)$ and $P(B)$ respectively. What follows is a concise explanation of the model. A more detailed analysis can be found in [Wilhemsson & Milstein, 1999] and [Sharma et al., 1996].

To obtain the relation between the physical quantities and the parameters of the model, Rayleigh fading was considered. The amplitude α of the received signal is therefore

$$f(\alpha) = \frac{2\alpha}{\bar{\gamma}} e^{-\alpha^2/\bar{\gamma}}, \ \alpha \geq 0 \tag{17.2}$$

and the SNR is exponentially distributed, given by

$$f(\gamma) = \frac{1}{\bar{\gamma}} e^{-\gamma/\bar{\gamma}}, \ \gamma \geq 0 \tag{17.3}$$

where $\bar{\gamma}$ is the average SNR of the received signal. Since, we have two states in the GE Channel, let γ_t be the threshold for the SNR, where the channel changes the state. The stationary probabilities for the two states are given by

$$P^{\text{stat}}(B) = 1 - e^{-\rho^2} \tag{17.4}$$

$$P^{\text{stat}}(G) = e^{-\rho^2} \tag{17.5}$$

where $\rho^2 = -\gamma_t/\bar{\gamma}$. From these we arrive at the channel transition probabilities given by the following equations,

$$g = \frac{\rho f_D T_s \sqrt{2\pi}}{e^{\rho^2} - 1} \tag{17.6}$$

$$b = \rho f_D T_s \sqrt{2\pi} \tag{17.7}$$

where $f_D = \frac{v f_c}{c}$. Here v is the relative speed of the objects communicating, f_c is the frequency of the carrier and c is the velocity of light. T_s is the symbol duration. f_D indicates the Doppler frequency, while $f_D T_s$ signifies the normalized Doppler frequency. The error probabilities in different states can be computed as follows:

$$P_e(B) = \frac{1}{P^{stat}(B)} \int_0^{\gamma_t} f(\gamma) P_e(\gamma) d\gamma \tag{17.8}$$

and

$$P_e(G) = \frac{1}{P^{stat}(G)} \int_{\gamma_t}^{\infty} f(\gamma) P_e(\gamma) d\gamma \tag{17.9}$$

where $P_e(\gamma)$ is the symbol error probability given the value of γ and $f(\gamma)$ is as defined above. $P_e(\gamma)$ depends on the type of modulation used, but for BPSK (Binary Phase Shift Keying) – one of the common modulation schemes, we have $P_e(\gamma) = Q(\sqrt{r2\gamma})$, where [Proakis, 2001]

$$Q(x) = \frac{1}{\sqrt{2\pi}} \int_x^{\infty} e^{-t^2/2} dt. \tag{17.10}$$

17.3.1 Simulation

Following were the parameters set for the simulation of the Ultra Wide Band channel:

- carrier frequency = 4.0 GHz,

- information rate = 480 Mbps.

Two sets of simulation were run for different threshold reading. The threshold here signifies the SNR level at which the channel changes states. The first set was with the threshold set to 5 dB lower than the average SNR and the other with 10 dB less than the average. Due to the very high data bit rate involved the transition probability is very small. Therefore, channel transition become very rare events, and simulations determined the error probabilities for code words beginning in a certain state. These were then weighted by the steady state probability of the corresponding state and added together to obtain the overall probability rate. Two measures, the bit error rate and the symbol error rate are computed and plotted. The simulation was run for 10,000 code words to get a good estimate for each state. Mathematica software was used to solve the complex mathematical equations and obtain the channel model parameters for the physical quantities under consideration.

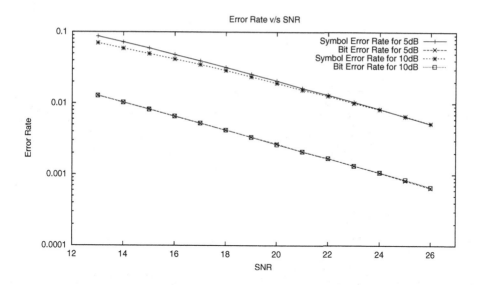

Figure 17.4.   The symbol error rate and bit error rates for different thresholds.

17.3.2   Simulation results

As can be seen from the Figure 17.4, the error probabilities decrease with increase in SNR as expected. The figure shows the symbol and the bit error probabilities observed. As expected the error rates follow a linear relationship with the increasing SNR on the logarithmic scale. We notice that around 20 dB average SNR for both the thresholds, the symbol error rate is about 0.02, which corresponds to an average of 5 symbol errors in a code word of 255 symbols. From the results, an error correction capability of 8 is seen as a good choice, as when the SNR is above 20 dB, the likelihood of more than 8 errors in a codeword of 255 is very low.

17.4   Architecture design options

Having decided on the codeword, investigation was carried out to determine appropriate algorithm and architecture. Figure 17.5 shows the various architectures available. Table 17.1 shows the hardware requirements of computational elements used in various architectures. Estimates have been made from the figures drawn in the papers when actual counts could not be obtained for any architecture. It should be noted that this is only the estimate of computational elements and, therefore, additional hardware will be needed for control.

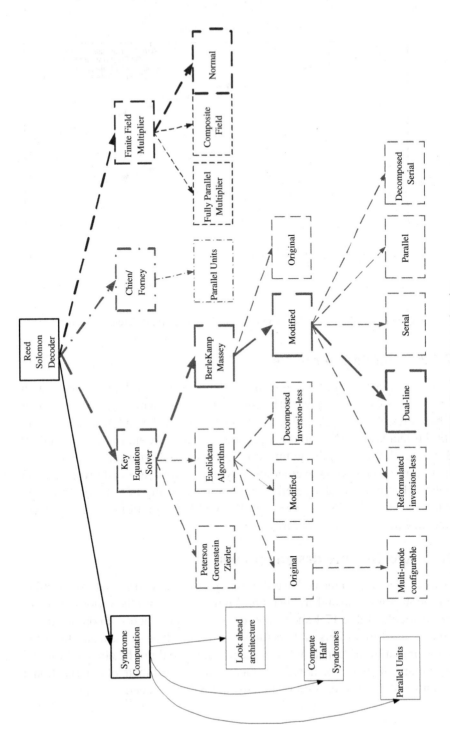

Figure 17.5. Design space exploration.

Table 17.1. Summary of hardware utilization of various architectures.

Architecture	Blocks	Adders	Multipliers	Muxes	Latches	Latency	Critical path delay
Syndrome Computation [Lee, 2003a]	2t	1	1	1	2		
Total		2t	2t	2t	4t	n	Mul + Add
Look ahead architecture (x units)	2t	x	x	1	2		
Total		2xt	2xt	2t	4t	n/x	Mul + Add
Original Euclidean [Lee et al., 2000]							
Divider Block	2t	1	1	3	2		
Multiply Block	t	2	1	3	3		
Total (Estimates)		4t	3t	9t	7t		
Actual [Lee, 2003b]		4t + 1	3t + 1	11t + 4	14t + 6	4t - 3	ROM+2×Mul+Add+2×Mux
Modified Euclidean [Lee et al., 2000]							
Degree Computation Block	2t	2	0	7	7		
Polynomial Arithmetic Block	2t	2	4	8	19		
Total (Estimates)		8t	8t	30t	52t		
Actual [Lee, 2003b]		8t	8t	40t + 2	78t + 4	10t + 8	Mul + Add + Mux
Decomposed inversion-less [Chang & Lee, 2001]		1	3	1	3t + 1	2t×(t+1)	Mul + Add + Mux
Modified BerleKamp Massey							
Serial		1	3	4	3t + 2	2t×(2t+2)	Mul + Add + Mux
Decomposed inversion-less [Chang et al., 1998]		2	3	2	5	2t×(t+1)	Mul + Add + Mux
Parallel		t	3t + 2	t	3t + 1	2t	2×Mul + 2×Add + Mux
Dual-line [Kang & Park, 2002]		2t	4t + 1	2t	4t + 1	3t + 1	Mul + Add
Reformulated inversion-less [Sarwate et al., 2001]		3t + 1	6t + 2	3t + 1	6t + 2	2t	Mul + Add
Chien/Forney		2t	2t + 2	2t + 2	2t + 10	4	max(Mul + Add, ROM)

17.4.1 Design decisions

In order to choose a good architecture for the application, various factors have to be considered.

- *Gate count.* This determines the silicon area to be used for development. It is a one-time production cost, but can be critical if it is too high.

- *Latency.* Latency is defined as the delay between the received code word and the corresponding decoded code word. The lower the latency, the smaller is the FIFO buffer size required and therefore, it also determines the silicon area to a large extent.

- *Critical path delay.* This determines the minimum clock period, i.e. maximum frequency that the system can be operated at.

Table 17.1 shows a summary of all the above mentioned parameters. For our intended UWB application, speed is of prime concern as it has to be able to support data rates as high as 1.0 Gbps. At the same time, power has to be kept low, as it is to be used in portable devices as well. This implies that the active hardware at any time should be kept low. Also, the overall latency and gate count of computational elements should be low since that would determine the total silicon area of the design.

Key equation solver. Reformulated inversion-less and dual line implementation of the modified Berlekamp Massey have the smallest critical path delay among all the alternatives of the Key Equation Solver. Inversion-less and dual-line architectures are explained in [Chang et al., 1998] and [Kang & Park, 2002] respectively. When comparing inversion-less and dual-line implementations, dual line is a good compromise in latency and computational elements needed. The latency is one of the lowest and it has the least critical path delay of all the architectures summarized. Thus, dual-line implementation of the BM algorithm is chosen for the key-equation solver. Another benefit of this architecture is that the design is very regular and hence easy to implement.

RS code. As we can see from Table 17.1, the hardware requirement for the entire block is a function of t, the error correction capability, and the latency is a function of both n and t. Thus, while we want to have a code with high error correction capability, we can not have a very high value of t as the hardware needed is proportional to it. The value of n determines the bit-width of the symbol and therefore the hardware needed, but only logarithmically. However, one would want to have a value of $n = 2^m - 1$, to derive maximum benefit out of the hardware. The value of t is often chosen to be a power of 2 in order to maximize the hardware utilized in design. Taking into account the results of Channel Model Simulation, $RS(255, 239)$ is chosen, since it has an error correction capability of 8.

17.4.2 Highlights

Table 17.2 shows the various parameters for choosing dual line architecture with $n = 255$, $k = 239$, and $t = 8$. The overall critical path delay is hence Mul + Add.

Table 17.2. Summary of hardware utilization for dual-line architecture.

Architecture	Adders	Multipliers	Muxes	Latches	Latency
Syndrome Computation	2t	2t	2t	4t	n
Dual-line	2t	4t + 1	2t	4t + 1	3t + 1
Chien/Forney	2t	2t + 2	2t + 2	2t + 10	4
Total	6t	8t + 3	6t + 2	10t + 11	3t + n + 5
For RS(255, 239)	48	67	50	91	284

17.5 Design flow

The first step was to develop a C-model for the decoder. 'Gcc' compiler was used to compile the code and to check if the code worked correctly. Output of each intermediate stage was compared with the expected output according to the algorithm with the aid of an example.

Once the algorithm was fully developed and tested in C, VHDL-code was developed. The VHDL code was structured such that it could be easily synthesized. A wrapper class was written around it, in order to test it. This VHDL code was compiled and tested using Cadence tools. 'Ncsim' was used to simulate the system and generate the output stream for the same input tests as were used for testing C code. The output stream from VHDL and C were then compared.

When this output was found to be matched for various input test cases, synthesis experiments were started. Ambit from Cadence was used to analyse the hardware usage and frequency of operation after various optimisation settings.

The design flow needed for verification of synthesized design and power estimation has been explained in Figure 17.6. As shown in the figure the core VHDL modules were optimised and synthesized using *ambit*. The synthesized model was written out into a verilog netlist using *ambit* itself. Once the netlist was obtained, it was compiled using *ncvlog* into the work library together with the technology library. The library used was for the same technology as the

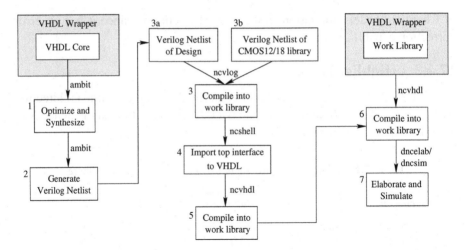

Figure 17.6. Design flow for design verification and estimation of power.

one used for synthesis. As can be seen, the wrapper modules were actually written in VHDL, while the compiled core was from the verilog. Thus, to allow interaction between the two, the top interface of the work library, was extracted into a VHDL file and then compiled into the work library. This was done using *ncshell* and *ncvhdl* respectively. This being done, the wrapper modules were compiled into the work library.

From this point onwards, two approaches were used. *Ncelab* and *ncsim* were used purely for simulating the synthesized design, and *dncelab* and *dncsim* were used to obtain power estimate, which were essentially the same tools, but included the *DIESEL* routines for estimating the power dissipated in the design. *Diesel* is an internal tool developed within Philips which estimates the power for the simulated design, and hence the accuracy of the results depends on the input provided.

17.6 Results

This section covers the results of various synthesis experiments conducted. Resource utilization, timing analysis and the power consumption were used as benchmarking parameters.

17.6.1 Area analysis

Ambit was run for 0.12 μm and 0.18 μm CMOS technology. The silicon area required was analysed for various timing constraints. A comparison for area of the decoder is shown in Table 17.3. This table shows the area requirement when the constraint was set to 5 ns, which can support 200 MHz frequency, i.e.

1.6 Gbps. The total number of design cells used, including the memory, were 12,768 and 12,613 for 0.12 μm and 0.18 μm respectively.

Table 17.3.   Resource utilization for the decoder.

Module	Module Area (μm^2)	
	CMOS12	CMOS18
Chien	7,663	15,675
FIFO	83,183	148,684
Forney	21,608	52,936
Gen_elp_eep	89,602	186,404
Gen_syndromes	17,828	34,754
top_view	219,913	438,472

17.6.2   Power analysis

The power estimates provided in this section are for design operation at 125 MHz, which translates to data rate of 1Gbps.

Variation with number of errors.   Figure 17.7 shows the variation of power with the number of errors found in the codeword for 0.12 μm technology. As can be seen from the graph obtained, the power dissipated for the FIFO and syndrome computation block is independent of the number of errors as expected. For the block that computes the Error Locator Polynomial (ELP) and Error Evaluator Polynomial (EEP), it is clearly seen that the power dissipated increases linearly with the number of errors. The Chien search block also shows a linear increase in the power dissipated.

The behaviour of Forney evaluator is a bit different from the other modules. We see that the power dissipated for the codeword with an even number of errors is not significantly larger to the one with the previous number of errors. The reason lies in the fact that the degree of EEP for codeword with one error is often the same as the one with two errors, and so on and so forth. However, as a general rule, there is still an increase in the power dissipation, because of some computation that is done for each error found.

Distribution of power in different modules.   Figure 17.8 shows a distribution of power when there are maximum number of errors correctable in the received code word, while Figure 17.9 shows the distribution when the code word is received intact. As can be seen, in the case of no errors, bulk of the power is consumed in computing syndromes, apart from the memory. In the event of maximum errors detected, the Forney block consumes the maximum power.

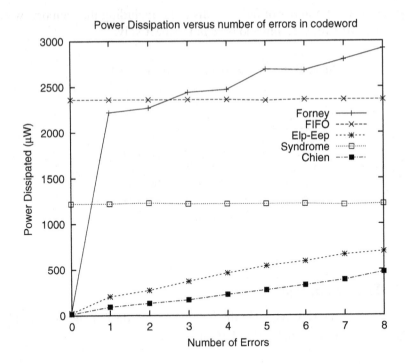

Figure 17.7. Graph showing variation of power dissipated with number of errors for different modules.

The total power consumption varies from 14 mW to 17 mW with the former corresponding to no-error case, while the latter corresponding to maximum errors.

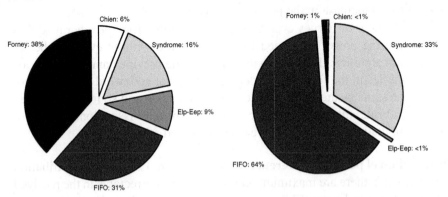

Figure 17.8. Power consumed by various blocks when 8 errors are found.

Figure 17.9. Power consumed by various blocks when no errors are found.

17.7   Benchmarking

Please note that for all the designs $RS(255, 239)$ code has been used for benchmarking. The design using modified Euclidean Algorithm is very hardware intensive. The design proposed in [Lee, 2003a] uses roughly 115K gates for 0.12 μm CMOS technology operating at 6.0 Gbps excluding memory. The proposed design only uses 12K cells including memory in both 0.12 μm and 0.18 μm technology. The results are better even when they are normalised for throughput and technology. The latency of the design is only 284 cycles when compared to 355 cycles in [Lee, 2003a].

In terms of power, a design was proposed in [Chang et al., 2002] for low power. In that design, 62 mW of power is used in the best case, including memory, using 0.25 μm CMOS technology, and 100 mW are consumed in the worst case. In our design, only 17 mW of power is used in the worst case using 0.12 μm technology. The area of the chip proposed in [Chang et al., 2002] using 0.25 μm CMOS technology is 5 mm^2, while the area of the proposed design is 0.22 mm^2 with 0.12 μm technology.

17.8   Optimisations to design

From the results, it was observed that the FIFO and the Forney block consumed most of the power. These blocks were investigated further and redesigned to improve the performance. The original design of FIFO involved a serial arrangement of shift-registers. This design was the most compact in terms of area but consumed more power since at every cycle all the elements were shifted by one. The design was hence, modified to have only one read and write every clock cycle. This increased the design area, but significantly reduced the power. Area of the new design of FIFO is now 109,000 μm^2 (with 0.12 μm technology), while the power consumed is only 970 μW, 60% lower than the earlier design. This results in power savings even in the case when no errors are found.

For the Forney block, design was optimised by combining two table lookups into one for computing the inverse of elements. This led to a better circuit in terms of area and also decreased the power significantly. The optimised design for Forney now occupies an area of 13,400 μm^2, about 38% lower than original design. The power consumption is lower by at least 1.5 mW for all cases. Table 17.4 shows the new area distribution of the decoder.

Power analysis was repeated for the optimised design. Figure 17.10 shows the power distribution in various modules when there are 8 errors in the received codeword, while Figure 17.11 shows the distribution when the codeword is received intact. As we can see, the FIFO now takes less than half the power in no-error case, as compared to two-thirds in the original design. In the

Table 17.4.   Resource utilization for the decoder in CMOS12 in optimised design.

Module	Module Area (μm^2)
Chien	7,655
FIFO	108,906
Forney	13,408
Key Equation	89,587
Syndromes	17,719
Top View	237,414

case of 8-errors, the power consumption of Forney has now reduced to about a quarter as compared to one-third in the original design.

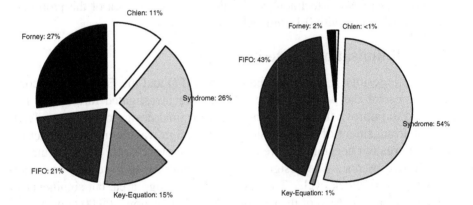

Figure 17.10.   Power consumed when 8 errors are found in optimised design.

Figure 17.11.   Power consumed when no errors are found in optimised design.

Figure 17.12 shows the variation of power with the number of errors. The trend in the power consumption of Forney is the same as before the optimisation. The total power consumption of the design now lies between 12 mW to 14 mW depending upon the no-error case to when maximum errors are found. It should be noted that 9.5 mW of power is consumed in driving the input. Thus, only about 2.5 mW to 4.5 mW is actually consumed in the transitions in the design. Voltage scaling measures can also be applied on the design to further lower the power consumption.

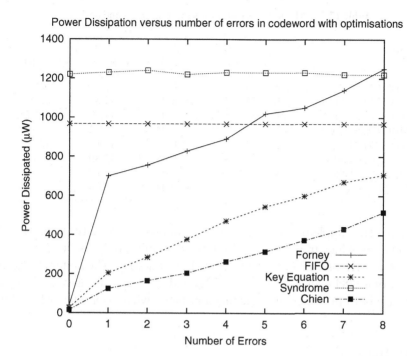

Figure 17.12.   Variation of power dissipated with number of errors for different modules with modifications in the design.

17.8.1   Embedded memory for FIFO

The option of using embedded memory for FIFO was also explored, since in the final design embedded memory shall be used. Table 17.5 shows an estimate of the area and power consumption for various libraries. In the table, x refers to the number of words in a row, y to the number of rows per block and z to the number of blocks. The total number of words in memory is $x \times y \times z$. As can be seen *AMDC C12ESRAM* is the only one that provides a lower power than our simulated design. However, all of them seem to have a better area than the synthesized design. When *AMDC C12ESRAM* is used for the final design, the area would be reduced by 0.08 mm^2, that is 80% lower than the synthesized design. The power consumption, however, remains the same.

17.9   Conclusions

A uniform comparison was drawn for various algorithms that have been proposed in literature. This helped in selecting the appropriate architecture for the intended application. The modified Berlekamp Massey algorithm is chosen for the VHDL implementation. Dual line architecture is used, which is as fast as serial and has low latency as that of a parallel approach.

Table 17.5. Memory Estimates for various libraries and designs.

Library	Size (bits)	Words x	y	z	Area (mm²)	Power (μW/MHz)	at 125 MHz
AMDC C12XSRAM	2.2K	36	8		0.02	13.03	1628.75
LTG C12FSRAM	2.2K	72	4		0.03	11.32	1415
LTG C12FSRAM	2.2K	36	8		0.02	11.57	1446.25
AMDC C12ESRAM	2.2K	72	2	2	0.02	7.2	900
LTG C12FDSRAM	2.2K	72	4		0.06	11.86	1482.5
LTG C12FDSRAM	2.2K	36	8		0.05	12.88	1610
AMDC C12EDSRAM	2.2K	72	4		0.04	10.7	1337.5
LTG C12FTSRAM	2.2K	72	4		0.03	13.9	1737.5

The decoder implemented is capable of running at 200 MHz in ASIC implementation, which translates to 1.6 Gbps and requires only about 12K design cells and an area of 0.22 mm² with CMOS12 technology. The system has a latency of only 284 cycles for RS(255, 239) code. The power dissipated in the worst case is 14 mW including the memory block when operating at 1.0 Gbps data rate.

References

Ahlin, L. [1985]. Coding methods for the mobile radio channel. In *Nordic Seminar on Digital Land Mobile Communications*.

Blahut, R.E. [1983]. *Theory and Practice of Error Control Codes*. Addison-Wesley.

Chang, H.C., and C.B. Shung [1998]. A (208,192;8) Reed-Solomon decoder for DVD application. In *IEEE International Conference on Communications*.

Chang, H.C., and C.Y. Lee [2001]. An area-efficient architecture for Reed-Solomon decoder using the inversion less decomposed Euclidean algorithm. In *IEEE International Symposium on Circuits and Systems*.

Chang, H.C., C.C. Lin, and C.Y. Lee [2002]. A low-power Reed-Solomon decoder for STM-16 optical communications. In *IEEE Asia-Pacific Conference on ASIC*.

Kang, H.J., and I.C. Park [2002]. A high-speed and low-latency Reed-Solomon decoder based on a dual-line structure. In *IEEE International Conference on Acoustics, Speech, and Signal Processing*.

Lee, H., M.L. Yu, and L. Song [2000]. VLSI design of Reed-Solomon decoder architectures. In *IEEE International Symposium on Circuits and Systems*.

Lee, H. [2003a]. High-speed VLSI architecture for parallel Reed-Solomon decoder. In *IEEE transactions on VLSI Systems*.

Lee, H. [2003b]. An area-efficient Euclidean algorithm block for Reed-Solomon decoder. In *IEEE Computer Society Annual Symposium on VLSI*.

Proakis, J.G. [2001]. *Digital Communications*. New York: Mc Graw Hill.

Sarwate, D.V., and N.R. Shanbhag [2001]. High-speed architectures for Reed-Solomon decoders. In *IEEE transactions on VLSI Systems*.

Sharma, G., A. Dholakia, and A. Hassan [1996]. Simulation of error trapping decoders on a fading channel. In *IEEE Transaction on Vehicular Technology*.

Wang, H.S., and N. Moayeri [1995]. Finite-state Markov channel – A useful model for radio communication channels. In *IEEE Transaction on Vehicular Technology*.

Wicker, S.B., and V.K. Bhargava [1994]. *Reed Solomon Codes and Their Applications*. Piscataway, NJ: IEEE Press.

Wilhemsson, L., and L.B. Milstein [1999]. On the effect of imperfect interleaving for the Gilbert-Elliott channel. In *IEEE Transactions on Communications*.

Index

Philips Research Book Series

1. H.J. Bergveld, W.S. Kruijt and P.H.L. Notten: *Battery Management Systems.* 2002   ISBN 1-4020-0832-5
2. W. Verhaegh, E. Aarts and J. Korst (eds.): *Algorithms in Ambient Intelligence.* 2004   ISBN 1-4020-1757-X
3. P. van der Stok (ed.): *Dynamic and Robust Streaming in and between Connected Consumer-Electronic Devices.* 2005   ISBN 1-4020-3453-9
4. E. Meinders, A.V. Mijritskii, L. van Pieterson and M. Wuttig: *Phase-Change Optical Recording Media.* 2006   ISBN 1-4020-4216-7
5. S. Mukherjee, E. Aarts, R. Roovers, F. Widdershoven and M. Ouwerkerk (eds.): *AmIware.* Hardware Technology Drivers of Ambient Intelligence. 2006
                                                                  ISBN 1-4020-4197-7
6. G. Spekowius and T. Wendler (eds.): *Advances in Healthcare Technology.* Shaping the Future of Medical Care. 2006   ISBN 1-4020-4383-X
7. W.F. J. Verhaegh, E. Aarts and J. Korst (eds.): *Intelligent Algorithms in Ambient and Biomedical Computing.* 2006   ISBN 1-4020-4953-8

springer.com